# Chromatography: Concepts and Applications

# Chromatography: Concepts and Applications

Edited by **Carlos Dayton**

New York

Published by Callisto Reference,
106 Park Avenue, Suite 200,
New York, NY 10016, USA
www.callistoreference.com

**Chromatography: Concepts and Applications**
Edited by Carlos Dayton

International Standard Book Number: 978-1-63239-112-4 (Hardback)

Printed in the United States of America.

# Contents

# Preface

The term chromatography was introduced in 1906 by a Russian botanist Mikhail Tswett. This technique is an effective separation device that is used in all branches of science. It is usually the only means of separating elements from complicated mixtures. The first systematic use of chromatography was explained by James and Martin in the year 1952, for the use of gas chromatography in the analysis of fatty acid mixtures. Vast areas of chromatographic procedures make use of differences in size, charge and other properties. There are several types of chromatography that have been introduced in the book. Some of them are column chromatography, high performance liquid chromatography (HPLC), gas chromatography, size exclusion chromatography, ion exchange chromatography etc. This book includes detailed descriptions about the applications of chromatography by different research findings. Every issue discussed in this book has lists of references tagged at the bottom to provide students and researchers with opening pointers for independent chromatography investigations.

This book unites the global concepts and researches in an organized manner for a comprehensive understanding of the subject. It is a ripe text for all researchers, students, scientists or anyone else who is interested in acquiring a better knowledge of this dynamic field.

I extend my sincere thanks to the contributors for such eloquent research chapters. Finally, I thank my family for being a source of support and help.

**Editor**

# Column Liquid Chromatography

Changming Zhang, Zhanggen Huang and Xiaohang Zhang
*State Key Laboratory of Coal Conversion,*
*Institute of Coal Chemistry,*
*Chinese Academy of Sciences, Taiyuan,*
*China*

## 1. Introduction

In the processing of coal and petroleum, there are many products produced such as gas and lighter liquid which is easy to use. At the same time, there is heavy material produced which is difficult to use. Such as, in crude oil refine processing, oil thermal cracking and catalytic cracking of petroleum, many residua oils, asphalts, and heaviest "waste" residual will be produced. The quantity of heavy oils is often large. So, it is important to study the property of heavy oils.

The column liquid chromatography (CLC) is an important and indispensable analysis method to study heavy oils. It is not only a separation means, but is also analysis means, especially for analysis of hydrocarbon group type.

Hydrocarbon group type analysis means the determination of the following classes of compounds:

1.  Saturated compounds, including paraffinic and naphthenic hydrocarbons.
2.  Aromatic compounds, (containing at least one benzene ring). Their molecules containing one benzene ring are classified as mono-aromatics, those with two aromatic rings as di-aromatics, etc.
3.  Resins, including polar substances containing elements other than C and H in the molecule (nitrogen, sulphur and oxygen in particular)
4.  Asphaltenes, including polar substances and asphaltenes only soluble in one or two polar solvents such as quinoline , which have large molecular weight and high aromatic ring number.

Now analysis methods existed have some deficiencies. Such as GC method can not be used to analyze compounds having high boiling point. The application of high performance liquid chromatography (HPLC) to hydrocarbon group-type analysis is characteristic with its high efficiency, high speed, and high sensitivity. But HPLC is only suitable for analysis of substances soluble in $n$-pentane [1].

TLC-FID [2-3] method can be also used to analysis the THF-soluble party in asphalt-samples and show great advantages. But, the components were combusted during TLC-FID analysis process and this lack made it not suitable for other analysis with preparation fraction. It should be pointed that the conventional method such as ASTM method use amount of solvent is large and some solvents has high toxicity [4, 5]. Moreover, there are too troublesome for some operation in traditional method. Hence, the separation of products containing heavy components remains a difficult task up to now.

Refereeing the literatures [4-10], the authors of this paper establish an optimum CLC method to analyze group-type of heavy oils through a series of studies. This paper detail introduces this method and its many applications which include preparation of high-level road asphalt, the characterization of molecular weight distributions (MWDs) and analysis of heterocyclic aromatic components of heavy oils.

## 2. The establish of CLC method

### 2.1 Column, support and heating apparatus

The dimension of glass chromatographic column is 90 mm length and 6 mm I. D. Silica gel with particle size range from 100 to 200 meshes was provided by marine chemical plant of Qingdao China. Silica gel was active under temperature of 180°C for 4 hours before use. Oxide of alumna 0.047-0.147 mm used was purchased from chemical and medical reagent company in Shanghai China. Muffle furnace (50°C-1000°C) and oven was used for sample preparation and heating.

### 2.2 Reagents

N-heptane, dichloromethane, trichloromethane as eluent solvents all were analytical grade reagents produced by Tianjin Chemical Reagent Factory (China). Pure reagents as model compounds were supplied by Aldrich Chemical Company (USA), including tetracosane (99.5% pure), dibenz[ah]anthracen ( 98%, pure), and acetanilide ( 99% pure), etc.

### 2.3 Analytical instruments

Fourier transforms FT-IR spectra were measured by a Bio-Rad Excalibur Series FTS 3000 spectrometer in the range of 4000-400 cm$^{-1}$ using KBr pellets. $^{1}$H NMR measurements were made with a Bruker Avance 500 spectrometer operating at 500.1 MHz.

## 3. The establish of group-type analysis method by CLC

### 3.1 Optimum chromatographic condition

As a base line, some pure reagents were chosen as model components prepared for CLC. These model compounds were tetracosane for saturates, dibenz[ah]anthracen for aromatics and acetanilide for resins. There is no appropriate pure reagent used for asphaltene fraction, so the insoluble fraction of tetrahydrofuran in one asphalt sample was used for asphaltene fraction.

Through a series of investigations,the optimum chromatographic operation was performed. The final optimum conditions were obtained as follows: Chromatographic column was glass column being 90 mm length, 6 mm i.d. The amount of silica gel used was from 1 to 1.5 gram.

The amount of alumina was from 1.5 to 1.8 gram. Total sample used was about 0.1 gram. The solvent of heptanes, mixture of heptanes/ dichloromethane (1/2.5, V/V) and mixture of dichloromethane/ trichloromethane (1/3, V/V) were as elutes corresponding to saturated hydrocarbon, aromatic hydrocarbon and resin respectively. The amount of heptanes, heptanes/ dichloromethane, and dichloromethane/ trichloromethane was 20ml, 35ml and 30ml respectively. Each fraction collected was dried in vacuum under 60°C until the weight keep constant.

Through above group analysis, the experimental deviation and recovery of CLC method are summarized in Table 1. From Table, it can be seen that the average of deviation and recover are -1.546% and 100.681% respectively; the results are good.

| Pure Reagents | Weight of preparation | Content W% | Determination W% | Deviation W% | Recover % | No. |
|---|---|---|---|---|---|---|
| Tetracosane | 0.0547(g) | 100 | 98.095 | -1.905 | 98.095 | 1017 |
| Tetracosane | 0.0508 | 100 | 97.964 | -2.036 | 97.964 | 1020 |
| Tetracosane | 0.0311 | 27.154 | 26.658 | -0.496 | 98.173 | 1201 |
| Tetracosane | 0.0285 | 26.571 | 25.638 | -0.933 | 96.489 | 1202 |
| Average | | | | -1.343 | 97.680 | |
| Dibenz(ah)anthracen | 0.042 | 100 | 98.095 | -1.805 | 98.095 | 1030 |
| Dibenz(ah)anthracen | 0.0442 | 100 | 97.964 | -2.036 | 97.964 | 1103 |
| Dibenz(ah)anthracen | 0.0259 | 22.629 | 20.909 | -1.720 | 92.399 | 1201 |
| Dibenz(ah)anthracen | 0.0294 | 27.402 | 26.843 | -0.559 | 97.960 | 1202 |
| Average | | | | -1.530 | 96.605 | |
| Acetanilide | 0.0875 | 100 | 97.600 | -2.400 | 97.600 | 1024 |
| Acetanilide | 0.0247 | 23.024 | 22.642 | -0.382 | 98.341 | 1202 |
| Acetanilide | 0.0365 | 100 | 96.438 | -3.562 | 96.438 | 1222 |
| Acetanilide | | 27.402 | 26.843 | -0.559 | 97.853 | 1201 |
| Average | | | | -1.726 | 97.558 | |
| Asphltene | 0.0261 | 22.815 | 24.790 | 1.975 | 108.657 | 1201 |
| Asphltene | 0.0395 | 100 | 94.937 | -5.063 | 94.937 | 1211 |
| Asphltene | 0.0387 | 100 | 98.450 | -1.550 | 98.450 | 1301 |
| Average | | | | -1.546 | 100.681 | |

Table 1. Experimental deviation and recovery of model compound.

### 3.2 Check of chromatographic resolution rate by FT-IR

The result of CLC method was checked by Fourier transform infrared (FT-IR) method .The spectra IR were acquired in the transmission mode as 64 scan in the IR range from 4000 to 500cm-1 at a resolution of 4cm-1. KBr standard pellets were used, and the samples were dried and then mixed with KBr, ground, and palletized.

IR spectrums of pure reagents including tetracosane, dibenz(ah)anthracen and acetanilide were obtained and used for standards. The IR spectrums of different fractions collected from flow out separated of the mixture reagents, and spectrums were compared with above standard spectrums. The results were shown in Figure 1.

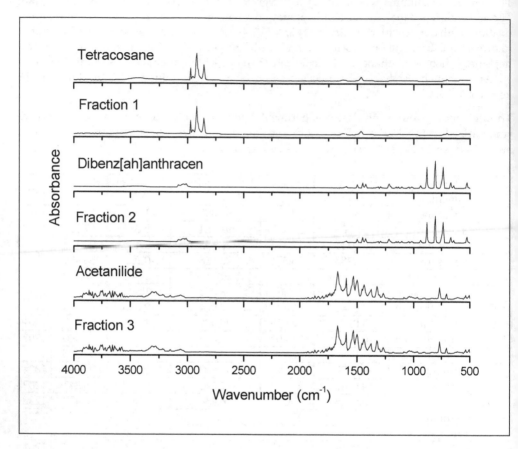

Fig. 1. Infrared spectrum for pure reagents and different fraction.

It is important to indicate that the IR spectra of fraction 1 collected (from 1202# sample) show similarity with pure tetracosane reagent. IR spectra for fraction 2 and fraction 3 show accordant results with dibenz(ah)anthracen and acetanilide respectively.

### 3.3 Check of chromatographic resolution rate by [1]H NMR

The CLC method was checked also by [1]H NMR. It measured different fractions collected from flow out separated of the mixture reagents and spectrums were compared with above standard spectrums. The high resolution [1]H NMR spectra of pure model compounds and fraction 1-3 are shown in Figure 2.

It is difficult to separate complex and heavy sample, however the IR and [1]H NMR analysis of the prepared fractions from CLC were all good agreement with pure reagents. This observation indicate the optimum CLC parameter in this work guarantee a good qualitative results.

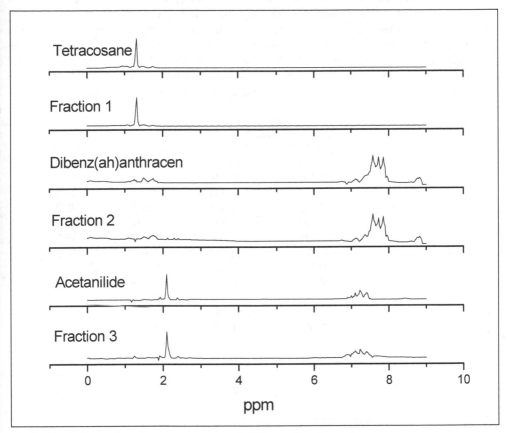

Fig. 2. ¹H NMR results of pure reagents and different fraction.

### 3.4 Evaluation of analysis of group composition by CLC

The recover rate and experiment deviations for model compounds were summarized in Table 1. It can be seen that the experiment result are fine.

Compared with routine ASTM method, these optimum chromatographic conditions show many advantages. First, the reagent and sample consumed was fewer than total solvent of 300 ml of classic ASTM method. Second, the dichloromethane and trichloromethane used in present study, compared with toluene and benzotrichloride used, has lower toxicity.

## 4. Applications of group type analysis by CLC

### 4.1 The application in making high grade road asphalt

Coal is used as the main source of energy in China. The crude oil produced in China is paraffinic; therefore, it is not suitable for road asphalt. China is trying to produce high grade road asphalt from the mixture of coal and petroleum [11, 12].

Three asphalt samples from petroleum and coal processing for high grade paving asphalt were characterized by established method. Sample NE-6, NE-9, NE-11 were the heavy products by co-processing of Shijiazhuang oil (a petroleum factory in China) and Yanzhou coal (a typical coal in China). The coal and oil ratio was 1:1. Among asphalt samples, the preparation of NE-6 sample was under the role of Fe catalyst during co-processing. NE-9 sample was related to Mo catalyst. The sample TLA is from Trindid Lake Asphalt. The results of group type analysis for four asphalt samples were shown in Table 2.

| Name | Test | Saturates | Aromatics | Resins | Asphaltenes |
|------|------|-----------|-----------|--------|-------------|
| NE-6 | (1) | 5.496 | 60.154 | 17.327 | 17.023 |
|  | (2) | 4.986 | 58.230 | 20.128 | 15.855 |
|  | Average | 5.241 | 59.191 | 18.727 | 16.439 |
|  | Deviation | 0.255 | 0.961 | 1.401 | 0.584 |
| NE-9 | (1) | 9.950 | 21.379 | 52.906 | 15.765 |
|  | (2) | 8.001 | 22.087 | 52.348 | 17.554 |
|  | Average | 8.975 | 21.733 | 52.627 | 16.659 |
|  | Deviation | 0.974 | 0.354 | 0.279 | 0.895 |
| NE-11 | (1) | 7.375 | 66.379 | 23.659 | 2.586 |
|  | (2) | 8.497 | 67.984 | 20.850 | 2.668 |
|  | Average | 7.936 | 67.182 | 22.254 | 2.627 |
|  | Deviation | 0.561 | 0.802 | 1.404 | 0.041 |
| TLA | (1) | 5.496 | 60.154 | 17.327 | 17.023 |
|  | (2) | 4.986 | 58.230 | 20.928 | 15.855 |
|  | Average | 5.241 | 59.191 | 19.128 | 16.439 |
|  | Deviation | 0.255 | 0.961 | 1.800 | 0.584 |

Table 2. Results of groups composition of asphalts (W%).

From Table 2 it can be seen that the application of established method to real asphalt samples show good results. Different samples have different group composition characterize. The experiment deviations of contents(W%) are in the ranges from 0.255% to 1.800%.

FTIR experiments were performed to check the qualitative ability of established method. IR spectra of saturated fraction, aromatic fraction and resin fraction for sample NE-9 were shown in Figure from 3 to 5. It is important to note intense absorption peaks for saturated fraction (Fig.3). Based the standard IR handbook, the absorption peaks around 719.45cm$^{-1}$, 1377.17 cm$^{-1}$, 2850.78 cm$^{-1}$, 2918.29 cm$^{-1}$ and 2959.79 cm$^{-1}$ was attributed to characteristics peak for $\delta$ (CH$_2$)$_N$ N>6, $\delta$ (CH$_3$), $\upsilon$ $_s$CH$_3$, $\upsilon$ $_{as}$ (CH$_2$) and $\upsilon_{as}$ CH$_3$ respectively. These data show that the prepared saturated fraction has a high purity.

As Figure 4 show, the absorption peaks around 748.38 cm$^{-1}$, 812.03 cm$^{-1}$, 877.61 cm$^{-1}$ and 3049.45 cm$^{-1}$ belong to character peak of aromatic C-H absorption. The peaks at 1602.84 cm$^{-1}$,1580 cm$^{-1}$ and 1410 cm$^{-1}$ were characteristics absorption peak of aromatic carbon. Obviously, the obtained aromatic hydrocarbon fraction has a good purity.

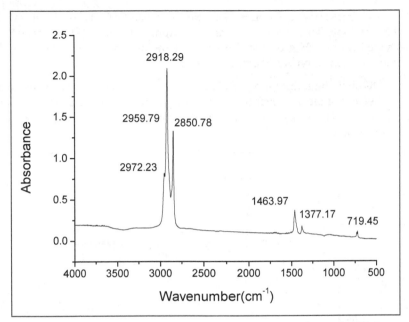

Fig. 3. Infrared spectrum of the saturated hydrocarbon fraction of sample NE-9.

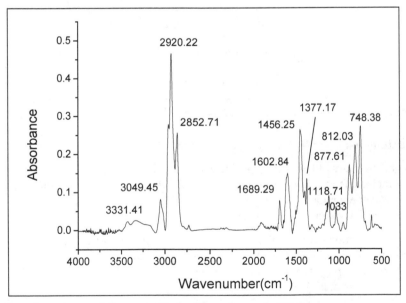

Fig. 4. Infrared spectrum of the aromatic fraction of sample NE-9.

The results from Figure 5 show that the resin fractions concentrate some oxygen-containing compounds. This conclusion can be approved by the appearing peak around 1215.15 cm-1, which is characteristics absorption peak for phenol compounds, and peak around 3649.31

cm-1, which is characteristics absorption peak for dissociate OH. The peaks at 1033.84 cm-1 and 1608.63 cm-1 attribute to the absorption from OH and C-O-C group. This is comprehensible because OH group in the structure the phenol connects to the aryl group, which may induce some aromatic absorption peaks.

The FTIR results show high resolution of CLC method established. It is difficult to separate complex and heavy sample, however the IR analysis of the prepared fractions from the CLC chow all good results This observation indicate that chromatographic parameter guarantee a good qualitative results.

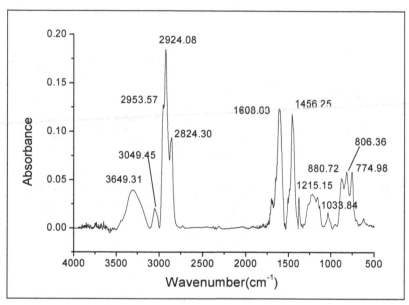

Fig. 5. Infrared spectrum of the resin fraction of sample NE-9.

## 4.2 The determination of MWDs by CLC coupled with SEC

Among characteristics of heavy oil, the size exchange chromatography (SEC) can be used to determine molecular weight distributions (MWDs), weight average molecular weight (Mw) and number average molecular weight (Mn), etc. With heavy oil of a group as example, the conditions of SEC are summarized as follows.

The analysis conditions are: a Shimadzu LC-10A high performance liquid chromatograph with an SPD-10AUP UV detector, the chromatographic column of SHIMPACK -801 (30 cm length, 0.8 cm i.d., polystyrene 6 μm), mobile phase of THF; flow rat with 1.2 ml/min; column temperature at 25°C.

The SEC chromatograms are shown in Figure 6, MWDs results are listed in Table 3.

In Figure 6, the sources of coal asphalt , KP petroleum asphalt, ethylene residue oil and vacuum residue oil  are from Shanxi coking plant in China, Korea refining, Xinjiang oil refinery in China and Saudi Arabia's oil refining, respectively.

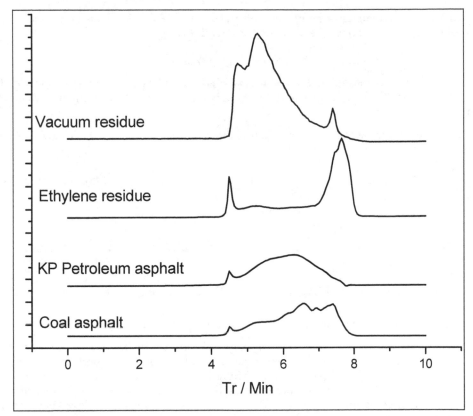

Fig. 6. The SEC chromatograms of typical heavy oils.

| Samples | Mw | W% | | | | | |
|---|---|---|---|---|---|---|---|
| | | M>5000 | M5000 -3000 | M3000 -1000 | M1000 -500 | M500 -300 | M<300 |
| Coal asphalt | 1032.200 | 1.658 | 2.802 | 21.199 | 29.730 | 22.745 | 21.863 |
| KP Petroleum asphalt | 1905.674 | 0.804 | 17.169 | 59.899 | 12.274 | 3.979 | 5.873 |
| "Ethylene" residue oil | 764.788 | 0.191 | 7.481 | 15.201 | 9.648 | 9.666 | 57.810 |
| Vacuum residue oil | 1886.698 | 3.683 | 1.490 | 61.166 | 20.566 | 6.399 | 6.693 |

Table 3. The MWDs of typical heavy oils

How much is the "representative" characteristics of this SEC method? This is an important problem to need know to treating these spectra and data of SEC. The so-called "representative" refers that extent which could be determined out of sample. Because most present SEC method is only suitable to compounds having UV adsorbent and soluble of THF, so, it is needed to know representative of whole sample. This problem will be completed only by CLC. Because the four groups: saturates, aromatics, resins and asphaltenes quantitatively could be obtained by CLC determination, then the "representative "(R index) will be calculated as the following.

$$R = 100\ \% - W_{asp}\ \% - W_{alk}\ \% \tag{1}$$

Which R represents the representation index; $W_{asph}$ % and $W_{alk}$ % represent the weight percent of asphaltene in sample and the weight percent of saturated hydrocarbons in sample, respectively.

With samples of Figure 6 as example, their R indexes from this CLC analysis are listed in Table 4.

| Name of samples | (1) | (2) | (3) | Average | Max of deviation % |
|---|---|---|---|---|---|
| Coal asphalt | 73.64 | 72.89 | 73.86 | 73.46 | -0.78 |
| KP petroleum asphalt, | 99.33 | 99.28 | 98.76 | 99.12 | -0.36 |
| Ethylene residue oil | 90.36 | 89.87 | 90.56 | 90.26 | -0.43 |
| Vacuum residue oil | 96.17 | 96.21 | 95.76 | 96.05 | 0.33 |

Table 4. The R indicators.

These results show that the CLC coupled with SEC is an effective mean to analyze MWDs.

## 4.3 Analysis of resin component by CLC coupled with HPLC

As components of resin of heavy oil are very complicated, so to analyze them is very difficult by only one method. However, CLC coupled with high performance liquid chromatography (HPLC) can separate successfully, quality and quantity these compositions. Because the resin fraction got concentrate oxygen-containing compounds and other hetero-atom-containing compounds by CLC separation, then the analysis of these hetero-atom-containing compounds became easy to by HPLC. With slurry oil (Tianjing Refinery of China) as an example, the analysis of components in resin was summarized as follows.

The preparation of resin fraction was same as that of above description of CLC; the HPLC was performed on a Shimadzu LC-3A chromatogram with a SPD-1 UV detector, operated at 254 nm. Two ODS (4.6x20 cm) columns in series were operated at 40 ºC with methanol /water=78:22(V/V) as the mobile phase, flowing at a rate of 0.8 ml/min. Typical separation chromatogram is shown in Figure 7.

From Figure 7 it can be seen the high resolution separation rate of complex compositions, these confirmed that the CLC preparation is successful and HPLC analysis is better.

The three qualitative methods of HPLC were selected to determine compositions of resin fraction. The three methods [13] are follows.

1.   The qualitative method of relative retention time (RRT).
2.   The qualitative method of stop- flow UV scanning.
3.   The qualitative method of UV characteristic index V'.

The quantitative determination of compositions was by the method of external standard (E-X) and the calculation formula uses the following.

$$W_x\ \% = (R_{ex}/C_x) * (S_x/S_{ex}) * (V_{ex}/V_x) * R_{es}\ \% \tag{2}$$

where $W_x$ % is the weight content percent of x composition in heavy oil sample, $R_x$ % and $C_{ex}$ % are the concentration of preparation solution of resin fraction and external standard solution, respectively, $S_x$ and $S_{ex}$ are the peak areas of component x and external standard, respectively, $V_{ex}$ and $V_x$ are the injection volumes of external standard solution and resin solution, respectively, $R_{es}$% is the weight percent of resin fraction in heavy oil sample. The qualitative and quantitative results are in Table 5.

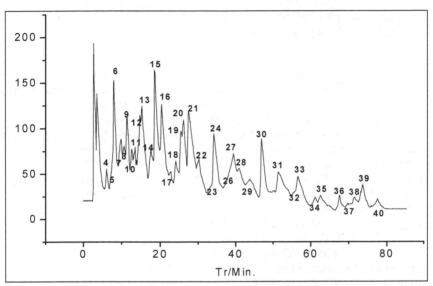

Fig. 7. HPLC chromatogram of resin fraction.

| Number of peak | Component | Quantitative results | Number of peak | Component | Quantitative results (ppm) |
|---|---|---|---|---|---|
| 9 | 3-Methyl indole | 610 | 15 | N-phenyl pyrrole | 800 |
| 10 | Quinoline | 240 | 16 | 7,8-Benzoquinoline | 1000 |
| 11 | Phenanthrene-quinone | 240 | 18 | p-phenyl phenol | 500 |
| 12 | Carbazole | 500 | 20 | N-phenyl indole | 600 |
| 13 | 4-Methylquinoline | 640 | 24 | Dibenzofuran | 630 |
| 14 | 2-Amino-phenol | 570 | 33 | N-Ethyl carbozole | 370 |
| Note | | ppm | 35 | 2,2'-Biquinoline | 290 |

Table 5. Results of components of resin (ppm).

## 5. Conclusion

A modified method for group type analysis of asphalt using CLC was established. The small-type CLC technique shows many advantages, such as high resolution rate, rapid operation, and requires minimal quantities of sample and solvent. The both of IR and $^1$H NMR results check the high resolution of this method.

The CLC method compared with routine ASTM method, the reagents used in this method are small amount and lower toxicity. These are beneficial to environmental protection and human health. This is very important for modern analysis.

The CLC method of this paper is an important and indispensable analysis method to study heavy oils. It is not only a separation means, but is also analysis means.This method was successfully applied to many analysis aspects, such as making high grade road asphalt, characterizing MWDs and analysis heterocyclic of aromatic compositions of heavy oils.

The analysis of heavy oil is a long and difficult task. We systematically summarized these studies and hope that these will help our colleagues.

## 6. References

[1] Changming Zhang, Aiying Li, Yongji Li, Zengmin Sen. Analysis of the class composition of some residual oils and asphalts by HPLC. *Preprints*, Division of petroleum chemistry[C].INC. American Chemical Society, 1989, 34(2):240-246.

[2] Zhe Wang, Changming Zhang. A study on the relationship between the composition and the usage of asphaltic heavy oil. *Preprints*, Division of petroleum chemistry[C].INC. American Chemical Society, 1992, 37(3):933-936.

[3] Zhang Changming, Li Aiying, Li Ying, Zhang linmei. Instrumental analysis and systematic investigation on heavy oils from coal. *Chinese journal of Chromatography* , 1999, 17(4):372-375.

[4] Copyright by the ASTM international. *Standard test method for separation of asphalt into four fractions.* 2002, Thu Dec 05 15; 56; 14.

[5] Shu-an Qian, Peng-zhou Zhang, Bai-ling Li, Structural characterization of pitch feedstocks for coke making. *Fuel*, 1995, 64(8): 1085-1091.

[6] Standard of geologic office of the People's Republic of China, *Analytical method of class composition for crude oil and extract organic*,1987,05-23.

[7] Justin D, Fair, Chad M.Kormos, Flash column chromatograms estimated from thin-layer chromatography data. *Journal of Chromatography A*, 2008, 1211:49-54.

[8] Davies, Don R., Johnson, Todd M. Isolation of three components from spearmint oil: An exercise in column and thin-layer chromatography. *J. Chem. Educ.* 2007,84(2):318-320.

[9] B. Concho-Grande, M. Rodriguez-Comesafia, J.Simal-Gandara, Sample HPLC determination of colistin in modicated feeds by pre-column derivatization and fluorescence detection. *Chromatographia* , 2001,54(7/8):481-484.

[10] B Liawruangrath, S. Liawruangrath, High performance thin layer chromatographic determination of erythromycin in pharmaceutical preparation. *Chromatographia*, 2001,54(5/6):405-408.

[11] Yongbing Xue, Jianli Yang, Zhenyu Liu, Zhiyu Wang, Zengnou Liu,Yunmei Li,Yuzhen Zhang, Paving asphalt modifier from co-processing of FCC slurry with coal. *Catalysis Today* 2004,98:333-338.

[12] Aroon Shenoy, Prediction of high temperature rheological properties of aged asphalts from the flow data of the original unaged samples. *Construction and Building Materials*, 2002,16 (8):509-517.

[13] Changming Zhang, Xiaohang Zhang, Jianli Yang, Zhenyu Liu, Analysis of polynuclear aromatic hydrocarbons in heavy ducts derived from coal and petroleum by high performance liquid chromatography. *J. chromatogr. A*, 2007, 167, 171–177.

# Chromatographic Separation and Identification of Sildenafil and Yohimbine Analogues Illegally Added in Herbal Supplements

Hakan Göker*, Maksut Coşkun and Gülgün Ayhan-Kılcıgil
*Central Instrumental Analysis II Laboratory, Faculty of Pharmacy, Ankara University,
Tandogan, Ankara,
Turkey*

## 1. Introduction

Herbal medicines are major source of aphrodisiacs and have been used worldwide for thousands of years by different cultures and civilizations. Recently, consumption of dietary supplements has been becoming more popular around the world. Unfortunately, the adulteration of dietary supplements with undeclared synthetic chemical compounds is steadily increasing according to the literature. Some herbal products advertised as "all natural" have in contrast been found to contain synthetic PDE-5 inhibitors. There are currently three PDE5 inhibitors **Sildenafil** (Langtry & Markham, 1999) (Viagra; Pfizer, New York, US), **Tadalafil** (Meuleman, 2003) (Cialis; Eli Lilly, Indianapolis, US), and **Vardenafil** (Keating & Scott, 2003) (Levitra; Bayer Pharmaceuticals Co, Wuppertal, Germany), approved worldwide for the treatment of male erectile dysfunction, further two agents **Udenafil** (Salem et al., 2006) (Zydena; Dong-A PharmTech Co, Korean), **Mirodenafil** (Jung, 2008) (Mvix, Life Science R&D Center of SK chemical, Beijing, Tianjin, Shanghai) were licensed only in Korea. They produce vascular smooth muscle relaxation, promote penile blood flow, and hence, induce erection. These kinds of commercially available herbal aphrodisiac products have been spiked with the above-mentioned legal drugs, but also with their analogues, which have not been subjected to formal pharmacokinetic or other pharmacological testing in either humans or animals.

The practice of self-medication by an increasing number of patients, the incessant aggressive advertising of these herbal aphrodisiacs, the invasion of the medicinal market with uncontrolled dietary supplements and the absence of real directives amplifies the potential health hazards to the community. Since the sildenafil is an chemical, it must not been found in any foodstuffs, but an increasing number of sildenafil analogues have been discovered in dietary and herbal supplements even in soft drinks, this number is steadily increasing and some time their types are unknown and necrosis not observed instead of paracetamol. Hence, it is prudent to test the safety and efficacy thus might have unknown and harmful side-effects. Structural analogues are also synthetic chemicals with slightly altered chemical structures and have similar erectile effects on the body. Nevertheless, it is not uncommon

---

* Corresponding Author

for chemicals with similar structures to possess slightly or entirely different before any new chemical is licensed as drug for human use. This testing process is lengthy and costly; on average, it takes 9.5 years and costs US$800 million to license a new drug. Many drug analogues, without the aforementioned drug testing process, are available for human consumption properties. Phenacetin, structurally similar to paracetamol, has been associated with renal papillary necrosis not observed with paracetamol (Poon, 2007). Many drug analogues, without the aforementioned drug testing process, are avaliable for human comsuption via different channels. Examples include analogues of psychoactive drugs, anabolic steroids, and Sibutramine which was one of the most abused compound as anti-obesity drugs.

The most commonly reported side effects of sildenafil are headaches, flushing of the face, upset stomach and nasal congestion. We met a dieatery supplement having combination of sildenafil and paracetamol, probably in order to prevent headeache caused by sildenafil. Other side effects include sensitivity to light, blurred vision, urinary tract infection, diarrhea and dizziness. The main problem with sildenafil and analogues are that they interact with many other medications. They can rapidly decrease blood pressure up to dangerously low. They can interact with nitrates such as nitrogliserin, which are often prescribed to heart patients. Sildenafil has not to be administrated to patients with heart problems taking nitrate medications because of the severe potentiation of vasodilatory effects.

It is well known, the first developed and consequently the most famous phosphodiesterase inhibitor is Sildenafil (Langtry & Markham, 1999) approved by the FDA in early April 1998. Novel PDE5 inhibitor, **Lodenafil carbonate,** breaks down in the body to form two molecules of the active drug lodenafil. This formulation has higher oral bioavailability than the parent drug (Toque et al., 2008) Fig 1.

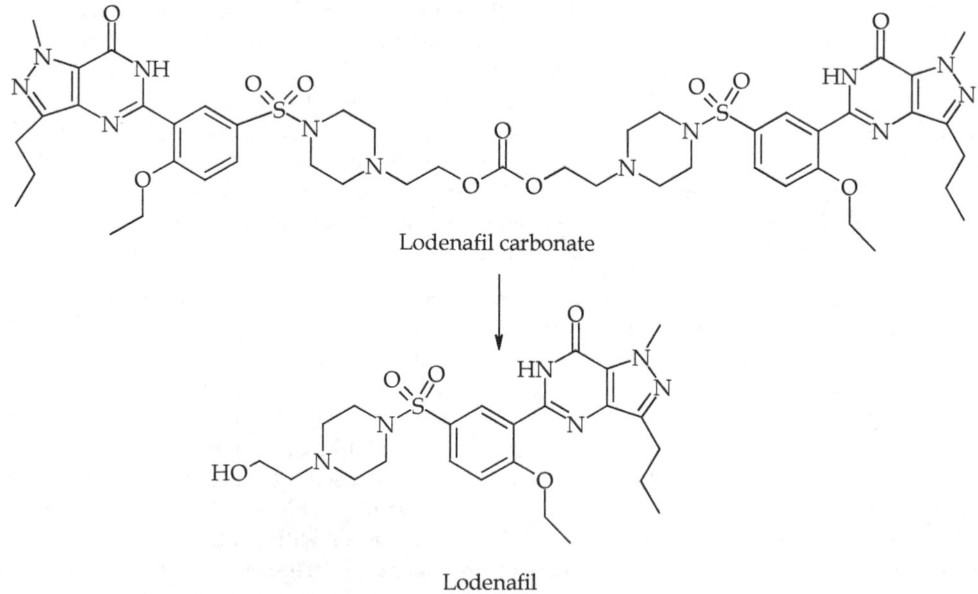

Lodenafil carbonate

Lodenafil

Fig. 1. Prodrug lodenafil carbonate.

**Avanafil** was discovered by pharmaceutical developer "Vivus" as PDE5 inhibitors, the safety and efficacy for erectile dysfunction, a possible lunch in 2012. It has a particular advantage over its potential competitors, the effects are demonstrable very quickly (in 15 minutes or less) (Bell & Palmer, 2011) Fig 2.

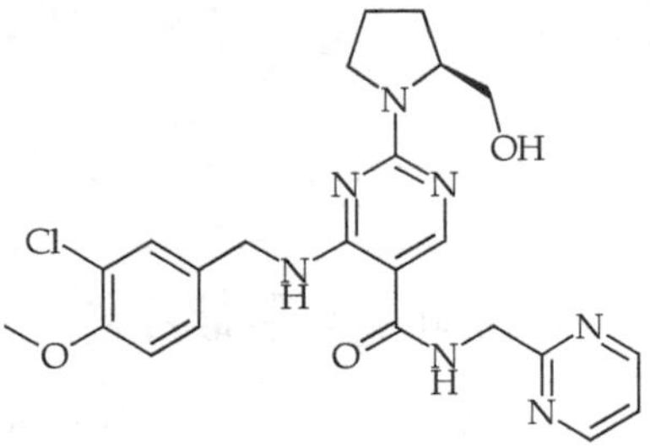

Fig. 2. Avanafil.

The one of the last discovered illegal PDE-5 inhibitor is called as "**Acetylvardenafil**" which was found in dieatery supplement known as MEGATON® in USA (Lee et al., 2011). Sulfonyl group of Vardenafil was substituted by an acetyl group Fig 3(a) .

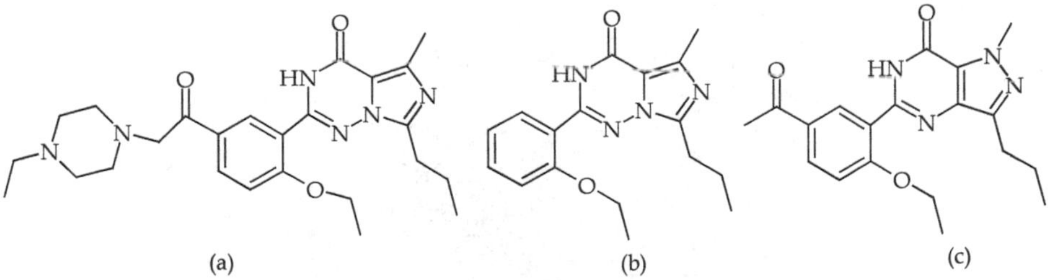

(a)                                    (b)                                    (c)

Fig. 3. (a) Acetylvardenafil; (b) Desulfovardenafil; (c) Gendenafil.

Another new analogue of vardenafil, **Desulfovardenafil**, in which the $N$-ethylpiperazine ring and the sulphonyl group were removed from the vardenafil structure, was identified for enhancing erectile function in herbal health product marketed, namely Power58 Platinum (Lai et al., 2007, Lam, 2007) Fig 3(b). New sildenafil analogues **Gendenafil** had an acetyl group instead of sulfonyl-$N$-methylpiperazine moiety was determined as 5-[2-ethoxy-5-acetyl-phenyl]-1-methyl-3-$n$-propyl-1,6-dihydro-7$H$-pyrazolo[4,3-$d$]pyrimidin-7-one (Lin et al., 2008) Fig3(c). Other vardenafil analogue was found to be added illegally into a dietary supplement marketed for male erectile dysfunction (MED). Its structure was determined as 2-(2-ethoxyphenyl)-5-methyl-7-propyl-imidazo[5,1-$f$][1,2,4]triazin-4(3$H$)-one and called "**Piperidenafil** or **Piperidino vardenafil** or trivial name **Pseudovardenafil**" (Park et al., 2007; Lai et al., 2007b) Fig 4(a).

Very recently, two new analogues of sildenafil in which the piperazine ring and the sulfonyl group were replaced by a piperazinone and a hydroxyethyl structure, respectively were isolated from a herbal product in Germany. Based on the piperazinone structure, the compounds were named **Piperazinonafil** in Fig 4(b) and **Isopiperazinonafil** (Wollein et al., 2011) in Fig 4(c).

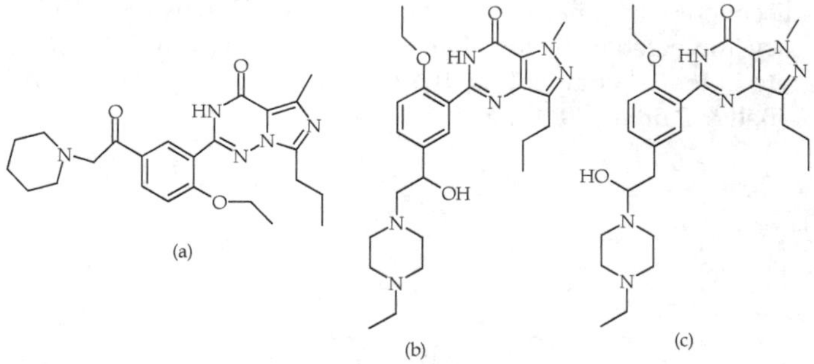

Fig. 4. (a) Piperidenafil; (b) Piperazinonafil; (c) Isopiperazinonafil.

Another sildenafil analogue was detected from a health supplement claimed for human MED, the structure of this new analogue was characterized as **dithio-desmethylcarbodenafil** containing 2 thiocarbonyl groups instead of 2 carbonyl groups, and 4-methyl substitution on the piperazine ring, rather than 4-ethyl substitution when compared to sildenafil (Ge et al., 2011) Fig. 5(a).

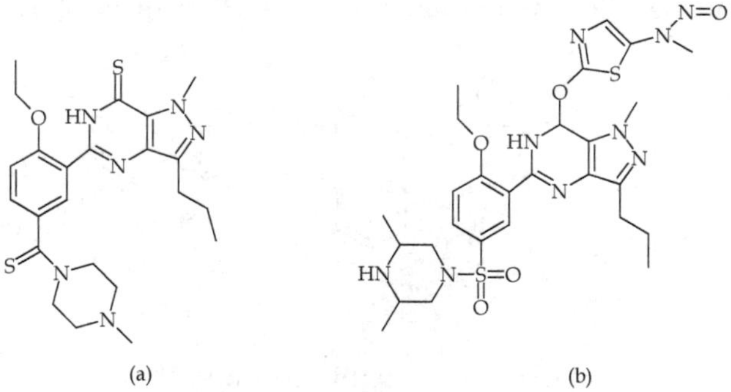

Fig. 5. (a) Dithio-desmethylcarbodenafil; (b) **Nitroso-prodenafil.**

Other new unapproved analogue of sildenafil was detected in capsules of a herbal dietary supplement promoted as a libido enhancing product. This is the first time a PDE-5 inhibitor and a potential NO donor were identified in one molecule. A hydrolysis experiment showed that the new analogue was a prodrug of aildenafil and was therefore named **nitroso-prodenafil** (Venhuis et al., 2011) Fig 5(b). Both PDE-5 inhibitors and nitrosamines cause vasodilatation by increasing levels of NO. To their coincidental use is warned against because it may cause a fatal drop in blood pressure. In addition, nitrosamines are known carcinogens. The findings indicate the dangerous level of advancement in medicinal chemistry by producers of unapproved drugs.

Tadalafil (Cialis®) was approved in 2003 by the FDA as the third phosphodiesterase type 5 enzyme (PDE-5) inhibitor to treat MED (Meuleman, 2003). Then, different tadalafil analogues have been found as adulterants in illegal products. Hasegawa et al. detected N-octyl-nortadalafil Fig 6(a) together with cyclopentynafil **Fig** 6(b) in dietary supplement (Hasegawa et al., 2008). Both of them are the first compounds reported to be new tadalafil and sildenafil analogues.

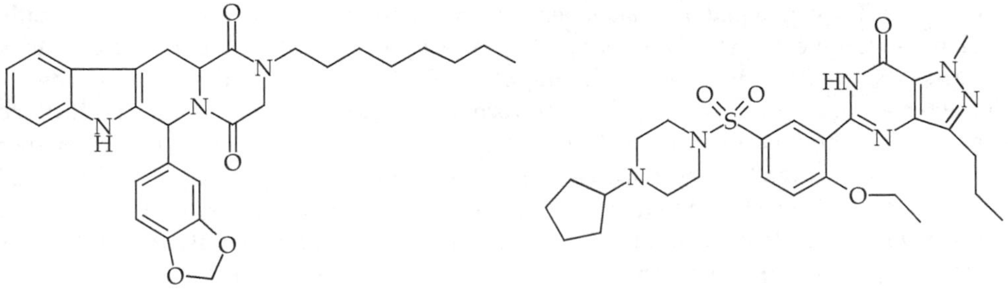

(a)

(b)

Fig. 6. (a) N-octylnor-tadalafil; (b) Cyclopentynafil.

In Taiwan, one of the dietary supplement which was claimed on the treatment of male erectile dysfunction was firstly screened in 2009 and Tadalafil and its doctored version was newly identified. Since it is having amino group instead of methyl in tadalafil it was named as **aminotadalafil** (Zou et al., 2006; Lin et al., 2009) Table-1. This compound has two asymmetric carbons, theoritically two pairs of enantiomers exist. The chromatographic separation of its stereoisomers was reported by using chiral LC-MS (Kurita et al., 2008). Using this method, RR-Aminotadalafil and SR-Aminotadalafil were detected in some health food. In addition, an interaction product of aminotadalafil was isolated from an illegal health food product. The structure of the interaction product was elucidated and unknown compound was characterized as condensation product of aminotadalafil and hydroxy-methylfuraldehyde and is probably the result of a drug-excipient incompatibility (Häberli et al., 2010) Fig 7.

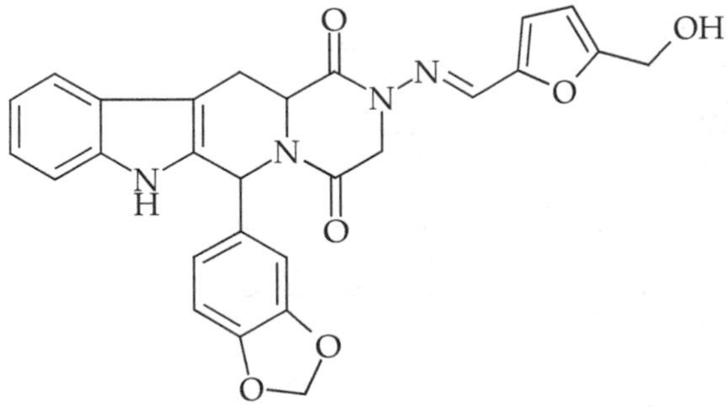

Fig. 7. Condensation product of aminotadalafil and hydroxymethylfuraldehyde.

Last flash development for the treatment of MED is discovering of **Zoraxel** (RX-10100) by Rexahn Pharmaceutical company (Albersen et al., 2010). Zoraxel is containing clavulanic acid that is centrally acting in the CNS and may be a more effective MED treatment for patients who are responsive or unresponsive to PDE-5 inhibitors. It is being developed as an orally administered, on-demand tablet to treat sexual dysfunction, and has extensive and well-established safety in humans. For the future, it is being expected, Zoraxel will be on to worldwide best-selling drug.

In our central instrumental analysis laboratory, we also try to detect these commercially available supplements (which are sent by the Ministry of Food Agriculture and Livestock of Turkey before it grants a license for import to Turkey)  whether they possess new or old sildenafil analogues by using high-performance liquid chromatography with diode array detection and mass spectrometry (HPLC-DAD–MS) and nuclear magnetic resonance (NMR) analyses, NMR is the only analytical technique which provides full structural information from novel compounds, acquisition of MS data or comparison of retention times may not be sufficient, in this case, LC-MS/NMR allowed to identify the adulterants without any need for references (Kesting et al., 2010) this means that these kind of analogues are not easy to detect by ordinary laboratory methods,  we have identified the listed analogues are given in Table 1-2  up to date. With the aim of evaluating the potential risks of commercialized aphrodisiac products on consumer health, the aim of present work is to investigate simple HPLC-MS method and NMR data of synthetic and natural analogues of aphrodisiacs.

**Table-1** shows the determination of some sildenafil and tadalafil analogues and Dapoxetin HCl, in commercially available health supplements in Turkey. Their formulas, $^1H$-$^{13}C$-NMR spectra, ESI(+) m/e values and their LC chromatograms are given in Table 1 and Fig 8, 9, respectively. Fig 9 shows the good separation of caffeine, some sildenafil analogues and

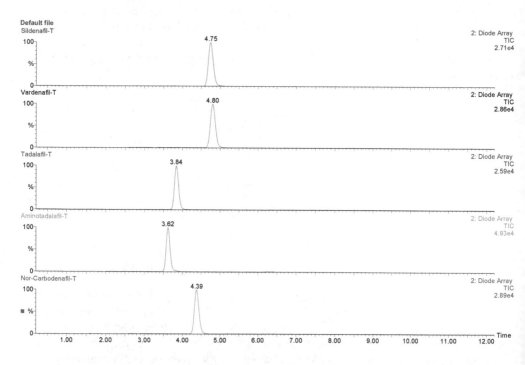

Fig. 8. HPLC chromatogram of *4.75* (Sildenafil), *4.80* (Vardenafil), *3.84* (Tadalafil), *3.62* (Aminotadalafil),  *4.39* (Nor-carbodenafil) (*RT* : Retention times as min).

Chromatographic Separation and Identification of Sildenafil and Yohimbine Analogues Illegally Added in Herbal Supplements

19

| Compound name | M.w. ESI(+) m/z | Formulas | $^1$H-NMR δ ppm |
|---|---|---|---|
| Sildenafil citrate (Langtry & Markham, 1999) $C_{22}H_{30}N_6O_4S$. $C_6H_8O_7$ | 474 475(M+H) | | (CD$_3$OD): 0.99 (t, 3H), 1.46 (t, 3H), 1.79 (m, 2H), 2.59 (s, 3H), 2.64, 2.72, 2.78, 2.82 (citrate prot.), 2.87 (t, 2H), 2.97 (br.t, 4H), 3.22 (br.s, 4H), 4.23 (s, 3H), 4.29 (q, 2H), 7.37 (d, J=8.8Hz, 1H), 7.9 (dd, J=8.8, 2.4Hz, 1H), 8.14 (d, J=2.4Hz, 1H) |
| Tadalafil (Meuleman, 2003) $C_{22}H_{19}N_3O_4$ | 389 390(M+H) | | (CD$_3$OD): 2.93 (s, 3H), 3.05 & 3.1 (dd, 11.6, 1.2Hz & 11.6, 1.2Hz, 1H), 3.66 (dd, J=16.0, 4.4Hz, 1H), 3.97 (d, J=17.6Hz, 1H), 4.20 (dd, J=17.6, 1.6Hz, 1H), 4.43 (dd, J=11.2, 4Hz, 1H), 5.84 (dd, J=5.6, 1.2Hz, 2H), 6.17 (s,1H), 6.67 (d, J=7.6Hz, 1H), 6.79 (d, J=1.6Hz, 1H), 6.82 (dd, J=8.0, 1.6Hz, 1H), 7.01 (td, J=6.8,0.8Hz, 1H), 7.06 (td, J=7.6, 1.2Hz, 1H), 7.25 (d, J=8Hz, 1H), 7.51 (d, J=8Hz, 1H). |
| Vardenafil HCl (Keating & Scott, 2003) $C_{23}H_{32}N_6O_4S$. HCl | 488 489(M+H) | | (CD$_3$OD) : 0.98 (t, 3H), 1.32 (t, 3H), 1.45 (t, 3H), 1.82 (m, 2H), 2.52 (s, 3H), 2.96 (t, 2H), 3.19 (q, 2H), 3.3 (br.s, 8H), 4.29 (q, 2H), 7.41 (d, J=8.8Hz, 1H), 7.98 (td, J=9.2, 1.6Hz, 1H), 8.05 (d, J=2Hz, 1H). |
| Udenafil (Salem et al., 2006) $C_{25}H_{36}N_6O_4S$ | 516 517(M+H) | | (CD$_3$OD) : 0.99 (t, 3H), 1.04 (t, 3H), 1.36 (m, 2H), 1.69-2.05 (m, 8H), 2.18 (m, 2H), 2.27 (s, 3H), 2.85-3.05 (m, 6H), 4.18 (t, 2H), 4.23 (s, 3H), 7.34 (d, J=8.4Hz, 1H), 7.96 (dd, J=8.4,2.4 Hz, 1H), 8.25 (dd, J=2.4, 0.8Hz, 1H). |
| Aminotadalafil (Zou et al., 2006; Kurita, 2008, Lin et al., 2009) $C_{21}H_{18}N_4O_4$ The mixture of isomers | 390 391(M+H) | | (CD$_3$OD): 3.11 & 3.14 (dd, 14.4, 1.2Hz & 11.6, 0.8Hz, 1H), 3.69 (dd, J=16.0, 4.4Hz, 1H), 4.11 (d, J=17.6Hz, 1H), 4.29 (dd, J=17.6, 2Hz, 1H), 4.43 (dd, J=11.2, 3.6Hz, 1H), 5.84 (dd, J=5.6, 1.2Hz, 2H), 6.14 (s,1H), 6.67 (d, J=7.6Hz, 1H), 6.80 (d, J=2Hz, 1H), 6.84 (dd, J=8.2, 1.6Hz, 1H), 7.01 (td, J=6.8, 0.8Hz, 1H), 7.06 (td, J=7.6, 1.6Hz, 1H), 7.24 (d, J=8Hz, 1H), 7.51 (d, J=8Hz, 1H). |
| Thiosildenafil (Zou et al., 2008) $C_{22}H_{30}N_6O_3S_2$ | 490 491(M+H) | | (CD$_3$OD) : $^1$H-NMR 1.0 (t, 3H), 1.54 (t, 3H), 1.82 (m, 2H), 2.26 (s, 3H), 2.51 (br.t, 4H), 2.9 (t, 2H), 3.06 (br.s, 4H), 4.34 (q, 2H), 4.48 (s, 3H), 7.39 (d, J=8.4Hz, 1H), 7.9 (dd, J=8.4, 2.4Hz, 1H), 8.33 (d, J=2.4Hz, 1H). |
| Hydroxythiohomosildenafil (Li et al., 2009b) $C_{23}H_{32}N_6O_4S_2$ | 520 521(M+H) | | (CD$_3$OD) : $^1$H-NMR 0.99 (t, 3H), 1.54 (t, 3H), 1.82 (m, 2H), 2.52 (t, 2H), 2.61 (br.t, 4H), 2.89 (t, 2H), 3.07 (br.t, 4H), 3.61 (t, 2H), 4.34 (q, 2H), 4.47 (s, 3H), 7.39 (d, J=8.4Hz, 1H), 7.89 (dd, J=8.4, 2.4Hz, 1H), 8.35 (d, J=2.4Hz, 1H). |
| Thiodimethylsildenafil (Thiomethisosildenafil) (Dimethylthinsildenafil) (Sulfoaildenafil) (Reepmeyer & Avignon, 2009; Gratz et al., 2009) $C_{23}H_{32}N_6O_3S_2$ | 504 505(M+H) | | (CD$_3$OD): $^1$H-NMR 0.99 (t, 3H), 1.04 (d, 6H), 1.54 (t, 3H), 1.86 (m, 4H), 2.87 (m, 2H), 2.91 (t, 2H), 3.61 (dd, 2H), 4.34 (q, 2H), 4.49 (s, 3H), 7.39 (d, J=8.4Hz, 1H), 7.89 (dd, J=8.4, 2.4Hz, 1H), 8.33 (d, J=2.4Hz, 1H). |
| Homothiomethisosildenafil (Homosulfoaildenafil) (Li et al., 2007) $C_{24}H_{34}N_6O_3S_2$ 518 Published in US Patent, however it is detected first time in herbal supplement. | 518 519(M+H) | | (CDCl$_3$) : 1.01 (t, 3H), 1.04 (d, 6H), 1.21 (t, 3H), 1.86 (m, 4H), 2.11 (m, 2H), 2.95 (t, 2H), 2.99 (m, 2H), 3.66 (dd, 2H), 4.28 (t, 2H), 4.53 (s, 3H), 7.19 (d, J=8.4Hz, 1H), 7.86 (dd, J=8.4, 2.4Hz, 1H), 8.82 (d, J=2.4Hz, 1H). $^{13}$C-NMR (CDCl$_3$) : 11.03, 14.2, 19.5, 22.4, 22.58, 27.8, 39.6, 50.5, 52.4, 72.5, 113.3, 120.2, 129.6, 130.8, 132.2, 132.58, 134.1, 146.3, 146.7, 159.76, 172.06. |
| Dimethylsildenafil (Aildenafil) (Methisosildenafil) (Wang et al., 2007) $C_{23}H_{32}N_6O_4S$ | 488 489(M+H) | | (CDOD$_3$): 0.99 (t, 3H), 1.04 (d, 6H), 1.47 (t, 3H), 1.81 (m, 2H), 1.89 (t, 2H), 2.87 (m, 4H), 3.61 (dd, 2H), 4.23 (s, 3H), 4.3 (q, 2H), 7.35 (d, J=8.8Hz, 1H), 7.87 (dd, J=8.8, 2.4Hz, 1H), 8.17 (d, J=2.4Hz, 1H). |
| Dimethylhomosildenafil* (Homomethisosildenafil) $C_{24}H_{34}N_6O_4S$ | 502 503(M+H) | | (CDCl$_3$): 1.01(t,3H), 1.05(d,6H), 1.19(t,3H), 1.84(m,2H), 1.93(t,2H) 2.05(m,2H), 2.93(t,2H), 3.03(m, 2H), 3.68(dd,2H), 4.26(t,2H), 4.27 (s,3H), 7.16(d,J=8.8Hz,1H), 7.84 (dd,J=8.8,2Hz,1H), 8.81(d,J=2Hz, 1H), 10.84(br.s,1H) (in Table 2) |
| Nor-carbodenafil Desmethylcarbodenafil (Piazza & Pamukçu, 2001) $C_{23}H_{30}N_6O_3$ Published in US Patent, however, it is detected first time in herbal supplement | 438 439(M+H) | | (CDCl$_3$): 1.01 (t, 3H), 1.59 (t, 3H), 1.84 (m, 2H), 2.34 (s, 3H), 2.47 (br.s, 4H), 2.9 (t, 2H), 3.56 & 3.78 (br.s, 4H), 4.25 (s, 3H), 4.32 (q, 2H), 7.05 (d, J=8.4Hz, 1H), 7.55 (dd, J=8.4, 2.4Hz, 1H), 8.55 (d, J=2.4Hz, 1H), 10.9 (s, 1H). $^{13}$C-NMR 14.3, 14.8, 22.6, 28.1, 38.4, 42.5(br. s), 46.3, 48(br.s), 55.3(br.s), 65.87, 113.2, 120.2, 124,7, 129.16, 130,7, 132.1, 138.7, 146.9, 147.6, 153.9, 157.6, 169.4. |
| Nor-acetildenafil Demethylhongdenafil (Reepmeyer & Woodruff, 2007) $C_{24}H_{32}N_6O_3$ | 452 453(M+H) | | (CD$_3$OD): 0.99 (t, 3H), 1.46 (t, 3H), 1.8 (m, 2H), 2.31 (s, 3H), 2.57 & 2.67 (br s, 8H), 2.86 (t, 2H), 3.93 (s, 2H), 4.21 (s, 3H), 4.28 (q, 2H), 7.23 (d, J=8.8 Hz, 1H), 8.16 (dd, J=8.8, 2.2 Hz, 1H), 8.46 (d, J=2.2 Hz, 1H) |
| Dimethylacetildenafil* (Goker et al., 2010) $C_{25}H_{34}N_6O_3$ | 466 467(M+H) | | (CDCl$_3$): 1.04 (t, 3H), 1.05 (d, 6H), 1.63 (t, 3H), 1.83 (t, 2H), 1.89 (m, 2H), 2.9 (m, 2H), 2.93 (t, 2H), 3.02 (m, 2H), 3.79 (s, 2H), 4.28 (s, 3H), 4.37 (q, 2H), 7.09 (d, J=8.8Hz, 1H), 8.16 (dd, J=8.8, 2Hz, 1H), 9.15 (d, J=2Hz, 1H), 10.85 (br.s, 1H). |
| Dapoxetine HCl (Li et al., 2009b) $C_{21}H_{23}NO$. HCl | 305 306(M+H) | | (CDCl$_3$): 2.2 (s, 6H), 2.34-2.45 (m, 1H), 2.59-2.7 (m, 1H), 3.55-3.63 (m, 1H), 3.86-3.9 (m, 1H), 4.02-4.4 (m, 1H), 6.62 (d, J=7.6Hz, 1H), 7.19-7.5 (m, 9H), 7.74-7.77 (m, 1H), 8.2-8.24 (m, 1H). |

*Detected in our lab. for the first time by us.

Table 1. Some synthetic PDE-5 inhibitors as analogues sildenafil and tadalafil were isolated by us.

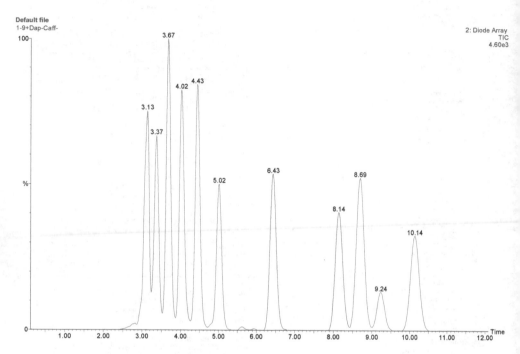

Fig. 9. HPLC chromotogram of *3.13* (Caffeine), *3.37* (Dimethylacetildenafil), *3.67* (Udenafil), *4.02* (Noracetildenafil), *4.43* (Dimethylsildenafil), *5.02* (Dimethylhomosildenafil), *6.43* (Hydroxythiohomosildenafil), *8.14* (Thiomethisosildenafil), *8.69* (Thiosildenafil), *9.24* (Dapoxetine HCl), *10.14* (Thiohomomethisosildenafil) (*RT* : Retention times as min).

Dapoxetin HCl. **Dapoxetine HCl**, is reported to be a selective serotonin reuptake inhibitor under investigation for the treatment of premature ejaculation (PE) (Li et al., 2009a). Since we have met this compound in one of the herbal drinks for adults, we have added it to our list. As it is well known, caffeine is very common for this kind of health supplements, it was added to chromatogram as well.

## 1.1 Natural aphrodisiacs

Formulas, $^1$H-$^{13}$C-NMR spectra, ESI(+) m/e values and LC-MS ion chromatograms of natural aphrodisiacs are given in Table 2 and Fig. 10.

L- **Arginine** is not herb, but a nonessential amino acid, it is found naturally in foods such as meat, dairy, poultry and fish, it also may be synthesized in the laboratory, in spite of there is insufficient evidence to rate effectiveness for male fertility and female sexual problem, but it is possible to see the ingredients of many aphrodisiacs, recently. It is also available as oral L-arginine supplements, which some product manufacturers market as a "natural Viagra" (Stanislavov & Nikolova, 2003).

Chromatographic Separation and Identification of Sildenafil and Yohimbine Analogues Illegally Added in Herbal Supplements

21

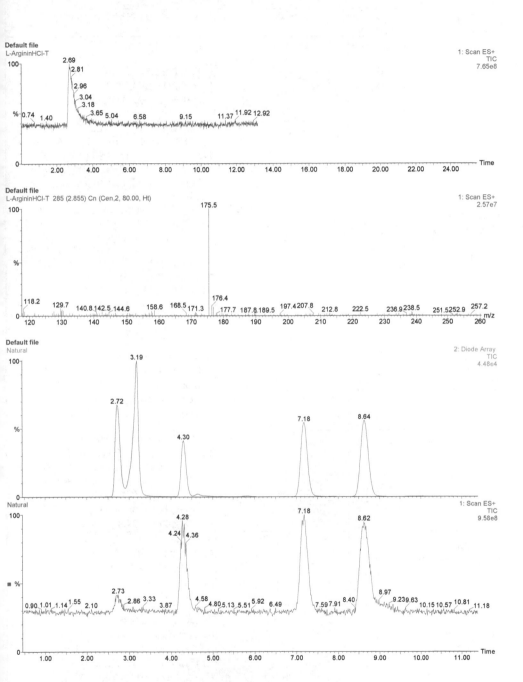

Fig. 10. HPLC-MS ion chromatogram of  2.69 (L-Arginin HCl), 2.72 (Inosine), 3.19 (Icariin), 4.3 (Yohimbin), 7.18 (Imperatonin), 8.64 (Osthole). (RT : Retention times as min).

| Compound name | M.w. ESI(+)m/z | Formulas | ¹H-NMR δ ppm |
|---|---|---|---|
| **L-Arginin HCl** (Stanislavov & Nikolova, 2003) $C_6H_{14}N_4O_2$ | 174 *175(M+H)* | | **(CD₃OD):** 1.27 (m, J=7.2Hz, 2H), 1.85 (q, J=7.2Hz, 2H), 3.21 (m, J=2.8Hz, 1H), 3.57 (t, J=6.4Hz, 1H), 4.6(s, 1H). |
| **Yohimbine** (Saini et al., 2010; Melnyk et al., 2011) $C_{21}H_{26}N_2O_3$ | 355 *356(M+H)* | | **(CD₃OD):** 1.16 (q, 12Hz, 1H), 1.35-1.42 (m, 1H), 1.46-1.56 (m, 2H), 1.65-1.69 (m, 1H), 1.90 (qd, 1H), 1.99 (dq, J=10Hz, 1H), 2.23 (t, J=11.2Hz, 1H), 2.34 (dd, J=12.0, 2.8Hz, 1H), 2.46 (dt, 1H), 2.62 (td, J=11.2, 4.4, 1H), 2.72 (dd, J=15.6, 4.8Hz, 1H), 2.90 (dd, J=11.2, 3Hz, 1H), 2.93-3.02 (m, 1H), 3.1 (dd, J=11.2, 4.8Hz, 1H), 3.4 (br.d, J=11.8, 1H), 3.76 (s, 3H), 4.23 (m, 1H), 6.95 (t, J=7.6Hz, 1H), 7.02 (t, 7.6Hz, 1H), 7.27 (d, J=7.6Hz, 1H), 7.36 (d, J=7.4Hz, 1H). **¹³C-NMR(CDCl₃):** 21.7, 23.3, 31.4, 34.3, 36.7, 40.7, 51.9, 52.3, 52.9, 59.8, 61.3, 66.9, 108.3, 110.7, 118.1, 119.4, 121.4, 127.4, 134.4, 135.9, 175.6. |
| **Inosine** (Deuster & Simmons, 2004) $C_{10}H_{12}N_4O_5$ | 268 *269(M+H)* | | **(DMSO-d₆):** 3.51 & 3.62 (m, 2H), 3.91 (q, J=4Hz, 1H), 4.09 (t, J=4Hz, 1H), 4.46 (t, J=5Hz, 1H), 5.1 (br.s, 1H), 5.15 (br.s, 1H), 5.4 (br.s, 1H), 5.83 (d, J=4.4Hz, 1H), 8.04 (s, 1H), 8.30 (s, 1H), 12.35 (br.s, 1H). |
| **Icariin** (Liu et al., 2005; Dell'Agli, 2008) $C_{33}H_{40}O_{15}$ | 676 *677(M+H)* | | **(CD₃OD):** 0.9 (d, J=6Hz, 3H), 1.63 (s, 3H), 1.72 (s, 3H), 3.15-3.75 (benzylic, ethylenic and sugar protons), 3.89 (s, 3H), 3.9-4.22 (sugar protons), 5.06 (d, J=7.2 Hz, 1H), 5.24 (t, 1H), 5.41 (d, J=1.6Hz, H), 6.65 (s, 1H), 7.01 (d, J=8.8Hz, 2H), 7.88 (d, J=8.8Hz, 2H) |
| **Osthole** (Liao et al., 2010) $C_{15}H_{16}O_3$ | 244 *245(M+H)* | | **(CDCl₃):** 1.67 (s, 3H), 1.84 (s, 3H), 3.53 (d, J=7.2Hz, 2H), 3.92 (s, 3H), 5.22 (t, J=7.2Hz, 1H), 6.23 (d, J=9.2Hz, 1H), 6.83 (d, J=8.6Hz, 1H), 7.29 (d, J=8.7Hz, 1H), 7.61 (d, J=9.2 Hz, 1H) |
| **Imperatorin** (Liao et al., 2010) $C_{16}H_{14}O_4$ | 270 *271(M+H)* | | **(CDCl₃):** 1.72 (s, 3H), 1.74 (s, 3H), 5.01 (d, 2H), 5.61 (t, J=7Hz, 1H), 6.37 (d, J=9.2Hz, 1H), 6.81 (d, J=2.4Hz, 1H), 7.36 (s, 1H), 7.69 (d, J=2.4Hz, 1H), 7.76 (d,J=9.2Hz, 1H) |

Table 2. Natural aphrodisiacs were isolated by us from some herbal supplements.

Chromatographic Separation and Identification of Sildenafil and Yohimbine Analogues Illegally Added in Herbal Supplements

23

**Yohimbine** is an alkaloid which are found naturally in *Pausinystalia yohimbe*, a classical aphrodisiac which has been recently revalued for its pro-sexual properties and extensively commercialized without control in some countries (Saini et al., 2010, Melnyk et al., 2011) . However, there is little evidence on its efficacy in the treatment of ED and it is, therefore, not currently recommended (Albersen et al., 2010).

**Inosine** is a nucleoside consisting of ribose and hypoxanthine, most commonly found in supplements that claim to be "energy promoters", naturally it is found in brewer's yeast and organ meats, however it can be synthesized in laboratory (Deuster & Simmons, 2004). Their nerve-stimulating action probably enhances sexual functions or help to remedy sexual functions when these are reduced due to degeneration of nerve tissue.

**Icariin** is one of the primary active component of *Epimedium* extracts, which has been used to treat impotence and improve sexual function by acting as PDE5 inhibitor as nafil derivatives (Liu et al., 2005; Dell'Agli, 2008). Since the illegally using of synthetic nafil derivatives have been prohibited, recently icariin has become very attractive by the drug manufacturers and suppliers, because of its natural character. Nowadays, the percentage ratio of Icariin in the commercial *Epimedium* extracts has been increased over 80-90. However there is no sufficient information about its pharmacological profiles and safety. Further investigation needs to be done to examine any benefits that could occur from supplements.

**Osthole** and **imperatorin,** coumarin compounds have been reported to exhibit various biological activities (Liao et al., 2010). It was reported that, both of them were found to help relax the corpus cavernosa of the penis, which would potentially help with blood flow, in phenylephrine-precontracted endothelium-intact rabbit corpus cavernosum (Chen et al., 2000; Chiou et al., 2001), however there is no information for human uses.

Herbal medications are being progressively utilized all over the world and it is believed that herbal remedies are not hazards, however some adverse reactions have been increased. Tribulus terrestris is frequently used because of its aphrodisiac effect. But there is an article (Talasaz, 2010), which reports a case of *T-terrestris*-induced hepatotoxicity, nephrotoxicity and neurotoxicity in an Iranian male patient.

In this text, we also present a new sildenafil analogue was found to have been added illegally to a herbal drinks marketed for the enhancement of sexual function. This analogue has never been found. Therefore, it prompted us to elucidate its structure. The structure was determined as 5-[5-[[(3,5-dimethyl-1-piperazinyl]sulfonyl]-2-propoxyphenyl]-1,6-dihydro-1-methyl-3-propyl-7*H*-pyrazolo [4,3-d]pyrimidin-7-one. Owing to the inclusion of a methylene group in dimethylsildenafil, the detected compound was called **Dimethyhomosildenafil**. The sample was purified with column chromatography. The IR, HPLC-/MS (ESI+), and completely assigned NMR data of **dimethylhomosildenafil** have been observed.

## 2. Material and methods

### 2.1 Equipments

Uncorrected melting points were measured on an Büchi B-540 capillary melting point apparatus. $^1$H (400 MHz) and $^{13}$C(100 MHz) NMR spectra were recorded employing a

VARIAN MERCURY 400 MHz FT spectrometer, with $CDCl_3$ as solvent. Chemical shifts ($\delta$) are in ppm relative to TMS. The LC/MS were taken on a Waters Micromass ZQ connected with Waters Alliance HPLC, using ESI(+) method, with C-18 column. Elemental analyses were performed by Leco CHNS-932. The infrared spectrum was recorded in the 600-3600 cm-1 range using a Jasco FT-IR-420 spectrometer and KBr pellets.

## 2.2 Extraction and isolation

The water contents of the alüminyum can (250 ml) were extracted with the mixture of dichloromethane-methanol (95:5) and evaporated, residue was directly carried out to a open column with silica gel 60 (0.04-0.063 mm) and eluted with dichloromethane-isopropanol (97:3). Fractions were collected and analyzed by TLC. All of the fractions having the target compound were collected and the solvent was evaporated and crystallization of the residue from ethanol gave 0.021g of white powder compound was obtained, m.p: 189-190°C, Anal. Calcd. for $C_{24}H_{34}N_6O_4S$. 0.5 HOH: C 56.34, H 6.90, N 16.42, S 6.27. Found: C 56.47, H 6.95, N 16.21, S 6.28.

## 2.3 Structure identification

### 2.3.1 NMR correlation data of dimethylhomosildenafil

Dimethylhomosildenafil was dissolved in $CDCl_3$ and subjected to 1D and 2D NMR spectroscopic analysis ([1]H, [13]C, DEPT, homo-COSY, HSQC and HMBC). The data are shown in Table 2.

### 2.3.2 Analysis condition of HPLC/MS

LC-MS coupled with positive and negative (ESI+) Electro Spray method was used to determine its molecular weight. The HPLC of LC/MS was carried out on a column XTerra® MS C-18 (4.6 X250 mm,5 µm) with Acetonitrile: Methanol:0.05 M Ammonium acetate in water (55:20:25) as mobile phase. The flow rate was 0.9 mL/min, the injection volume was 5 µL and the appropriate running time (at least 15 min). The eluate was monitored by a photo-diode array detector at 254 nm. The analytical condition of mass was as follows: capillary voltage :3.41 kV, cone voltage : 26 V, source temperature : 100 °C : desolvation temperature : 350°C. The HPLC chromatogram of dimethylhomosildenafil is given in Fig 2 with 5.02 min. $r_t$ values. This method was carried out all the given HPLC-MS analysis in this text.

**Table 3** shows the [1]H-NMR, [13]C-NMR, DEPT, COSY, HSQC and HMBC spectral data of isolated compound **1**, which were similar to that of dimethylsildenafil. The spectroscopic numbering used is given in Table 3. The difference of this compound, than dimethylsildenafil is related with ether protons connected to the C-19. Here is one more metyhlene group as propoxy. The 3,5-dimethyl protons of piperazine were observed at $\delta H$ 1.05 (d, 6H) as expected, in the HMBC spectrum the correlation of H-28,29/H-24,26, DEPT and HSQC results indicated that dimethyl group attached to C24-26. Since the 2,6-diequatorial methyl groups would be lower energetic form, so the configuration is established as a cis diequatorial methyl configuration as shown in Table 1 as it is in the

Chromatographic Separation and Identification of Sildenafil and Yohimbine Analogues Illegally Added in Herbal
Supplements

25

methisosildenafil (Reepmeyer & Avignon, 2009). The IR spectrum of dimethylhomosildenafil is given in Fig 11. This compound also must be put on the inspection list for illegal health-related substances because of the unknown safety and toxicity profile.

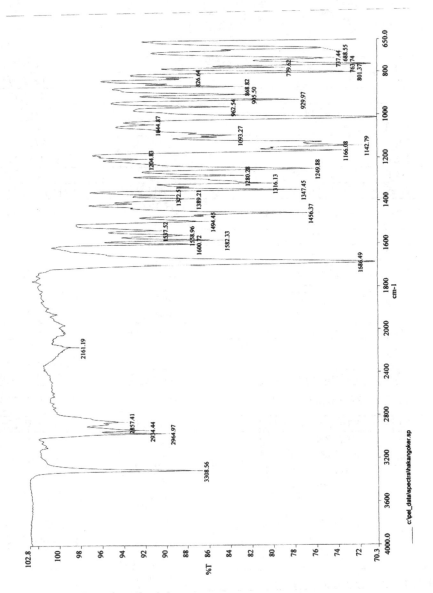

Fig. 11. IR spectrum of Dimethylhomosildenafil.

| No | $^{13}$C | $^1$H | DEPT[a] | COSY | HMBC |
|---|---|---|---|---|---|
| 3 | 146.7 | --- | | | C-3/H-12 C-3/H-11 |
| 5 | 147.2 | --- | | | C-5/H-15 |
| 6 | -- | 10.84(br.s,1H) | | | --- |
| 7 | 153.8 | --- | | | --- |
| 8 | 124.7 | --- | | | C-8/H-10 |
| 9 | 138.6 | --- | | | C-9/H-11 |
| 10 | 38.4 | 4.27(s,3H) | 3 | | --- |
| 11 | 27.99 | 2.93(t,2H,J=7.2 Hz) | 2 | H-11/H-12 | C-11/H-13, C-11/H-12 |
| 12 | 22.5 | 1.84(m,2H,J=7.6 Hz) | 2 | H-12/H-11, H-12/H-13 | C-12/H-13, C-12/H-11 |
| 13 | 14.3 | 1.01(t,3H,J=7.6 Hz) | 3 | H-13/H-12 | C-13/H-12, C-13/H-11 |
| 14 | 129.4 | --- | | -- | C-14/H-18 |
| 15 | 131.1 | 8.81(d,1H, $J_m$=2 Hz) | 1 | H-15/H-17 | C-15/H-17 |
| 16 | 121.2 | --- | | | C-16/H-18 |
| 17 | 131.8 | 7.84(dd,1H,$J_o$=8.8 Hz, $J_m$=2 Hz) | 1 | H-17/H-18, H-17/H-15 | C-17/H-15 |
| 18 | 113.2 | 7.16(d,1H,$J_o$=8.8 Hz) | 1 | H-18/H-17 | ---- |
| 19 | 159.7 | --- | | | C-19/H-15, C-19/H-17 C-19/H-18, C-19/H-20 |
| 20 | 72.1 | 4.26(t,2H,J=7.2 Hz) | 2 | H-20/H-21 | C-20/H-21, C20/H-22 |
| 21 | 22.54 | 2.05(m,2H,J=6.8 Hz) | 2 | H-21/H-20, H-21/H-22 | C-21/H-20, C-21/H-22 |
| 22 | 10.85 | 1.19(t,3H,J= 7.6 Hz) | 3 | H-22/H-21 | C-22/H-21, C-22/H-20 |
| 23,27 | 52.3 | 1.93(t,2H, J=10.4 Hz, axial) & 3.68(dd,2H, J=10.4 Hz, J=1.6 Hz equatorial) | 2 | H-23,H-27 axial/ H-24,H-26 | C-23,27/H-28,29 |
| 24,26 | 50.5 | 3.03(m, 2H, J=3.6 Hz, axial) | 1 | H-24,H-26/ H-23,H-27 axial H-24,H-26/ H-28,H-29 | C-24,26/H-28,29 C-24,26/H-23,27 |
| 28,29 | 19.5 | 1.05(d,6H,J=6.8 Hz) | 3 | H-28,H-29/ H-24,H-26 | ---- |

δ ppm in CDCl₃, $J$ in Hz  a) Number in DEPT is the number of attached protons.

Table 3. NMR data of Dimethylhomosildenafil.

Chromatographic Separation and Identification of Sildenafil and Yohimbine Analogues Illegally Added in Herbal Supplements

27

## 3. Conclusion

Oral PDE5 inhibitors are the treatment of choice for MED. The physiological mechanism of erection involves release of nitric oxide (NO) in the corpus cavernosum as a result of sexual stimulation. NO then activates the enzyme guanylate cyclase, which results in increased levels of cyclic guanosine monophosphate (cGMP), leading to smooth muscle relaxation in blood vessels supplying the corpus cavernosum and allowing inflow of blood. Nafil derivatives have no direct relaxant effect on isolated human corpus cavernosum, but enhance the effect of NO by inhibiting PDE5, which is responsible for degredation of cGMP in the corpus cavernosum. When sexual stimulation causes local release of NO, inhibition of PDE5 by nafil analogues causes increased levels of cGMP in the corpus cavernosum, resulting in smooth muscle relaxation and inflow of blood to the corpus cavernosum. This mode of action means that PDE5 inhibitors are ineffective without sexual stimulation. The PDE-5 inhibitors have helped many men with MED, and the FDA indicate that Sildenafil citrate, Vardenafil HCl and Tadalafil are safe and well-tolerated when taken as directed by men who have gotten approval from their doctors. These drugs must not been used without medical examination or prescription. PDE-5 inhibitors also increase the risk of a variety of cardiovascular diseases, including heart attack, myocardial infarction, and sudden death. The medication may interact with other drugs which should be mortal, e.g. synergic effect with alpha-blockers. Other side effects associated with PDE-5 drugs, such as priapism, severe hypotension, increased intraocular pressure and sudden hearing loss and blidness. The PDE-5 inhibitors must not buy over the internet or other non-standart source, otherwise, the men run several risk, such as it should be counterfeit product that does not have the legal structural compound which has been untested for safety or no same purity as the real drug.

## 4. Acknowledgment

We thank Prof. Dr. Erden Banoğlu (Gazi University, Faculty of Pharmacy, Ankara) for providing sample of Homothiomethisosildenafil which was also isolated from the herbal dietary supplement by him. Central Instrumental Analysis Lab. of Pharmacy, Faculty of Ankara University provided support for acquisition of the IR, NMR, HPLC-MS spectrometers and elemental analyzer used in this work.

## 5. References

Albersen, M.; Shindel, A. W.; Mwamukonda, K. B. & Lue, T. F. (2010). The future is today: emerging drugs for the treatment of erectile dysfunction, *Expert Opin. Emerging Drugs* 15:467-480.

Bell, A. S. & Palmer, M. J. (2011). Novel phosphodiesterase type 5 modulators: a patent survey (2008-2010), *Expert Opinion on Therapeutic Patents* 21:1631-1641.

Chen, J.; Chiou, W. F.; Chen, C. C. & Chen, C. F. (2000). Effect of the plant-extract osthole on the relaxation of rabbit corpus cavernosum tissue *in vitro*. *J. Urol.* 163:1975-80.

Chiou, W. F.; Huang, Y. L.; Chen, C. F. & Chen, C. C. (2001). Vasorelaxing Effect of Coumarins from *Cnidium monnieri* on Rabbit Corpus Cavernosum. *Planta Med.* 67:282-284.

Dell'Agli, M. D.; Galli, G. V.; Cero, E. D.;  Belluti,  F.; Matera, R. & Zironi, E. (2008). Potent inhibition of human phosphodiesterase-5 by Icariin Derivatives, *J. Nat. Prod.* 71: 1513-1517.

Deuster, P. A. & Simmons, R. G. (2004) Dietary Supplements and Military Divers. A Synopsis for Undersea Medical Officers, Available from: www.usuhs.mil/mem/hpl/DietarySupplementUMO.pdf.

Ge, X.; Li, L.; Koh, H. L. & Low, M.Y. (2011). Identification of a new sildenafil analogue in a health supplement, *J. Pharm. Biomed. Anal.* 56:491-496.

Göker, H.; Coskun, M. & Alp, M. (2010). Isolation and identification of a new acetildenafil analogue used to adulterate a dietary supplement: dimethylacetildenafil, *Turk J Chem.* 157-163.

Gratz, S. R.; Zeller, M.; Mincey, D. W. & Flurer, C. L. (2009). Structural characterization of sulfoaildenafil, an analog of sildenafil, *J. Pharm. Biomed. Anal.* 50:228-231.

Häberli, A.; Girard, P.; Low, M. Y. & Ge, X. (2010). Isolation and structure elucidation of an interaction product of aminotadalafil found in an illegal health food product, *J. Pharm. Biomed. Anal.* 53:24-28.

Hasegawa, T.; Takahashi, K.; Saijo, M.; Ishii, T.; Nagata, T.; Kurihara, M.; Haishima, Y.; Goda, Y. & Kawahara, N. (2009). Isolation and Structural Elucidation of Cyclopentynafil and *N*-Octylnortadalafil Found in a Dietary Supplement, *Chem Pharm. Bull.* 57:185-189.

Jung, J. Y.; Kim, S. K.; Kim, B. S.; Lee, S. H.; Park, Y. S.; Kim, S. J.; Choi, C.; Yoon, S. I.; Kim, J. S.; Cho, S. D.; Im,  G. J.; Lee, S. M.; Jung, J. W. & Lee, Y. S. (2008). The penile erection efficacy of a new phosphodiesterase type 5 inhibitor, mirodenafil (SK3530), in rabbits with acute spinal cord injury. *J. Vet. Med. Sci.*, 70:1199-204.

Keating, G. M. & Scott, L. J. (2003). Vardenafil: a review of its use in erectile dysfunction, *Drugs*, 63:2673-703.

Kesting, J. R.; Huang, J. & Sørensen, D. (2010). Identification of adulterants in a Chinese herbal medicine by LC-HRMS and LC-MS-SPE/NMR and comparative *in vivo* study standarts in a hypertensive rat model, *J. Pharm. Biomed. Anal.* 51:705-711.

Kurita, H.; Mizuno, K.; Kuromi, K.; Suzuki, N.; Ueno, C.; Kamimura, M.; Fujiwara, A.; Owada, K.; Ogo, N. & Yamamoto, M. (2008). Identification of Aminotadalafil and its Stereoisomers Contained in Health Foods Using Chiral Liquid Chromatography-Mass Spectrometry, *J. Health Science*, 54:310-314.

Lai, K. C.; Liu, Y. C.; Tseng, M. C.; Lin, Y. L. & Lin, J. H. (2007a). Isolation and Identification of a Vardenafil Analogue in a Dietary Supplement, *J. Food Drug Anal.* 15:220-227.

Lai, K. C.; Liu, Y. C.; Tseng, M. C.; Lin, Y. L. & Lin, J. H. (2007b). Isolation and Identification of a Vardenafil Analogue in a Functional Food Marketed for Penile Erectile Dysfunction, *J. Food Drug Anal.* 15:133-138.

Lam, Y. H.; Poon, W. T.; Lai, C. K.; Chan, A. Y. W. & Mak, T. W. L. (2008). Identification of a novel vardenafil analogue in herbal product, *J. Pharm. Biomed. Anal.* 46:804-807.

Langtry, H. D. & Markham, A. (1999). Sildenafil: a review of its use in erectile dysfunction, *Drugs*, 57:967-89.

Lee, H. M.; Kim, S. C.; Jang, Y. M.; Kwon, S. W. & Lee, B. J. (2011). Separation and structural elucidation of a novel analogue of vardenafil included as an adulterant in a dietary supplement by liquid chromatography-electrospray ionization mass spectrometry, infrared spectroscopy and nuclear magnetic resonance spectroscopy, *J. Pharm. Biomed. Anal.* 54:491-496.

Chromatographic Separation and Identification of Sildenafil and Yohimbine Analogues Illegally Added in Herbal
Supplements

29

Li, L.; Low, M. Y.; Ge, X.; Bloodworth, B. C. & Koh, H. L. (2009a). Isolation and structural elucidation of dapoxetine as an adulterant in a health supplement used for sexual performance enhancement, *J. Pharm. Biomed. Anal.* 50:724-728.

Li, L.; Low, M. Y.; Aliwarga, F.; Teo, J.; Ge, X. W.; Zeng, Y.; Bloodworth, B. C. & Koh, H. L. (2009b). Isolation and identification of hydroxythiohomosildenafil in herbal dietary supplements sold as sexual performance enhancement products, *Food Addit. Contam.* 26:145-151.

Li, S.; Ren, J.; Zhao, Y.; Lv, Q. & Guo, J. (2007). Pyrazolopyrimidinethione Derivatives, Salts and Solvates, thereof, preparation methods and use thereof, US 2007/0219220.

Liao, P. C.; Chien, S. C.; Ho, C. L.; Wang, E. I. C.; Lee, S. C.; Kuo, Y. H.; Jeyashoke, N.; Chen, J.; Dong, W. C.; Chao, L. K. & Hua, K. F. (2010). Osthole Regulates Inflammatory Mediator Expression through Modulating NF-kB, Mitogen-Activated Protein Kinases, Protein Kinase C, and Reactive Oxygen Species, *J. Agric. Food Chem*, 58: 10445-10451.

Lin, M. C.; Liu, Y. C.; Lin, Y. L. & Lin, J. H. (2008). Isolation and Identification of a Novel Sildenafil Analogue Adulterated in Dietary Supplements, *J Food Drug Anal.* 16:15-20.

Lin, M. C.; Liu, Y. C.; Lin, Y. L. & Lin, J. H. (2009). Identification of a Tadalafil Analogue Adulterated in a dietary Supplement, *J. Food Drug Anal.* 17:451-458.

Liu, R.; Li, A.; Sun, A.; Cui, J. & Kong, L. (2005). Preparative isolation and purification of three flavonoids from the Chinese medicinal plant *Epimedium koreamum* Nakai by high-speed counter-current chromatography, *J Chromatog. A, 1064*:53-57.

Melnyk, J. P. & Marcone, M. F. (2011). Aphrodisiacs from plant and animal sources-A review of current scienfitic literature, *Food Research International.* 44:840-850.

Meuleman, E. J. (2003). Review of tadalafil in the treatment of erectile dysfunction, *Expert Opin. Pharmacother,* 4:2049-56.

Park, H. J.; Jeong, H. K.; Chang, M. I.; Im, M. H.; Jeong, J. Y.; Choi, D. M.; Park, K.; Hong, M. K.; Youm, J.; Han, S. B; Kim, D. J.; Park, J. H. & Kwon, S. W. (2007). Structure determination of new analogues of vardenafil and sildenafil in dietary supplements, *Food Addit. Contam.* 24:122-129.

Piazza, G. A. & Pamukçu, R. (2001). Method of treating a patient having precancerous lesions with phenyl purinone derivatives, US 6,200,980.

Poon, W. T.; Lam, Y. H.; Lai, C. K.; Chan, A. Y. W. & Mak, T. W. L. (2007). Analogues of erectile dysfunction drugs: an under-recognised threat, *Hong Kong Med J.* 13:359-63.

Reepmeyer, J. C. & Woodruff, J. T. (2007). Use of liquid chromatography-mass spectrometry and a chemical cleavage reaction for the structure elucidation of a new sildenafil analogue detected as an adulterant in an herbal dietary supplement, *J. Pharm. Biomed. Anal.* 44:887-893.

Reepmeyer, J. C. & D'Avignon, D. A. (2009). Structure elucidation of thioketone analogues of sildenafil detected as adulterants in herbal aphrodisiacs, *J. Pharm. Biomed. Anal.* 49:145-150.

Saini, N. K.; Singhal, M.; Srivastava, B. & Sharma, S. (2010). Natural Plants Effective in Treatmen of Sexual Dysfunction: A Review, *The Pharma Research*, 4:206-224.

Salem, E. A.; Kendirci, M. & Hellstrom, W. J. (2006). Udenafil, a long acting PDE5 inhibitor for erectile dysfunction *Curr Opin. Investig. Drugs*, 7:661-9.

Stanislavov, R. & Nikolova, V. (2003) Treatment of Erectile Dysfunction with Pycnogenol and L-Arginine, *J Sex Marital Therapy*, 29:207-213.

Talasaz, A. H.; Abbasi, M. R.; Abkhiz, S. & Dahti-Khavidaki, S. (2010). *Tribulus terrestris*-induced severe nephrotoxicity in a young healthy male, *Nephrol Dial. Transplant*. 25: 3792-3793.

Toque, H. A.; Teixeira, C. E.; Lorenzetti, R.; Okuyama, C. E.; Antunes, E. & De Nucci, G. (2008). Pharmacological characterization of a novel phosphodiesterase type 5 (PDE5) inhibitor lodenafil carbonate on human and rabbit corpus cavernosum, *European Journal of Pharmacology*, 591:189-195.

Venhuis, B. J.; Zomer, G.; Hamzink, M.; Meiring, H. D.; Aubin, Y. & Kaste, D. (2011). The identification of a nitrosated prodrug of the PDE-5 inhibitor aildenafil in a dietary supplement: A Viagra with a pop, *J. Pharm. Biomed. Anal*. 54:735-741.

Wang, J.; Jiang, Y.; Wang, Y.; Zhao, X.; Cui, Y. & Gu, J. (2007). Liquid chromatography tandem mass spectrometry assay to determine the pharmacokinetics of aildenafil in human plasma, *J. Pharm. Biomed. Anal*. 44:231-235.

Wollein, U.; Eisenreich, W. & Schramek, N. (2011). Identification of novel sildenafil-analogues in an adulterated herbal food supplement, *J. Pharm. Biomed. Anal*. 56:705-712.

Zou, P.; Hou, P.; Low, M. Y. & Koh, H. L. (2006). Structural elucidation of a tadalafil analogue found as an adulterant of a herbal product, *Food Addit. Contam*. 23:446-451.

Zou, P.; Hou, P.; Oh, S. S. Y.; Chong, Y. M.; Bloodworth, B. C.; Low, M. Y. & Koh, H. L. (2008). Isolation and identification of thiohomosildenafil and thiosildenafil in health supplements, *J. Pharm. Biomed. Anal*. 47:279-284.

# 3

# Column Chromatography for Terpenoids and Flavonoids

Gülçin Saltan Çitoğlu and Özlem Bahadır Acıkara
*Ankara University,*
*Turkey*

## 1. Introduction

Natural products have coming from various source materials including terrestrial plants, terrestrial microorganisms, marine organisms, terrestrial vertebrates and invertebrates have importance as they provide an amazing source of new drugs as well as new drug leads and new chemical entities for further drug development (McCurdy & Scully, 2005; Chin et al., 2006). Morphine, vincristine, codeine, digitoxin, quinine, galantamine and taxol are just some of the typical examples of drugs that have been introduced from natural sources (Heinrich et al., 2004; Balunas & Kinghorn, 2005).

Natural products can be mainly divided into three groups such as primary metabolites, secondary metabolites and high molecular weight polymeric materials (Hanson, 2003). Primary metabolites including nucleic acids, amino acids, sugars; occur in all cells and play a central role in the metabolism and reproduction of the cells. High molecular weight polymeric materials such as cellulose, lignins and proteins take a part in the cellular structure. Secondary metabolites, small molecules which are not essential for the growth and development of the producing organism have importance because of their biological activities on other organisms. Natural product term refers to any naturally occurring compounds but in most cases mean secondary metabolite (Hanson 2003; Sarker et al., 2005). Secondary metabolites mainly consist of these following groups:

- Terpenoids and steroids
- Fatty acid derivatives and polyketides
- Alkaloids
- Phenylpropanoids
- Nonribozomal polypeptides
- Enzyme cofactors (McMurry, 2010).

## 2. Isolation of terpenoids and flavonoids by column chromatography

### 2.1 Terpenoids

Terpenoids are the most widespread, chemically interesting groups of secondary metabolites with over 30,000 known compounds including steroids (Wang et al., 2005; Umlauf, 2004). Many terpenes have biological activities and are used for the treatment of human diseases. Among the pharmaceuticals, the anticancer drug Taxol® and the antimalarial drug Artimesinin are two of the most renowned terpene-based drugs. Terpenoids and steroids are originated from isoprenoit ($C_5$) units derived from isopentenyl (3-methyl-3-en-1-yl) pyrophosphate. These $C_5$ units are linked together in a head-to-tail manner. Based on the number of the isoprene units, terpenoids are classified as monoterpenes ($C_{10}$), sesquiterpenes ($C_{15}$), diterpenes ($C_{20}$), sesterpenes ($C_{25}$), triterpenes ($C_{30}$), tetraterpenes ($C_{40}$) and polyterpenes (Wang et al., 2005). Mono and sesquiterpenes are the main constituents of the essential oils. However di- and triterpenoids which are not volatile compounds, generally found in gums and resins. Tetraterpenoids constitute a group of terpenoids called as carotenoids. This group includes carotenes, xanthophylls and carotenoic acids and the most important polyterpenoid is the rubber (Sameeno, 2007; Raaman, 2006).

Terpenoids are chemically lipid-soluble compounds and they can be extracted with petroleum ether generally. Sesquiterpene lactones, diterpenes, sterols and less polar triterpenoids extraction can be also performed by using benzene, ether and chloroform. Ethyl acetate and acetone extracts contain oxygenated diterpenoids, sterols and triterpenoids. Ethanol, methanol and water led to the extraction of highly oxygenated namely polar triterpenes as well as triterpenoid and sterol glycosides. Total extraction of the material carried out by any polar solvents such as acetone, aqueous methanol (%80) and aqueous ethanol and then re-extraction with hexane, chloroform and ethyl acetate is also leads to successive extraction of terpenoids and sterols (Harborne, 1998; Bhat, 2005).

Gas-Liquid Chromatography (GLC) is known as the best method for analyses of terpenoids especially mono- and sesquiterpenoids. Isolation of the mono- and sesquiterpenoids is also achieved by preparative GLC currently. Thin layer chromatography (TLC) can be used as another rapid, useful method for terpenoids and sterols detection with concentrated $H_2SO_4$ and heating due to all terpenoids and steroids (except carotenoids) are colourless compounds. TLC is also allowing to the isolation of various classes of terpenoids on silica gel and silver nitrate impregnated silica gel coated plates (Harborne, 1998; Bhat, 2005).

For isolation of various terpenoids especially sesqui-, di-, tri- and tetraterpenoids as well as sterols column chromatography is convenient method. As stationary phase silica gel, alumina, cellulose, sephadex, polyamid are used for the separation of different types of secondary metabolites but of this silica gel is the most extensively used adsorbent for particularly nonpolar and medium polar compounds including terpenoids and sterols. Silver nitrate impregnated silica gel is also provide separation of terpenoids containing unsaturation (Bhat, 2005; Sarker et al., 2006). Terpenoids are generally alicyclic compounds and isomerism is common. Due to the twisted cyclohexane ring, in chair form, different geometric conformations are possible depending on the substitution around the ring. Therefore, stereochemistry is commonly found in terpenoids. These structural features may cause artifact formation during isolation procedure (Harborne, 1998).

## 2.1.1 Monoterpenoids

The monoterpenoids which are composed of the condensation of two isoprene units are important components of essential oils (Gould, 1997). They are widely distributed in nature, most of which have been found in higher plants. However a number of halogenated derivatives have been isolated from marine organisms and have been found in defense and pheromonal secretions of insects. Monoterpenes have intensely purgent odors and they are the most common volatile compounds in plants responsible for fragrance and flavor. Therefore monoterpenes have a great commercial interest for food industry as well as perfume and fragrance industry (Robbers et al., 1996). Geraniol, a major component of geranium oil *(Pelargunium graveolens)* and its isomer, linalool; citral a major constituent of lemon oil, is obtained commercially from lemon grass oil *(Cymbopogon flexuosus)*, menthol is found in the essential oil of the field mint, *Mentha arvensis,* and possesses useful physiological properties including local anaesthetic and refreshing effects, terpineol and α-pinene are found in pine oil (turpentine), camphor, which was isolated from the camphor tree, *Cinnamomum camphora* are some of the typical examples of monoterpenoids (Hanson, 2003).

Isolation for mono- as well as sesquiterpenoids the classic procedure is obtaining essential oils by steam distillation. However extraction with non-polar solvents such as petroleum ether, ether and hexane can be preferred due to artifact formation at the raised temperatures (Harborne, 1998). Adsorbtion chromatography on silica gel is the simplest and most effective method for separation of terpenoids and GLC is used commonly for identification as well as isolation of the monoterpenoids. Column chromatography is also a valid method for fractionation of monoterpenoids. Isocratic elutions with solvents such as pentane, petroleum ether, hexane or gradient elution with mixtures of solvents in increasing polarity leads to successive isolation (Sur, 1991). Additionally, faster techniques of column chromatography such as flash chromatography may be preferred due to conventional column chromatography for separating procedure is time-consuming and frequently gives poor recovery owing to band tailing (Ikan, 1991).

The genus *Tagetes* belongs to the Asteraceae family. *Tagetes minuta* has essential oil and ocimenone which was reported to have mosquito larvicidal activity is the major constituent of this oil. Separation of the essential oil of *T. minuta* on silica gel column eluting with $Et_2O$ resulted in 10 fractions which the first four of these led to the isolation of (Z)-β-ocimene, dihydrotagetone, (Z)-tagetone (Z)-ocimenone and (E)-ocimenone. Additionally, 3,7-dimethyloct-1-en-6-one, 3,7-dimethyl-5-hydroxyoct-1-en-6-one and 3,7-dimethyloct-1,7-dien-6-one were obtained by rechromatography of fraction V respectively (Garg & Mehta, 1998). *Tagetes patula* L. another species from this genus allows to the isolation of acyclic monoterpene glycosides. Methanolic extract of the flowers was separated on silica gel column chromatography using $CHCl_3$-MeOH mixtures to yield 2-methyl-6-methylen-2,7-octadiene 1-O-β-D-glucopyranoside (Garg et al., 1999).

R=H        3,7-dimethyloct-1-en-6-one
R=OH       3,7-dimethyl-5-hydroxyoct-1-en-6-one
R=H, Δ⁷    3,7-dimethyloct-1,7-dien-6-one

2-methyl-6-methylen-2,7-octadiene 1-O-β-D-glucopyranoside

*Artemisia tridentata* ssp. *vaseyana*, *Artemisia cana* ssp. *viscidula* and *Artemisia tridentata* ssp. *spiciform* led to the isolation of monoterpenoids. For each plant sample, air-dried ground leaves and flower heads were extracted with pentane in soxlet extractor. The extracts were concentrated in vacuo, and vacuum short path distilled to yield yellowish oils. The each oil isolated from *A. tridentata* ssp. *vaseyana*, *A. cana* ssp. *viscidula* and *A. tridentata* ssp. *spiciformis* was separated by flash chromatography on silica gel using 19:1 hexane-EtOAc followed by 4:1 hexane-EtOAc except for the oil isolated from *A. tridentata* ssp. *spiciformis* which was flash chromatographed with 9:1 hexane-EtOAc as the second solvent system. *A. tridentata* ssp. *vaseyana* essential oil was separated into three major fractions by column chromatography. GC analysis of the first chromatographic fraction indicated the presence of four constituents. Two major compounds were isolated and identified by comparison of spectral data to literature values. The first was 1,8 cineole (eucalyptol) and the second was *trans*-3-(1-oxo-2-methyl-2-propenyl)-2,2-dimethylcyclopropylmethanol which is thermally unstable and isolated as its GC artifact 2,4-diisopropenyl-5*H*-furan. The third compound was 2,2-dimethyl-6-isopropenyl-2*H*-pyran and the fourth was 2,3-dimethyl-6-isopropyl-4*H*-pyran. Thujone was determined as the major components of the second fraction. In the third fraction sabinol, chrysanthemol, chrysanthemyl acetate, fraganyl acetate, fraganol and 2-isopropenyl-5-methylhexa-*trans*-3,5-diene-1-ol were identified as the major components. Four major constituents obtained from *Artemisia cana* ssp. *viscidula* chromatographic separation and the compounds were identified as santolina triene, α-pinene, rothrockene and artemisia trien was found to be in first fraction. The second of four chromatographic fractions gave five components; artemiseole, 1-8 cineole, trans-3-(1-pylmethanol) which is thermally unstable and isolated as its GC artifact 2,4-diisopropenyl-5*H*-furan, 2,2-dimethyl-6-isopropenyl-2*H*-pyran, 2-isopropenyl-5-methylhexa-*trans*-3,5-diene-1-ol. Crysanthemal as well as eight compounds identified as camphor, isolyratol, lyratol, chrysanthemol, chrysanthemyl acetate, fraganyl acetate, fraganol and 2-isopropenyl-5-methylhexa-*trans*-3,5-dien-1-ol eight compounds were isolated by preparative GC from the third and fourth chromatographic fraction of *A. cana* ssp. *viscidula* respectively. Volatile oils obtained from the neutral pentane extract of *A. tridentata* ssp. *spiciformis* were flash chromatographed into five separate fractions to give mainly known compounds. The first fraction containing hydrocarbons was analyzed by preparative GC and contained santolina triene, α-pinene, camphene and rothrockene. Fraction two contained artemiseole, 1,8-cineole and oxidosantolina triene, fraction three contained lyratal, thujone and camphor and fraction four contained sabinyl acetate and chrysanthemyl acetate. The final alcohol fraction contained α-santolina alcohol, sabinol, chrysanthemol, isolyratol, lyratol and lavandulol (Gunawardena et al., 2002).

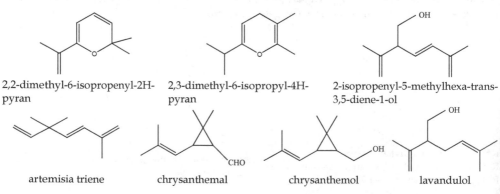

2,2-dimethyl-6-isopropenyl-2*H*-pyran

2,3-dimethyl-6-isopropyl-4*H*-pyran

2-isopropenyl-5-methylhexa-trans-3,5-diene-1-ol

artemisia triene

chrysanthemal

chrysanthemol

lavandulol

| | | | |
|---|---|---|---|
| 1,8-cineole | thujone | sabinol | santoline triene |
| α-pinene | artemiseole | camphor | camphene |

*Artemisia annua* L. (sweet wormwood; Compositae), the source of the potent anti-malarial drug artemisinin, has been the subject of extensive phytochemical investigations over the past two decades. Sesquiterpenoids are the most abundant compounds in this species. Additionally, monoterpenoids, diterpenoids and flavonoids have been isolated. The seeds of *A. annua* were frozen in liquid $N_2$ and converted into a powder by grinding with a pestle and mortar. The powder was repetitively extracted with $CH_2Cl_2$, dried ($MgSO_4$) and solvent removed under reduced pressure to yield an aromatic green gum which was subjected to gradient (hexane-EtOAc 5 to 100%) column chromatography yielding 32 crude fractions. The crude fractions from column chromatography were further purified by repeated prep HPLC, using *n*-hexane-EtOAc-HOAc in varying proportions, according to the polarity of the crude fraction which was under investigation. Three monoterpenoids which was identified as 4-hydroxy-2-isopropenyl-5-methylene-hexan-1-ol, 1,10-oxy-α-myrcene hydroxide and 1,10-oxy-β-myrcene hydroxide, was isolated together with sesquiterpenoids and diterpenoid (Brown et al., 2003).

| | | |
|---|---|---|
| 4-hydroxy-2-isopropenyl 5-methylene-hexan-1-ol | 1,10-oxy-α-myrcene hydroxide | 1,10-oxy-β-myrcene hydroxide |

*Artemisia judaica* (L.) is a perennial fragrant shrub which grows widely in the deserts and Sinai Peninsula of Egypt. Mixture of the dry leaves of *A. judaica*, *A. monosperma* and *A. hera alba* is very common anthelmintic drug in the most of North African and Middle East countries under Arabic name of Shih. It has been reported that *A. judaica* essential oil has two major constituents as piperitone and *trans*-ethyl cinnamate. Piperitone showed insecticidal activity against *Callosobruchus maculatus*. Piperitone was isolated from aerial parts of the plant. Dried and powdered aerial parts of *A. judaica* were hydrodistilled in a Clevenger-type apparatus. The essential oil, pale yellow, was obtained and was dried over anhydrous sodium sulphate. The essential oil was chromatographed on silica gel column using hexane, 2.5% acetone-hexane, 10% acetone-hexane and acetone solvent system to give 45 fractions of 200 ml of each. The resulting fractions were concentrated under reduced pressure and examined by TLC to offer two main fractions. Fractions 10–17 was subjected to silica gel column eluted with chloroform to offer of piperitone (Abdelgaleil et al., 2008).

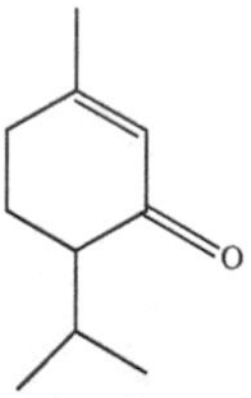

piperitone

α-Pinene type monoterpenoids have been isolated from the aerial parts of *Artemisia suksdorfii*. Extraction was performed by using $CH_2Cl_2$ at room temperature and after concentrated, subjected to column chromatography on silica gel. Gradient mixtures of hexane and $CH_2Cl_2$ and then $CH_2Cl_2$ and methanol were used for elution to obtain five fractions. Fraction 3 and 4 were separated on silica gel column and eluted with n-hexane-$CH_2Cl_2$ to yield Fraction 1-A and 1-B. Further purification with elution by using hexane: $CH_2Cl_2$-MeOH (5:7:0.5) of fraction 1-A on sephadex LH-20 column resulted in isolation of two α-pinene-type monoterpenoids; 7-hydroxymyrtenol and 7-hydroxymyrtenal (Mahmoud & Ahmed, 2006).

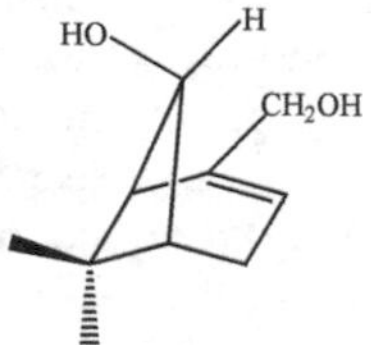

7-hydroxymyrtenol 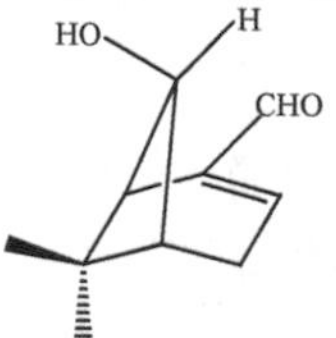                    7-hydroxymyrtenal

The *Mentha* genus (Labiatae) has importance as sources of essential oil production in the world. Additionally some members of this genus are used as herbal teas and spices. Menthone, mentol, menthyl acetate, neo-isomenthyl acetate, 1-menthyl-β-D-glucopyranosyl, 1-menthyl-6'-O-acetyl-β-D-glucopyranosyl have been identified mainly in various species. *Mentha longifolia* is widely distributed in Eurasia and tropical Asia. Longifone, a new chloro derivative of menthone was isolated from the aerial parts of the *M. longifolia*. After concentrated to dryness methanolic extract was re-diluted in water and then extracted with EtOAc. EtOAc soluble part subjected to silica gel column chromatography using hexane, hexane-CHCl₃, CHCl₃ and CHCl₃-MeOH as mobile phase. Fraction that eluted with 20% CHCl₃ in hexane yielded with longifone (M.S.Ali et al., 2002).

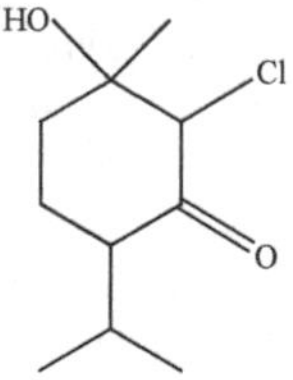

longifone

*Passiflora quadrangularis* L. (badea) is widely distributed in some regions of tropical America. Fruits of the plant are used locally to prepare different kinds of drinks with a pleasant and refreshing aroma. Two oxygenated monoterpenoids were isolated from the fruits extract whose odour strongly resembled the aroma of fresh fruit. After fruits were blended, pentane-$CH_2Cl_2$ (1:1) was used for extraction. Obtained organic extract was dried over $Na_2SO_4$ and concentrated. The concentrated extract was subjected to silica gel column chromatography with the following eluant solutions; pentane–$Et_2O$ (9:1), pentane–$Et_2O$ (2:1), pentane–$Et_2O$ (1:1), pentane–$Et_2O$ (1:2) and $Et_2O$ to obtain five fractions, fraction I to V, respectively. Fraction III and fraction V were further fractionated by column chromatography over silica gel using hexane–AcOEt (7:1 - 4:1) as eluents to yield (2E)-2,6-dimethyl-2,5-heptadienoic acid and (3S)-(5E)-2,6-dimethyl-5,7-octadiene-2,3-diol respectively. To obtain glycoside of (2E)-2,6-dimethyl-2,5-heptadienoic acid and (3E)-3,7-dimethyl-3-octene-1,2,6,7-tetrol fruits pulp was blended in a mixer with the pH adjusted to 7.0 with 5 N NaOH. After centrifugation supernatant was subjected to XAD-2 column chromatography and eluted with water then MeOH. The MeOH eluate was fractioned by multilayer coil counter current chromatography using $CHCl_3$-MeOH-$H_2O$ (7:13:18) to yield fifty fractions. Fractions 20-30 were rechromatographed on silica gel column chromatography using $CHCl_3$-MeOH (7:1, 5:1, 4:1, 3:1) mixtures. Fractions eluted with $CHCl_3$-MeOH (7:1) gave (2E)-2,6-dimethyl-2,5-heptadienoic acid-β-D-glucopyranosyl ester. (3E)-3,7-dimethyl-3-octene-1,2,6,7-tetrol was obtained from fractions eluted with $CHCl_3$-MeOH (5:1) after column chromatography on silica gel using EtOAc-BuOH-$H_2O$ (8:2:5) (Osorio et al., 2000).

(2E)-2,6-dimethyl-2,5-heptadienoic acid

(2E)-2,6-dimethyl-2,5-heptadienoic acid β-D-glucopyranosyl ester

(3S)-(5E)-2,6-dimethyl-5,7-octadiene-2,3-diol

(3E)-3,7-dimethyl-3-octene-1,2,6,7-tetrol

*Alpinia kadsumadai* Hayata is native to Hainan Island in Southern to China and has traditional usage in Chinese medicine as an antiemetic and for treatment of stomach disorders. Aerial parts of the *A. kadsumadai* contain monoterpenoids, sesquiterpenoids, diarylheptanoids, chalcones and flavonoids. $CH_2Cl_2$ extract of the aerial parts were subjected to column chromatography on silica gel and eluted with hexane-EtOAc mixture in increasing polarity. Fractions eluted with 15% EtOAc-hexane gives 1-terpinen-4-ol (Ngo &Brown, 1998).

1-terpinen-4-ol

*Carum carvi* L., Caraway (Umbelliferae) has been used as a popular aromatic herb and spice since antiquity and has been cultivated in Europe since the Middle Ages. Its fruit has been used for medicine and in cooking, and is listed in British, German and European pharmacopoeia. For medicinal purpose, it is used to relieve flatulent indigestion, colic and bronchitis. Studies on the fruits have revealed that the essential oil, and many monoterpenoids (d-carvone (main; 50–60%), l-limonene, carvacrol, trans-carveol, d-dihydrocarveol, l-dihydrocarveol, etc.) have been identified as the constituents. It was reported that monoterpeneoids have also been identified in the water soluble extracts of caraway. Commercial caraway was extracted with 70% methanol at room temperature. After evaporation of the solvent, the residue was partitioned into ether–water, EtOAc-water. Removal of the solvent from each phase gave the ether, EtOAc and aqueous extracts, The aqueous extract was chromatographed over Amberlite XAD-II ($H_2O$–MeOH). The methanol eluate was subjected to Sephadex LH-20 (MeOH) to give eight fractions (A–H). Fraction B was chromatographed over silica gel ($CHCl_3$–MeOH–$H_2O$ (17:3:0.2-4:1:0.1-7:3:0.5)-MeOH) to give 14 fractions ($B_1$–$B_{14}$). Fraction $B_3$ was passed through a Lobar RP-8 column (MeCN–$H_2O$ (3:17)) to give nine fractions ($B_{3-1}$–$B_{3-9}$), and fraction $B_{3-5}$ was subjected to HPLC (ODS, MeCN–$H_2O$ (3:37)). The main fraction was acetylated with $Ac_2O$ and pyridine, and the acetylated fraction was subjected to HPLC (ODS, MeCN–$H_2O$ (2:3)) to give two fractions. These two fractions were deacetylated by heating in a water bath with 5% $NH_4OH$–MeOH for 2 h, and passed through Sephadex LH-20 (MeOH) to give (1R, 2R, 4S)-p-menthane-1,2,8-triol and Rel-(1S, 2S, 4R, 8R)-p-menthane-1,2,8-triol. Fraction $B_{3-7}$ was subjected to HPLC (ODS, MeCN–$H_2O$ (1:9)) to give (1S, 2S, 4S, 8R)-p-menthane-2,8,9-triol; (1S, 2S, 4S, 8S)-p-menthane-2,8,9-triol; (1S, 2R, 4R, 8R)-p-menthane-2,8,9-triol and (1S, 2R, 4R, 8R)-p-menthane-2,8,9-triol. Fraction $B_{3-7}$ was subjected to HPLC (ODS, MeCN–$H_2O$ (1:9) to give Rel-(1R, 2S, 4R, 8S)-p-menthane-2,8,9-triol; Rel-(1R, 2S, 4R, 8R)-p-menthane-2,8,9-triol; Rel-(1S, 2S, 4R, 8S)-p-menthane-2,8,9-triol and Rel-(1R, 2S, 4R, 8R)-p-menthane-2,8,9-triol. From this mixture, Rel-(1R, 2S, 4R, 8R)-p-menthane-2,8,9-triol was isolated by silica gel column chromatography ($CHCl_3$–MeOH–$H_2O$ (9:1:0.1)). Fraction $B_9$ was subjected to a Lobar RP-8 column (MeCN–$H_2O$ (3:17)) and HPLC (CHA, MeCN–$H_2O$ (9:1)) to give (1S, 2R, 4R, 8S)-p-menthane-2,8,9-triol-9-O-β-D-glucopyranoside respectively. Fraction $B_{11}$ was also subjected to a Lobar RP-8 column (MeCN–$H_2O$ (3:17)) and HPLC (CHA, MeCN–$H_2O$ (9:1)) to give (1S, 2R, 4S)-p-menthane-1,2,8-triol-8-O-β-D-glucopyranoside respectively. Fraction $B_{10}$ was passed through a Lobar RP-8 column (MeCN–$H_2O$ (3:17)) to give eight fractions ($B_{10-1}$–$B_{10-8}$). Fraction $B_{10-4}$, fraction $B_{10-5}$ and $B_{10-7}$ were subjected to HPLC (CHA, MeCN–$H_2O$ (9:1)) to give (1S, 2S, 4R)-p-menthane-1,2,10-triol-2-O-β-D-glucopyranoside, (1S, 2S, 4R, 8R)-p-menthane-1,2,9-triol-2-O-β-D-glucopyranoside, and (1S, 2R, 4R, 8S)-p-menthane-2,8,9-triol-4-O-β-D-glucopyranoside respectively (Matsumura et al., 2001).

(1S, 2S, 4S, 8R)-p-menthane 2,8,9-triol

(1R, 2R, 4S)-p-menthane-1,2,8-triol

(1S, 2R, 4R, 8S)-p-menthane-2,8,9-triol-4-O-β-D-glucopyranoside

(1S, 2R, 4S)-p-menthane-1,2,8-triol-8-O-β-D-glucopyranoside

Carvacrol, one of the essential oil components of *Monarda punctata* was obtained as a lipase inhibitor. Lipase is an enzyme that hydrolyzes triacylglycerols (TGs). The digestion and absorption of natural lipids begins with hydrolysis by pancreatic lipase. The activity of this enzyme greatly affects the metabolism of fat and the concentration of TG in blood. Recently, inhibitors of lipase and lipid absorption have been isolated from natural sources with the aim of preventing and treating metabolic syndrome. *Monarda punctata* L. (Lamiaceae) is a traditional herbal medicine of North American Indians used as a remedy for colds and a treatment for nausea, vomiting, and rheumatic pains. Carvacrol was obtained from *M. punctata* essential oil. Powdered whole plants of *M. punctata* were extracted with acetone–$H_2O$ (80:20). The extract was suspended in $H_2O$, and extracted with $Et_2O$. The ether extract was suspended in EtOH–$H_2O$ (8:2), and extracted with hexane. The hexane soluble extract was passed through a silica gel column yielding 14 fractions, one of which, eluted with $CHCl_3$-MeOH (99:1) was an essential oil fraction whose major component was carvacrol. The $H_2O$ layer extract was a red-brown syrup. It was dissolved again in $H_2O$, and the aqueous solution was passed through a porous polymer gel column and eluted with $H_2O$, MeOH–$H_2O$ (80:20) and MeOH. The MeOH–$H_2O$ (80:20) eluate was subjected to on a reversed-phase column chromatography using ODS (Cosmosil 140$C_{18}$-OPN) and eluted with 20%, 30%, 40%, 50%, 60%, 80% MeOH in $H_2O$, and MeOH (fractions 1A–1G). Fraction 1C was subjected to YCCC and HPLC, yielding monoterpenoid glycosides monardins (A-F) together with flavonoids and some other phenolic compounds (Yamada et al., 2010).

| carvacrol | 2-methyl-5-(1-methylethyl) phenyl β-D-glucopyranoside | monardin C | monardin E |

## 2.1.2 Sesquiterpenoids

Sesquiterpenoids are generally synthesized by the mevalonate pathway and they are formed from three $C_5$ units (Dewick, 2009). The sesquiterpenoids which widely distributed in nature have similar properties to monoterpenoids and generally be less volatile than monoterpenoids (Robbers et al., 1996; Heinrich et al., 2004; Dewick, 2009). α-Bisabolol, a major component of matricaria (*Matricaria chamomilla*); γ-bisabolene which contributes to the aroma of ginger (*Zingiber officinale*); costunolide a bitter principle found in the roots of chicory (*Cichorium intybus*); parthenolide, an antimigraine agent in feverfew are some of the naturally occurring sesquiterpenoids (Dewick, 2009).

Artemisinin, antimalarial drug, is one of the most important sesquiterpene obtained from sweet wormwood, *Artemisia annua* L. (Asteraceae). This plant is known as Qinghao and has been used for the treatment of fevers and malaria in China for many centuries. The methyl ether of dihydroartemisinin that was developed for enhancing the solubility of the compound whilst retaining the biological activity is used clinically (Heinrich et al., 2004; Klayman et al., 1984). Artemisinin isolated from the leaves of *A. annua*. Petroleum ether extract of the plant was chromatographed on silica gel (70-230 mesh) using 7.5% EtOAc in

CHCl$_3$ solvent system. Artemisinin was isolated as fine white crystals in second fraction (Klayman et al., 1984).

artemisinin

*Curcuma zedoaria* Roscoe (Zingiberaceae), also known as white turmeric, zedoaria or gajutsu, has been used for menstrual disorders, dyspepsia, vomiting and for cancer traditionally. This plant has also been used for the treatment of cervical cancer in Chinese traditional medicine. *C. zedoaria* rich source of essential oils and many sesquiterpenoids as well as curcuminoids have been isolated (Syu et al., 1998; Lobo et al., 2009). Zedoarol, germacrone, curdione, β-elemene and curzeone are sesquiterpenoids which were isolated from *C. zedoaria* Shiobara et al (1986). CH$_2$Cl$_2$ extracts of the plant was chromatographed on silica gel using hexane-EtOAc gradient. Fraction 11 was rechromatographed after evaporation on Sephadex LH-20 using CHCl$_3$-MeOH (1:1) to afford curzeone. Zedoarol obtained from separation of fraction 21 on silica gel using hexane-EtOAc (97:3) and sephadex LH-20 (CHCl$_3$-MeOH, 1:1) respectively. Further separation of fraction 63 on silica gel (CH$_2$Cl$_2$) followed by on Sephadex LH-20 (CHCl$_3$-MeOH, 1:1) led to the isolation of germacrone (Shiobara et al., 1986). Ar-Turmerone and β-turmerone were obtained also from *C. zedoaria* rhizomes. Methanolic extract of the rhizomes were prepared and then was suspended in distilled water and partitioned with CHCl$_3$. After evaporation CHCl$_3$ extract was subjected to column chromatography on silica gel and eluted with gradient mixtures of CHCl$_3$ and MeOH (20:1 to 1:1) to afford eight fractions. Further separation was performed on fraction 2 on silica gel by column chromatography eluting with CHCl$_3$ and MeOH in increasing polarity (100:1 to 1:1) to obtain five subfractions. Subfraction 2 led to the isolation of ar-turmerone and β-turmerone after preparative TLC (hexane-EtOAc, 97:3) (Hong et al., 2001).

zedoarol                              germacrone                              curzeone

curdione                    β-elemene                    ar-turmerone                    β- turmerone

*Daucus carota* L. (Umbelliferae) is widely distributed in the world. Fruits of the plant have been used commonly as a medicine for the treatment of ancylostomiasis, dropsy, chronic kidney diseases and bladder afflictions in Chinese medicine. Flavonoids, anthocyanins, chromones, coumarins as well as sesqiterpenoids have been isolated from the *D. carota*. Sesquiterpenoids were isolated from the fruits of the plant. Fruits of the plant were extracted with 95% aqueous EtOH. Partition of the EtOH extract was performed with petroleum ether, $CHCl_3$, EtOAc and BuOH respectively, after suspended in $H_2O$. The $CHCl_3$ layer was fractionated on silica gel by column chromatography with gradient elution of petroleum ether-EtOAc (7:1-1:7) to yield 10 fractions. Fraction 6 was chromatographed on silica gel column chromatography petroleum ether-EtOAc (3:1-1:1) to give 5 subfractions. Subfraction 3 was separated by Sephadex LH-20 with MeOH followed by silica gel $CHCl_3$-$Et_2O$ (8:1) to obtain daucusol. Daucuside, a sesquiterpenoid glycoside was also isolated from the BuOH layer. Column chromatography on silica gel with eluting gradient of $CHCl_3$-$Et_2O$ (15:1-8:1) allows obtaining eight fractions. Repeated column chromatography on silica gel with $CHCl_3$-MeOH (9:1) provides five subfractions. Daucuside was obtained by purification of subfraction 2 using preparative HPLC (20% aqueous MeOH) (Fu et al., 2010).

daucusol                                                    daucuside

*Tanacetum parthenium* (L.) Schultz. Bip. (Asteraceae) known as feverfew, leaves have been used as antipyretic or febrifuge. Recent studies have revealed that feverfew effective in migraine by substantially reducing the frequency and severity of the headache. Responsible compound appears to be parthenolide, a germacranolide type sesquiterpenoid lactone. Parthenolide was reported to act as serotonin antagonist resulting in an inhibition of the release of serotonin from blood platelets. Parthenolide was isolated from the leaves of *T. parthenium*. Extraction of the plant material was done after exhaustive maceration in ethanol-water (90:10) at room temperature in the dark. The extract was filtered, evaporated under vacuum, and lyophilized. Subsequently, the hydroalcoholic extract was chromatographed on a silica gel column with hexane, $CH_2Cl_2$, EtOAc, MeOH, and MeOH-$H_2O$ (90:10). Next, the $CH_2Cl_2$ fraction was chromatographed on a silica gel column with different mixtures of solvents. The hexane- $CH_2Cl_2$ fraction resulted in isolation of parthenolide (Robbers et al., 1998; Tiuman et al., 2005).

parthenolide

*Valeriana officinalis* L. (Valerianaceae) known as valerian, is used in the treatment of conditions involving nervous excitability, such as hysterical states and hypochondriasis as well as insomnia. The main components of the valerian roots are the iridoids and volatile oil. Volatile oil contains numerous compounds including monoterpenoids, sesquiterpenoids (Heinrich et al., 2004). Valerenane sesquiterpeneoids were isolated from a $CH_2Cl_2$ extract of the Valeriana roots. Extract was concentrated and combined with 2% NaOH. Then aqueous layer were acidified and extracted with petroleum ether-$Et_2O$ (2:1) to obtain extract A . The remaining $CH_2Cl_2$ extract was washed with $H_2O$ and concentrated under vacuum. The residue was dissolved in petroleum ether and concentrated after filtration to yield extract B. Isolation procedure of extract B was performed on silica gel column using petroleum ether-$Et_2O$ mixture in increasing polarity. Fractions 26-34 contains Z-valerenyl acetate and E-/Z-valerenyl isovalerate which were isolated by means of preparative TLC (hexane-$Et_2O$, 4:1) followed by preparative GC. Extract A was dissolved in pentane and stored at -20 ºC after evaporation. Separation of the extract A was done using column chromatography on silica gel and eluted with petroleum ether-$Et_2O$ mixture from 10 upto 100%. Valerenic acid and hydroxyvalerenic acid were obtained from the 20 % $Et_2O$ and 100% $Et_2O$ fractions respectively by preparative TLC (hexane-$Et_2O$, 1:4). Acetoxy valerenic acid was also obtained from remaining pentane extracts by preparative TLC using hexane-$Et_2O$ (3:2) (Bos et al., 1986).

| R1 | R2 | |
|------|-----|------------------------|
| CHO | H | valerenal |
| COOH | H | valerenic acid |
| COOH | OH | hydroxyvalerenic acid |
| COOH | OAc | acetoxyvalerenic acid |

| R | |
|--------------------------|-------------------------|
| H | valerenol |
| Ac | E-valerenyl acetate |
| CO-CH₂-CH(Me)₂ | E-valerenyl isovalerate |

R
Ac                    Z-valerenyl acetate
CO-CH₂-CH(Me)₂ Z-valerenyl isovalerate

## 2.1.3 Diterpenoids

The diterpenoids are a large group of non-volatile terpenoids based on four isoprene units (Robbers et al., 1996).

Many of the diterpenoids are wood resin products. Abietic acid is the major component of colophony. Gibberellins are the best known plant hormones, taxol is used in the treatment of breast and ovarian cancer obtained from Pacific yew (*Taxus bravifolia*), are the known diterpenoids from nature (Hanson, 2003).

Taxol (Paclitaxel), a diterpenoid isolated from *Taxus brevifolia* Nutt. (Taxaceae), also known as the Pacific yew, used clinically in ovarian, breast, lung and prostate cancer effectively (Robbers et al., 1996; Wall & Wani, 1996; Heinrich et al., 2004). Taxol has been isolated from *T. brevifolia* using many different chromatographic techniquies and one of the way was described by Senihl et al. (1984) which employs normal phase chromatography columns for the separation procedures and includes multiple (seven) steps respectively as follows; 1. Extraction with alcohol and concentration, 2. Partition between water and dichloromethane, 3. Filtration chromatography, 4. Silica column chromatography, 5. Alumina chromatography. 6. Medium pressure silica column chromatography, 7. Preparative HPLC. For the other analogues, two or three other chromatographic columns, followed by preperative HPLC, were used.

taxol

*Ginkgo biloba*, (Ginkgoaceae) one of the oldest living plant species dating back more than 200 million years, is often reffered to as "living fossil". Medicinal uses of *G. biloba* was described in the Chinese Materia Medica more than 2.000 years ago and is used to treat memory and cognitive impairment, for which it has moderate efficacy with minimal side effects. The ginkgo leaves contain many active ingredients, including flavonoids, terpene trilactones (Jacobs & Browner, 2000). Triterpene lactones namely ginkgolides and flavonoids are believed to be associated with pharmacological activities of *G. biloba* extracts. While flavonoids can be obtained from many other plants, ginkgolides are unique compounents of the *G. biloba* extracts (Jaracz et al., 2004). It has been reported that flavone glycosides of the rutin type probably reduced the capillar fragility and reduce blood vessel which may prevent ischemic brain damage. Ginkgolides have been shown to inhibit platelet activating factor (PAF) as well as increasing blood fluidity and ciculation. In Europe ginkgo extract is sold as an approved drug (Robbers et al., 1996). It has been reported that many extraction methods have been developed for the extraction triterpene lactones efficiently such as using organic solvents, water, pressurized water or supercritical fluids. From these enriched extracts terpenic compounds can be separated by fractional recrystalization, repeated column chromatography, reversed phase HPLC, chromatography with Sephadex LH-20 or more efficiently by chromatography on NaOAc impregnated silica gel. In the following method was described by Jaracz et al. (2004) for the isolation of bilobalide and ginkgolides using column chromatography. The enriched triterpene trilactone extract was chromatographed on silica gel column. The column eluted with EtOAc-hexane solvent mixtures. The initial solvent system was EtOAc-hexane (3.5:6.5). Content of EtOAc in eluent was increased gradually in six steps to EtOAc-hexane (6.5:3.5). The fractions collected at EtOAc-hexane (4.5:5.5) contained bilobalide. Pure bilobalide was obtained as white powder after washing with $Et_2O$. The fractions collected at EtOAc/hexane (5:5) and (5.5:4.5) contained mixture of ginkgolide A/B and ginkgolide C/J, respectively. Ginkgolide mixtures were separated using further chromatographic methods to yield pure compounds (Jaracz et al., 2004). A simple preparative method for the isolation and purification of ginkgolides and bilobalide (ginkgo terpene trilactones) was also developed by Beek & Lelyveld (1997). *Ginkgo biloba* leaf extracts were used for extraction. After a partition step with EtOAc, the enriched intermediate extract was separated into the individual terpenes by medium-pressure liquid chromatography on silica impregnated with 6.5% NaOAc with a gradient from petroleum ether–EtOAc to EtOAc–MeOH. After recrystallization from $H_2O$–MeOH, all ginkgolides could be isolated in high purity. After a selective extraction with $H_2O$, leaves could also be used as a starting material (Beek & Lelyveld, 1997).

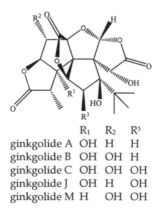

|              | R₁ | R₂ | R³ |
|--------------|----|----|----|
| ginkgolide A | OH | H  | H  |
| ginkgolide B | OH | OH | H  |
| ginkgolide C | OH | OH | OH |
| ginkgolide J | OH | H  | OH |
| ginkgolide M | H  | OH | OH |

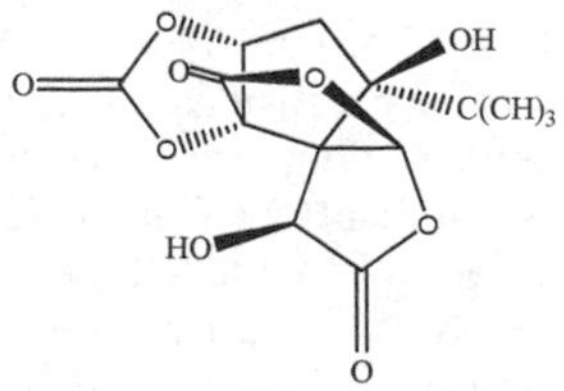

bilobalide

*Salvia divinorum* Epling & Jativa is known as hallucinogenic mint and traditionally used by Mazatec Indians of Oaxaca, Mexico in traditional medicine primarily for its psychoactive effects (Giroud et al., 2000; D.Y.W.Lee et al., 2005). Salvinorin A, a neoclerodane diterpenoid has been isolated and identified as the responsible compound for psychoactive effects. Additionally, salvinorin A have found to have high affinity and selectivity for the kappa opioid receptor is one of the three main types of opioid receptors (D.Y.W.Lee et al., 2005). The discovery of kappa opioid receptor as the molecular target of salvinorin A has opened up many opportunities for drug discovery and drug development for a number of psychiatric and non-psychiatric disorders (Vortherms & Roth, 2006; Li et al 2007). Salvinorin A isolated from the leaves of the *S. divinorum*. Dried leaves of the plant were sequentially extracted with hexane, acetone and MeOH. The acetone extract was fractioned by flash column chromatography with an equal mixture of activated carbon Celite 545. The column was eluted with acetone and hexane. The supernatant of the acetone extract was chromatographed on a silica gel column and eluted with $CHCl_3$-acetone to give five fractions. The fraction eluted with $CHCl_3$-acetone (20:1) was subjected to repeated silica gel column chromatography with a gradient of hexane and EtOAc (15:1-1:1) to afford subfractions. Combined subfractions were purified on silica gel column by $CHCl_3$-EtOAc (20:1-10:1) or hexane-EtOAc (5:1-2:1) solvent systems to yield salvinorin A together with other diterpenoids such as salvinorin B, -C, -D, -E, -F, -G, divinatorin C, -D, -E, hardwickiic acid (D.Y.W.Lee et al., 2005).

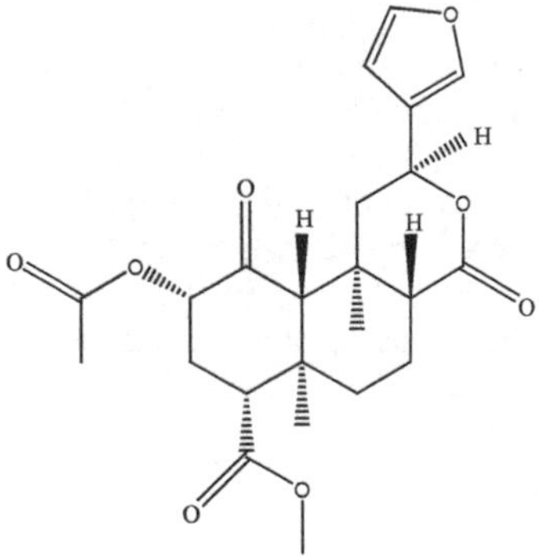

salvinorin A

## 2.1.4 Triterpenoids

The triterpenoids are formed from six isoprene units biosynthetically and widely distributed in nature including plants, microorganisms, animals and humans. Typical examples of the triterpenoids are steroids which have many important functions in mammals such as sex hormones (Robbers et al., 1996; Heinrich et al., 2004).

Oleanolic acid and its isomer ursolic acid are triterpenoids that exist widely in plants as well as in foods as their free forms or as their glycosides. Oleanolic acid and ursolic acid are well known for their hepatoprotective effects. They are used alone or in combination with other hepatoprotective ingredients as oral medications (Liu, 1995; 2005). It has also been reported that oleanolic acid and ursolic acid act at various stages of tumor development to inhibit tumor initiation and promotion, as well as to induce tumor cell differentiation and apoptosis (Liu, 2005). Oleanolic acid and ursolic acid have been isolated many natural sources. Oleanolic acid was obtained from grape as antimicrobial compound. Raisins were extracted with MeOH by maceration. The extract was concentrated and suspended in % 90 MeOH and then partitioned with hexane, $CHCl_3$ and EtOAc respectively. The hexane soluble extract was subjected to silica gel column chromatography and eluted with mixture of $CHCl_3$-MeOH  (1:0-0:1) to give nine fractions. Fraction 3 was separated on silica gel VLC column and eluted with hexane:isopropyl alcohol gradient mixtures (98:2 – 50:50) to yield oleanolic acid (Rivero-Cruz et al., 2008). Another example can be given for oleanolic acid isolation from *Salvia officinalis*. Leaves of the *Salvia officinalis* were extracted with MeOH and then extract was partitioned with EtOAc and n-BuOH respectively. EtOAc fraction was chromatographed on silica gel column chromatography using following solvent systems hexane-EtOAc (10:1-3:1-1:1)-$CHCl_3$-MeOH (10:1)-MeOH to give 4 fractions. Fractions 2 and 3 give diterpenes as well as oleanolic acid after  column chromatography on ODS (MeOH-$H_2O$ 60:40-90:10) followed by preparative HPLC (MeOH-$H_2O$, 85:15) (Ninomiya et al., 2004).

Ursolic acid was isolated from *Sambucus ebulus* L. (Elder) as anti-inflammatory agent. Isolation was carried out from ethanolic extract of the dwarf elder. Initial seperation was performed by means of liquid-liquid extraction of the crude extract with petroleum ether, diethyl ether, EtOAc and BuOH respectively. Diethyl ether fraction was subjected to silica gel column chromatography and petroleum ether and increasing amounts of ethyl acetate was used as mobile phase to afford eight fractions. Fraction 4 was divided in $CH_2Cl_2$ as soluble and insoluble part. $CH_2Cl_2$ insoluble part was subjected to liquid-liquid using petroleum ether, EtOAc, ACN and butyl-methyl ether (10:1:5:2). The lower layer was separated by high-speed counter current chromatography (HSCCC) using petroleum ether, EtOAc, ACN and butyl-methyl ether (10:1:5:2) to obtain three fractions and remained insoluble part. Insoluble part subjected to crystallization with mixture of ACN and tetrahydrofuran to afford ursolic acid as white platelets (Schwaiger et al., 2011). Ursolic acid was also obtained from many of the plants. One of them is *Orthosiphon stamineus* Benth., (Lamiaceae), a native plant to tropical Eastern Asia. Dried leaves of the plant were extracted with MeOH. After filtration and concentration, the crude extract was suspended in $H_2O$ and partitioned with hexane, $CHCl_3$, EtOAc and BuOH. The $CHCl_3$ soluble fraction was applied to silica gel column chromatography and eluted with EtOAc-hexane (7:3) to yield 5 fraction. Further purification by preparative TLC using EtOAc-hexane (3:2) led to the isolation of ursolic acid (Hossain & Ismail, 2010).

oleanolic acid                                    ursolic acid

*Curcurbita pepo* (pumpkin) belongs to Cucurbitaceae family is used as a vegetable for human consumption and also use in traditional medicine. *Cucurbita pepo* is used in the therapy of minor disorders of the prostate gland and the urinary bladder. *Cucurbita pepo* has received considerable attention in recent years because of the nutritional and health values of the seeds. The seeds are excellent source of protein and also pharmacological activity such as antidiabetic, anti fungal and antioxidant. Diets riched in pumpkin seeds have also been associated with lower levels of gastric, breast, lung and colorectal cancer. Seeds and fruit parts of cucurbits are reported to possess purgative, emetic and antihelmintic properties due to the secondary metabolite cucurbitacin content. Cucurbitacins are important functional component found in Cucurbitaceae and constitute a group of diverse triterpenoid substances which are well known for their bitterness and toxicity. They are highly oxygenated, tetracyclic triterpenes containing a cucurbitane skeleton and they are divided into twelve categories which range from cucurbitacins A to T. Specific forms of cucurbitacins are known to have varying potencies with regard to particular activities and effects. It is known that, for example, cucurbitacins B and D are the most potent feeding stimulants for diabroticite beetles, while cucurbitacin D exhibits anti-ovulatory activity in mice, and cucurbitacin B, D, and E all exhibit cytotoxic and anti-tumor effects. Several cucurbitane and hexanorcucurbitane glycosides and other types of triterpenoids have been isolated from the fruits of *Cucurbita pepo* (Gill & Bali, 2011). To obtain cucurbitacins, a liquid is obtained from cucurbitacin-containing plant material by compressing is extracted with a non-polar solvent to remove waxes, pigments, fatty acids, lipids and terpenes from the cucurbitacin-containing solution. For isolation and separation of cucurbitacins, retaining aqueous cucurbitacins-containing liquid is applied to a silica gel column chromatography, preferably the flash column chromatography. Elution is performed with a moderately polar solvent (e.g., CH$_3$Cl) firstly and then the column is eluted with a suitable mixture of solvents (e.g., CH$_3$Cl and acetone, toluene and acetone, EtOAc and acetone, or CH$_3$Cl and acetone), preferably in a ratio of about 95:5 by volume. This elution is collected essentially consists of the cucurbitacin B, which may then be additionally purified and dried. Then column is eluted with a second suitable mixture of solvents (e.g., CH$_3$Cl, acetone and MeOH; EtOAc, acetone and MeOH; or CH$_3$Cl, acetone and MeOH), preferably in a ratio of about 90:5:5 by volume. This elution is collected essentially contains cucurbitacin D. Finally, the silica gel column is eluted with a third suitable solvent mixture (e.g., CH$_3$Cl, acetone and MeOH; EtOAc, acetone and MeOH; or CH$_3$Cl, acetone and MeOH), preferably in a ratio of 80:5:15 by volume. This elution is collected mainly consists of the cucurbitacin E (Subbiah, 1999).

cucurbitacin D

*Centella asiatica* (L.) Urban (Umbelliferae) (Gotu kola), is widely cultivated as a spice or vegetable and is used in treatment of skin diseases, rheumatism, inflammation mental illness, epilepsy, diarrhea and wounds. Polyacetylenes, flavonoids and triterpenoids have been isolated from this plant and among them triterpenoids are major and the most important components of *C. asiatica*, regarded as a marker constituent in terms of quality control. The triterpenes obtained from *C. asiatica* are mainly pentacyclic triterpenic acids and their respective glycosides, belonging to ursane- or oleanane-type, including asiatic acid, asiaticoside, madecassic acid, madecassoside, brahmoside, brahmic acid, brahminoside, thankuniside, isothankuniside, centelloside, madasiatic acid, centic acid, cenellic acid, betulinic acid, indocentic acid, etc (Zeng & Qin, 2007; Nhiem et al., 2011). Chromatographic separation of the triterpenoids and their glycosides were performed from methanolic extract of the plant leaves. MeOH extract was suspended in $H_2O$ and partitioned with EtOAc. EtOAc soluble fraction was then subjected to column chromatography on silica gel and eluted with gradient of $CHCl_3$-MeOH (50:1-1:50) to yield five fraction. Fraction 1 was rechromatographed on silica gel using $CHCl_3$-MeOH (10:1) as an eluent to give four subfractions. Subfraction 3-4 give asiatic acid and quadranoside IV after purification on RP-18 column with MeOH-$H_2O$ (5:1) and MeOH-$H_2O$ (4:1) respectively. The $H_2O$ soluble fraction was chromatographed on Diaion HP-20P column eluted with step gradient of MeOH in $H_2O$ yielding the five fractions. Fraction 2 was rechromatographed on RP-18 column and eluted with acetone /$H_2O$ (2:1) to yield four subfractions. Subfraction 1 was separated on a silica gel column using $CHCl_3$/MeOH/$H_2O$ (30:10:1) as solvent system to afford asiaticoside G. Asiaticoside and asiaticoside F were obtained from subfraction 2 by means of further purification on silica gel column using $CHCl_3$-MeOH-$H_2O$ (35:10:1) (Nhiem et al., 2011).

| R₁ | R₂ | R₃ | |
|---|---|---|---|
| OH | H | H | asiatic acid |
| OH | OH | Sugar | asiaticoside G |
| OH | H | Sugar | asiaticoside |
| H | H | Sugar | asiaticoside F |
| OH | H | Glc | quadranoside IV |

*Calendula officinalis* L., (Asteraceae), (Marigold) is popular medicinal herb and cosmetic in Europe and in America. This plant has been recorded various national pharmacopoeias as well as European Pharmacopoeia. Marigold has been used for wound healing and topical anti-inflammation. The anti-inflammatory properties of the plant flowers have been attributed to triterpenoids some of which are lauryl, myristoyl and palmitoyl esters of faradiol. *Calendula* flowers were extracted using supercritical fluid extraction method under 500 bar pressure, 50 °C and 35kg h⁻¹ carbon dioxide flow. Prepared extract was separated on silica gel column chromatography and eluted with petroleum ether-CHCl₃-MeOH (50:49:1) to yield four fractions. Fraction 4, containing triterpenoid esters, was rechromatographed using low-pressure liquid chromatography on Lobar LiChroprep RP-18. Elution was carried out using MeOH to obtain nine subfractions. Fraction 3 gives faradiol-3-O-laurate. Faradiol-3-O-myristate and faradiol-3-O-palmitate was obtained from fraction 4 and maniladiol-3-O-myristate as well as maniladiol-3-O-palmitate were purified from fraction 5 by further separations in HPLC (Hamburger et al., 2003).

| R= laurate | faradiol-3-O-laurate |
| R= myristate | faradiol-3-O-myristate |
| R= palmitate | faradiol-3-O-palmitate |

| R= myristate | maniladiol-3-O-myristate |
| R= palmitate | maniladiol-3-O-palmitate |

*Panax ginseng* C.A. Meyer has been used as a traditional medicine in China for thousand of years. Ginseng root is one of the most important oriental medicines and is used worldwide to combat stress and disturbances of the central nervous system, for hypothermia, for its antioxidant and organ-protective actions, and for radio-protection. The name "ginseng" often leads to some confusion due to its use for different plants with different phytochemical constituents. True ginsengs are plants in the genus *Panax* from which Asian ginseng (*Panax ginseng*) and American ginseng (*Panax quinquefolium*) have received the most interest for phytomedicinal use. However, *Eleutherococcus senticosis*, a completely different plant not even in the genus *Panax*, is sometimes referred to as Russian or Siberian "ginseng" (Briksin 2000; Fukuda et al., 2000; Park et al., 2002; Ruan et al., 2010). There are two types of preparations from ginseng: white ginseng prepared by drying after peeling off and red ginseng prepared by steaming and drying (Shibata, 2001). Ginseng root contains dammarane and oleanane type saponins as well as polyacetylene derivatives and polysaccharides. Triterpene saponins called as ginsenosides are the well known chemical constituents of ginseng. More than 30 ginsenosides have been identified in ginseng. Ginsenosides Rb₁, Rb₂, Rc, Rd, Rg₁, Rg₂ and Re are the major constituents of white and red ginsengs. However Rg₃, Rg₅, Rg₆, Rs₁, Rs₂ and Rs₃ are known to be only compounds that have been isolated from red ginseng (Park et al., 2002). Some partly deglycosylated saponins such as Rh₁, Rh₂ and Rg₃ are obtained from red ginseng as artifacts produced during steaming (Shibata, 2001). Ginsenosides, Rb1, Rb2, Rc, Rd, Re, Rf, Rg1, and Rg2 are considered to be the most relevant for pharmacological activity (Briskin, 2000). Ruan et al. (2010) isolated a new ginsenosides from fresh roots of *Panax ginseng* together with known ginsenosides Rb₁, Rb₂, Rc and Rd. Ginseng roots was extracted with MeOH-H₂O (4:1) five times and then extract was concentrated to dryness under reduced pressure at 40 ºC. The crude extract was suspended in H₂O and subjected to D-101 resin column chromatography using MeOH-H₂O (0:1, 3:2) as eluents to afford total ginsenosides. Total ginsenosides was applied to silica gel column and eluted with CHCl₃-MeOH-H₂O (6:4:1) to yield three fractions. Fraction 1 was further chromatographed on preparative HPLC eluted with gradient $CH_3CN-H_2O$ 20% to 50%) to give the ginsenoside $Rb_1$, $Rb_2$, Rc and Rd and $Ra_3$ (Ruan et al. 2010). In another study ginsenosides were isolated from dried rootlet of ginseng which was steamed at 120 for 3 hours in an autoclave. Steamed ginseng was extracted with MeOH under reflux for 2 hr. The solvent was removed in vacuo to yield MeOH extract, which was suspended in water and extracted with $CH_2Cl_2$. The remaining aqueous layer was extracted with water-saturated n-BuOH The n-BuOH fraction was concentrated in vacuo to yield BuOH fraction, which was subjected to silica gel column chromatography. Five fractions were obtained using stepwise gradient elution (EtOAc-MeOH-$H_2O$, 40:1: 1 - 10:1: 1). Fraction 3 was chromatographed over silica gel using EtOAc-MeOH- $H_2O$ = 25:1:1 solvent. Ginsenosides $Rs_4$ and $Rs_5$ were obtained from fraction 3, which were further purified on Ag-impregnated preparative TLC using EtOAc-MeOH-$H_2O$, 15:1:1 solvent. They were further purified over semi-preparative HPLC using a reverse-phase column (LiChrospher 100 RP-18, 250 mm x 10 mm i.d.) with 60% $CH_3CN$ eluent to isolate ginsenoside $Rs_4$ and $Rs_5$. Fraction 2 was chromatographed over silica gel using hexane-isopropyl alcohol = 6:1 solvent to give $Rs_6$ and $Rs_7$ rich fractions. The fractions were further purified by semi-preparative HPLC using a reverse-phase column (LiChrospher 100 RP-18, 250 mm x10 mm i.d.) with 50% ACN eluent to yield ginsenosides $Rs_6$ and $Rs_7$ (Park et al., 2002).

R = -glc              ginsenoside $Rb_1$
R = -ara(p)           ginsenoside $Rb_2$
R = -ara(f)           ginsenoside Rc
R = -H                ginsenoside Rd

*Ganoderma lucidum* (Fr.) P. Karst. belongs to the family of Ganodermataceae (Basidiomyetes), has been used since ancient times. *G. lucidum* was believed to cure many kinds of diseases, and it was considered as an elixir that could revive the dead by the ancient people. *G. lucidum* possess important biological activities including anti-tumor, antimicrobial, antiviral (especially anti-HIV activities) and antiaging activities. It has been reported since the first triterpenoid ganoderic acid A was reported more than 150 compounds have been identified from *Ganoderma* spp. Triterpenoids are main components of this genus and have importance for their pharmacologic properties (Cheng et al., 2010). The air-dried and powdered fruit bodies of G. *lucidum* were extracted with EtOH-$H_2O$. The crude extract was washed with petroleum ether to remove fatty acids, then extracted with $CH_2Cl_2$ to give the total triterpenoids fraction. An aliquot of the $CH_2Cl_2$ extract was applied to a silica gel column eluted successively with $CHCl_3$–MeOH (200:1–1:1 gradient system) to obtain 5 fractions. Fraction 2 was subjected to a silica gel column eluted with petroleum ether–EtOAc (10:1–1:1 gradient system) to afford six fractions, F21–F26. F21 was subjected to Sephadex LH-20 column chromatography (petroleum ether–$CHCl_3$–MeOH, 2:1:1), then recrystallized to afford ganoderic acid DM. F22 was applied to a Sephadex LH-20 (petroleum ether–$CHCl_3$–MeOH, 2:1:1), and then further purified by semipreparative HPLC (MeOH–$H_2O$, 90:10) to afford ganodermanondiol and ganoderic acid T-Q. F23 was recrystallized to obtain lucidadiol, F24 was extensively subjected to silica gel column chromatography (petroleum ether–$CHCl_3$–$Me_2CO$, 8:1:1), Sephadex LH-20, and then further purified by semipreparative HPLC to give ganoderol B, lucidumol A and 15α-hydroxy-3-oxo-5α-lanosta-7,9,24(E)-trien-26-oic acid. F25 was purified by semipreparative HPLC (MeOH–$H_2O$, 70:30, detection wavelength, 252 nm) to give 3β-hydroxy-5α-lanosta-7,9,24(E)-triene-26-oic acid and 15α,26-dihydroxy-5α-lanosta-7,9,24(E)-trien-3-one. F26 was applied to a succession silica gel column (400–600 mesh, $CHCl_3$-MeOH, 100:1-10:1 gradient system) and Sephadex LH-20 (PE-$CHCl_3$-MeOH, 2:1:1) chromatography to afford 3β-hydroxy-7-oxo-5α-lanosta-8,24(E)-dien-26-oic acid, ganodermanontriol, 3β,7β-dihyroxy-12β-acetoxy-11,15,23-trioxo-5α -lanosta-8-en-26-oic acid methyl ester, ganoderiol F and lucideric acid A. In the same manner, F3 was applied to a silica gel column (400–600 mesh) eluted with PE (60–90 C)–$Me_2CO$ (7:1–3:1) to afford 6 fractions, F31–F36. Each fraction was subjected to silica gel column chromatography and repeated semipreparative HPLC to afford ganoderic acid D, 11α-hydroxy-3,7-dioxo-5α -lanosta-8,24(E)-dien-26-oic acid, 11β-hydroxy-3,7-dioxo-5α -lanosta-8,24(E)-dien-26-oic acid, lucidone A, ganolucidic acid E, 4,4,14, α-trimethyl- 3,7-dioxo-5α -chol-8-en-24-oic acid, ganoderic acid F ganoderenic acid D, ganoderic acid E, ganoderic acid J, ganoderenic acid F. F4 was separated by repeated column chromatography ($CHCl_3$-MeOH, 100:1-5:1 gradient system) and semipreparative HPLC (MeOH-$H_2O$, 40:60, detection wavelength, 252 nm) to afford ganoderic acid B, ganoderic acid A, 7β,12β-dihydroxy-3,11,15,23-tetraoxo-5α-lanosta-8-en-26-oic acid, 12β-hydroxy-3,7,11,15,23-pentaoxo-5α-lanosta-8-en-26-oic acid,

ganoderic acid A

ganoderenic acid B, methyl ganoderate H, 12β-acetoxy-7β-hydroxy-3,11,15,23-tetraoxo-5α -
lanosta-8,20-dien-26-oic acid, methyl ganoderic acid B, 12β-acetoxy-3β,7β-dihydroxy-
11,15,23-trioxo-5α-lanosta-8,20-dien-26-oic acid, ganolucidic acid A, methyl lucidenate C,
12β-acetoxy-3,7,11,15,23-pentaoxo-5α -lanosta-8-en-26-oic acid ethyl ester, ganoderic acid H
and ganoderic acid AM1. Whereas fraction F5 was separated by repeated column
chromatography (CHCl₃-MeOH, 100:1-1:1 gradient system) and semipreparative HPLC
(MeOH-H₂O, 30:70, detection wavelength, 252 nm) to give compounds 3β,7β,15α-
trihydroxy-11,23-dioxo-5α-lanosta-8-en-26-oic acid, ganoderic acid K, ganoderic acid G,
ganoderenic acid A (Cheng et al., 2010).

*Cimicifuga racemosa* or *Actaea racemosa* L. (Black cohosh), Ranunculaceae, is known as well-
known herbal medicine with health benefits in treating painful menstrual periods and
menopausal disorders. This plant is used primarily as a hormon-free phytomedicine in the
treatment of climacteric symptoms related to menopause. It has been clarified that main
components of the rhizoma are cycloartane type triterpenoids and their glycosides such as
actein, 27-deoxyactein and several cimicifugosides as well as isoflavones, alkaloids and
phenylpropanoids have been identified in Black cohosh (Chen et al., 2002; Watanabe et al.,
2002; Heinrich et al., 2004; Z.Ali et al., 2007; Li et al., 2007; Borelli & Ermst, 2008; Nian et al.,
2011). Chen et al. (2002) isolated actein, 26-deoxyactein and 23-epi-26-deoxyactein from the
rhizomes of *A. racemosa*. Rhizomes of the plant were extracted with MeOH and then
concentrated under vacuum to yield a syrup residue. A sample of the residue was
suspended in H₂O-MeOH (9:1) and fractionated by successive partitions with EtOAc and
BuOH respectively. The EtOAc-soluble fraction was subjected to column chromatography
on silica gel and eluted with CHCl₃, CHCl₃-MeOH (10, 20, 30, 40, 50, 75 %) and MeOH
respectively to afford 8 fractions. Fraction 5 was subjected to a normal phase silica gel
column chromatography by elution of petroleum ether-EtOAc mixture in increasing polarity
to give 10 subfractions. 26-deoxyactein was obtained by direct crystallization from
subfraction 5. The main liquor was sequentially subjected to RP-18 chromatographic column
separation eluted by ACN-H₂O followed by normal phase silica gel separation eluted by
petroleum ether-EtOAc-MeOH (10:7:0.5) to obtain actein. Subfraction 3 was subjected to a
RP-18 column separation eluted by ACN-H₂O and MeOH-H₂O followed by silica gel
column separation eluted with petroleum ether-EtOAc-MeOH (10:5:0.5) to yield 23-epi-26-
deoxyactein (Chen et al., 2002). Another isolation procedure was described by Watanabe et
al. (2002). Methanolic extract of the rhizomes of *C. racemosa* was concentrated under vacuum
and the viscous concentrate was passed through a Diaion HP-20 column and eluted with
30% MeOH, 50% MeOH, MeOH, EtOH and EtOAc successively. Fractions eluted with 50%
MeOH were subjected to column chromatography on silica gel eluting with stepwise
gradient mixtures of CHCl₃-MeOH- H₂O (19:1:0, 9:1:0, 40:10:1, 20:10:1) and finally with
MeOH to afford 7 fractions. Fraction 4 rechromatographed on silica gel column
chromatography eluting with CHCl₃-MeOH (30:1, 10:1, 5:1) and ODS silica gel with MeOH-
H₂O (8:5) to give 25-O-acetyl-12β-hydroxycimigenol 3-O-α-L-arabinopyranoside. Fraction 5
gave 12β-hydroxycimigenol 3-O-α-L-arabinopyranoside after separation on ODS silica gel
column chromatography eluted with MeOH- H₂O (8:5). Fraction 7 was chromatographed on
silica gel eluting with CHCl₃-MeOH (9:1) and ODS silica gel with MeOH- H₂O (4:3)
respectively to obtain 12β,21-dihydroxycimigenol 3-O-α-L-arabinopyranoside. The MeOH
eluate portion was separated on silica gel and eluted with CHCl₃-MeOH (19:1, 9:1, 4:1, 2:1)
mixture and MeOH finally to yield four fractions. Fraction 2 was subjected to column

chromatography on silica gel and eluted with $CHCl_3$-MeOH to obtain two subfractions (2a-2b). Subfraction 2a suspended in MeOH and after filtration rechromatographed on silica gel column chromatography eluting with $CHCl_3$-MeOH (30:1, 19:1) and ODS silica gel with ACN-$H_2O$ (1:1) and MeOH- $H_2O$ (8:3) as well as on Sephadex LH-20 with MeOH respectively to obtain cimicigenol 3-$O$-α-L-arabinopyranoside, 25-$O$-methoxycimicigenol 3-$O$-α-L-arabinopyranoside, 23-$O$-acetylshengmanol 3-$O$-α-L-arabinopyranoside, 27-deoxyactein and actein. The remain subfraction 2b led to the isolation of cimiracemoside F, cimiracemoside G, cimiracemoside H and 822R, 23R, 24R)-12β-acetyloxy-16β, 23:22,25-diepoxy-23,24-dihydroxy-9,19-cyclolanostan-3β-yl α-L-arabinopyranoside (Watanabe et al., 2002). *Actaea podocarpa* which is also known as *Cimicifuga americana* (Summer cohosh) led to the isolation of cyclolanostane type glycosides named podocarpasides (A-G) that was reported by Z.Ali et al. (2007).

R= H     26-deoxyactein
R=OH    actein

23-epi-26-deoxyactein

## 2.1.5 Tetraterpenoids

Members of this class also called as carotenes or carotenoids because of their occurrence in the carrot (*Daucus carota*). Carotenoids are naturally occurring pigments and they are responsible for the yellow, orange, red and purple colors of plants as well as bacteria and algae (Robbers et al., 1996; Heinrich et al., 2004). Fruits and vegetables are rich sources of carotenoids and it has been revealed that carotenoids are strong antioxidant compounds for the prevention of cancer and other human diseases (Burns et al., 2003). Numerous epidemiological studies have demonstrated that carotenoids may be responsible for the beneficial effects associated with the intake of green and yellow vegetables and fruits for cancer prevention in humans. It has been become clear that not only β-carotene but also α-carotene and lycopene, some xanthophylls such as lutein, canthaxanthin, fucoxanthin, halocynthiaxanthin, etc. possess significant cancer chemopreventive effects (Maoka et al., 2001).

The carotenoids can be divided into two groups. Hydrocarbons, soluble in petroleum ether is the first group, orange-red carotenoids α, β, γ-carotene, red-carotenoid lycopene are typical samples of hydrocarbon carotenoids. The second group called as xanthophylls which mainly contains oxygenated derivatives (alcohols, aldehydes, ketones, epoxides and acids) such as lutein, crytoxanthin, rhodoxanthin, violaxanthin, crocetin and ext. are soluble in ethanol (Ikan, 1991).

General isolation procedure of carotenoids was described by Kimura & Rodriguez-Amaya (2002) using column chromatography. Carotenoids are widely distributed in leafy vegetables; therefore they are providing good sources for isolation. Cold acetone exctract was prepared from lettuce, partitioned to petroleum ether, and then concentrated under vacuum to obtain crude extract. The crude extract was separated on MgO-Hyflosupercel (1:1 activated for 2 h at 110 °C) column chromatography using ether-petroleum ether (8%) and acetone-petroleum ether (10-15%, 15-18%, 25-40, 60-70%) as mobile phase to yield β-carotene, lactucaxanthin, violaxanthin, lutein, neoxanthin, chlorophylls respectively. All fractions eluted with petroleum ether containing acetone were washed four or three times with water in a separatory funnel to remove the acetone and then dried with $Na_2SO_4$. This method allows to efficiently and quickly separation of carotenoids (Kimure & Rodriguez-Amaya, 2002).

Lycopene, known as tomato pigment, is widely distributed in nature. It was isolated from *Tamus communis* firstly. Lycopene can be extracted from tomato paste with $MeOH$-$CH_2Cl_2$ mixture after dehydration of tomato paste by MeOH. The extract was concentrated, and then crystallized twice from benzene by the addition of MeOH to obtain lycopene of 98 to 99% purity. Further purification can be achieved by column chromatography on calcium hydroxide (Ikan, 1997).

lycopene

Palm oil contains carotenoids. Palm oil mill effluent that is remaining part of the palm oil industry was extracted with hexane. Hexane extract was evaporated and then subjected to column chromatography on silica gel. Elution was performed by hexane to obtain β-carotene successively. The best separation procedure was performed with the 1:6 ratio of extracted oil: silica gel and the 40 °C temperature (Ahmad et al., 2009).

β-carotene

Ripe fruits of paprika (red pepper), which are used widely as vegetables and food colorants, are good source of carotenoid pigments. The red carotenoids in paprika (*Capsicum annuum* L.) are mainly capsanthin, capsorubin and capsanthin 3,6-epoxide (Maoka et al., 2001). The methanol (MeOH) extract of the fruits of *C. annuum* L. was partitioned between *n*-hexane-$Et_2O$ (1:1) and 10% aqueous NaCl. The organic layer was concentrated to dryness. The residue was subjected to silica gel column chromatography using hexane, hexane-ether (8:2), hexane-ether (7:3), hexane-ether (5:5), ether, ether-acetone (2:8), ether-acetone (5:5) and acetone, successively. Each fraction was further purified by HPLC on a $C_{18}$ reversed phase column with $CH_2Cl_2$-$CH_3CN$ (2:8) as the eluent. Capsanthin obtained from fractions eluted with ether-acetone (5:5) from silica gel column, Capsanthin 3'-ester from eluted with hexane-ether (1:1) from silica gel column, Capsanthin 3,3'-diester eluted with hexane-ether (8:2) from silica gel column; Capsorubin eluted with ether-acetone (2:8) from silica gel column, Capsorubin 3,3'-diester eluted with hexane-ether (7:3) from silica gel column. Capsanthin 3,6-epoxide eluted with ether-acetone (2:8) from silica gel column. Cucurbitaxanthin A 3'-ester eluted with ether-hexane (3:7) from silica gel column (Maoka et al., 2001). Latoxanthin, a minor carotenoid was isolated from the fruits of *Capsicum annuum* var. *lycopersiciforme flavum*, yellow paprika. Exract of the yellow paprika was subjected to column chromatography and eluted with hexane-acetone mixture (3:7). Repeated chromatography on $CaCO_3$ column and then crystallization in benzene-hexane led to the isolation of latoxanthin as red crystals (Nagy et al., 2007).

Capsanthin

## 2.2 Flavonoids

Flavonoids are one of the largest groups of secondary metabolites and widely distributed in leaves, seeds, bark and flowers of plants with more than 4000 different structures which are classified according to their chemical structures as follows; flavones, flavonols, flavanones, dihydroflavonols, isoflavones, anthocyanins, catechins and calchones. Flavonoids which are part of human diet are thought to have positive effects on human health such as reducing risk of cardiovascular diseases and cancer. Most of the beneficial effects of flavonoids are attributed to their antioxidant and chelating abilities (Cook & Samman, 1996; Peterson & Dwyer, 1998; Heim et al., 2002; Rijke et al., 2006). Flavonoids are structurally related compounds with a chromane-type skeleton with a phenyl substituent in the $C_2$ or $C_3$ position (Rijke et al., 2006). They are consisting of phenylpropane ($C_6$-$C_3$) unit derived from shikimic acid pathway and $C_6$ unit derived from polyketide pathway biosynthetically (Heinrich et al., 2004).

Flavonoids are present generally as mixtures and it is very rare to find only one single flavonoid components in plants (Harborne, 1998). They are phenolic compounds and hydroxyl groups often located at positions 3, 5, 7, 3', 4' and/or 5'. One or more of these hydroxyl groups are frequently methylated, acetylated, prenylated or sulphated. Flavonoids are present in plants as aglycone or as their glycosides generally (Rijke et al., 2006).

Extraction of the flavonoids can be performed with solvents that are chosen according to their polarity. Less polar aglycones such as isoflavones, flavanones, dihydroflavonols, higly methylated flavones and flavonols can be extracted with $CH_2Cl_2$, $CHCl_3$, ether, EtOAC. However the more polar aglycones including hydroxylated flavones, flavonols, chalcones and flavonoid glycosides are generally extracted by polar solvents such as acetone, alcohol, water and their combinations (Harborne, 1975; 1998). Many different chromatographic techniques are employed for flavonoids isolation among them column chromatography remains the most useful technique for large scale isolation procedure. Conventional open-column chromatography is still widely used because of its simplicity and its value as an initial separation step. Preparative work on large quantities of flavonoids from crude plant extracts is also possible. Silica gel, polyamide, sephadex, cellulose are commonly used adsorbents. Silica gel is recommended mainly for separation of less polar flavonoids (isoflavones, flavanones, dihydroflavonols and highly methylated/acetylated flavones and flavonols). However, some flavonoid glycosides also can be purified on silica gel using more polar solvents as eluants. Cellulose can be considered as a saceld-up form of paper chromatography. Cellulose is suitable for separation of all class of flavonoids and their glycosides. However, cellulose has low capacity and limited resolving power. Sephadex led to the isolation of compounds on the basis of their molecular size. The hydroxypropylated dextran gel, Sephadex LH-20 is designed for use of with organic solvents or water/solvent mixtures. For Sephadex gels, as well as size exclusion, adsorption and partition mechanisms operate in the presence of organic solvents. Although methanol and ethanol can be used as eluents for proanthocyanidins, acetone is better for displacing the high molecular weight polyphenols. Slow flow rates are also recommended. Open-column chromatography with certain supports (silica gel, polyamide) suffers from a certain degree of irreversible adsorption of the solute on the column. Modifications of the method such as dry-column chromatography, vacuum liquid chromatography are also provide practical usage for the rapid fractionation of plant extracts. VLC with a polyamide support has been reported for succesive separation of flavonol glycosides. Medium-pressure liquid chromatography

flavone          flavonol          flavonone          dihydroflavonol

isoflavone          catechin          anthocyanin          chalcone

(MPLC) which is a closed column (generally glass) connected to a compressed air source or a reciprocating pump covers is also a simple alternative method to open-column chromatography or flash chromatography, with both higher resolution and shorter separation times. MPLC columns have a high loading capacity, up to a 1:25 sample-to-packing-material ratio, and are ideal for the separation of flavonoids. In MPLC, the columns are generally filled by the user. Particle sizes of 25 to 200 μm are usually advocated (15 to 25, 25 to 40, or 43 to 60 μm are the most common ranges) and both slurry packing and dry packing is possible. When compared a shorter column of larger internal diameter with a long column of small internal diameter (with the same amount of stationary phase) resolution is increased. Choice of solvent systems can be efficiently performed by TLC or by analytical HPLC (Harborne, 1975; 1998; Andersen & Markham, 2006).

*Vitex agnus-castus* (Verbenaceae), (Chasteberry or chaste-tree), is used for the treatment of management of female reproductive disorders including premenstrual problems (PMS), menopausal symptoms, and insufficient milk production. The German Commission E recommends it for menstrual problems, mastalgia, and premenstrual syndrome. The specific chemical components responsible for its clinical effects have not been determined but some iridoids, terpenoids as well as flavonoids have been isolated from the leaves or fruits (Hirobe et al., 1997; Hadju et al., 2007). Hirobe et al. (1997) was described the isolation of luteolin, artemetin, isorhamnetin, 4', 5-dihyroxy-3, 3', 6, 7-tetramethoxyflavone as well as four luteolin caffeoylglucoides. Fruits of the chasteberry were powdered and extracted with MeOH. The concentrated extract was partitioned between $H_2O$ and hexane, $CHCl_3$, BuOH respectively. BuOH fraction was subjected to HP-20 column chromatography and eluted with $H_2O$, 40%, 60% and 80% MeOH respectively to yield five fractions (Fr. A-E). Fr. D was subjected to further separation on Sephadex LH-20 and then silica gel column chromatography and elution was performed by $CHCl_3$-MeOH (1:1) and $CHCl_3$-MeOH-$H_2O$ (8:9:1-6.7:3:0.3) succesively. Further purification was performed by means of ODS MPLC and HPLC with MeOH-$H_2O$ and ACN-$H_2O$ to obtain luteolin 6-C-(4''-methyl-6''-O-trans-caffeoylglucoide), luteolin 6-C-(6''-O-trans-caffeoylglucoide), luteolin 6-C-(2''-O-trans-caffeoylglucoide) and luteolin 7-O-(6''-p-benzoylglucoide), luteolin and isorhamnetin. To obtain 4', 5-dihyroxy-3, 3', 6, 7-tetramethoxyflavone and artemetin hexane extract was subjected to column chromatography and eluted with hexane-EtOAc (10:0-0:10) followed with EtOAc-MeOH (1:1). The EtOAc-MeOH (1:1) eluate was subsequently separated using ODS MPLC and finally purified by means of ODS HPLC elution with 80% MeOH (Hirobe et al., 1997).

Dried and powdered flowering stems of *V. agnus castus* were extracted with MeOH and extracts were evaporated under reduced pressure to yield syrupy residue. The MeOH extract was dissolved in $H_2O$ and partitioned with $CHCl_3$ followed by BuOH. A part of the BuOH phase was fractionated on a silica gel column eluting with a gradient solvent system ($CHCl_3$-MeOH) to give nine main fractions (Frs. A-I). Fraction G was further chromatographed over silica gel column eluting with EtOAc-MeOH-$H_2O$ (100:5:2 to 100:17:13) to yield eight fractions (Frs. G1-8). Fr G6 was applied to repeated column chromatographies (CC) over Sephadex LH-20 eluted with MeOH to afford isoorientin (Luteolin 6-C-glucoside) and luteolin 7-O-glucoside. Fr. G2 was subjected to column chromatography on Sephadex LH-20 using MeOH to give compound (2''-O-trans-caffeoylisoorientin. Fractionation of Fr. F by open CC on silica gel using EtOAc-MeOH-$H_2O$ (100:17:13) yielded subfractions F1-6 . Fr. F2 was submitted to Sephadex LH-20 CC (MeOH) to afford pure 6''-O-transcaffeoylisoorientin (Kuruüzüm-Uz et al., 2008).

R₁ =H, R₂=CH₃, R₃=*trans* caffeoyl      luteolin 6-C-(4''-methyl-6''-O-trans-caffeoylglucoide)
R₁ =R₂=H, R₃=*trans* caffeoyl           luteolin 6-C-(6''-O-trans-caffeoylglucoide)
R₁ = *trans* caffeoyl, R₂=R₃=H         luteolin 6-C-(2''-O-trans-caffeoylglucoide)

R₁ =O-Glc(6''-p-hydroxybenzoyl) R₂=R₄=H, R₃=R₅=R₆=OH    luteolin 7-O-(6''-p-benzoylglucoide)
R₁ =R₂=R₄=R₅=OCH₃, R₃=R₆=OH                                4', 5-dihyroxy-3, 3', 6, 7-
                                                              tetramethoxyflavone
R₁ =R₃=R₅=R₆=OH, R₂=R₄=H                        luteolin
R₁ =R₂=R₄=R₅=R₆=OCH₃, R₃=OH                 artemetin
R₁ =R₃=R₄=R₆=OH, R₅=OCH₃, R₂=H         isorhamnetin

isoorientin                                   lutelin-7-glycoside

Isoflavonoids such as puerarin, daidzin and daidzein was isolated from *Pueraria lobata* (Willd) Ohwi roots. Pueraria roots are used in Chinese traditional medicine with "gegen" names for common cold. Daidzein, also known as soya isoflavon has spazmolytic activity. Pueraria roots were extracted with acetone to separate non-glycosidic flavonoids. Acetone extract was separated on silica gel column chromatography using hexane-EtOAc mixture to obtain daidzein, formononetin and puerarol. The residue was extracted against MeOH and then concentrated to dryness under vacuum. To obtain crude extract were dissolved in H₂O and extracted with BuOH to separate glycosidic compounds which were further fractitioned chromatographically on Sephadex LH-20 eluting with MeOH. The glycosidic mixture was chromatographed over silica gel column using CHCl₃-MeOH-H₂O (40:16:3) to separate puerarin, daidzin and other glycosides mixtures. Further purification of glycosides mixture was performed by HPLC (Ohshima et al., 1987).

daidzein                                                daidzin

*Crataegus* species (Hawthorn), (Rosaceae) is a traditional medicinal plant that possess beneficial effects on the heart and blood circulation. *Crataegus monogyna* and *C. laevigata* are the most often used species. Dried flowers, leaves and fruits are mainly used for mild to moderately severe heart failure and coronary heart diseases. Oligomeric procyanidins and (-)-epicatechin as well as flavonoids are known as the main active constituents (Svedström et al., 2002; Sözer et al., 2006). *C. laevigata* flowers and leaves methanolic water extract led to the isolation of procyanidins. Methanolic-water extract was prepared in an ultrasonic bath and then extract was concentrated to a smaller volume under vacuum. Concentrated extract was partitioned with petroleum ether and EtOAc respectively. The EtOAc soluble part evapoarated to the dryness and subjected to column chromatography on polyamide CC 6 column eluting with MeOH, MeOH-$H_2O$ (7:3) and $Et_2O$-$H_2O$ (7:3). Fractions 18-33 allow the isolation of (-)-epicatechin after purification by preparative HPTLC. Epicatechin dimmers; epicatechin-(4β→8)-epicatechin (procyanidin B-2); catechin-(4α→8)-epicatechin (procyanidin B-4); epicatechin-(4β→6)-epicatechin (procyanidin B-5) were obtained from the 43-75 and 117-121 fractions by column chromatography on Sephadex LH-20 uaing EtOH as eluents. Fractions 123-140 was separated with EtOH on Sephadex LH-20 to yield procyanidin trimers, epicatechin-(4β→8)-epicatechin-(4β→8)-epicatechin (procyanidin C-1); epicatechin-(4β→6)-epicatechin-(4β→8)-epicatechin and tetramer epicatechin-(4β→8)-epicatechin-(4β→8)-epicatechin-(4β→8)-epicatechin (procyanidin D-1) (Svedström et al., 2002). Many flavonoids such as hyperoside, rutin, and quercetin have also been isolated from *Crataegus* species. The dried and powdered leaves of *Crategus davisii* were extracted with petroleum ether and then with EtOH. The petroleum ether extract was concentrated and extracted with 60% EtOH. The aqueous extract was concentrated and extracted with $CHCl_3$ (Extract A). The EtOH extract was concentrated and extracted with toluene, $CHCl_3$ (Extract B) and EtOAc (Extract C) successively. Extract A was chromatographed first by vacuum liquid chromatography (VLC) on silica gel with Petroleum ether-$CHCl_3$-MeOH mixtures. Fractions 11-12 ($CHCl_3$-MeOH 80:20) were then chromatographed on a chromatotron (silica gel) again with PE-$CHCl_3$-MeOH mixtures and from fraction 42 (petroleum ether-$CHCl_3$, 20:80 ) crataequinone B were obtained. Extract B was chromatographed first by VLC on silica gel with toluene-$Et_2O$-MeOH mixtures. Fractions 6-8 (toluene-$Et_2O$, 75:25 to 60:40) were then chromatographed on a chromatotron (silica gel) again with toluene-$Et_2O$-MeOH mixtures and from fractions 40-48 ( toluene-$Et_2O$ 30 : 70 to 10:90 ) quercetin were obtained. Extract C was chromatographed first on a silica gel column with toluene-EtOH mixtures. Main fractions 153-304 (toluene-EtOH, 50:50) were then chromatographed on a chromatotron (silica gel) with toluene-$CHCl_3$-EtOH mixtures and from fractions 24-32 ($CHCl_3$-EtOH, 50:50 to 30:70) hyperoside, vitexin 4'-rhamnoside and rutin were obtained using preparative PC (acetic acid-$H_2O$, 15:85 ). Main fractions 305-380 (EtOH) were chromatographed on a polyamide column with $H_2O$-EtOH mixtures and from fractions 77-133 ($H_2O$-EtOH, 80:20) vitexin 2''-rhamnoside was obtained (Sözer et al., 2006). Apigenin, hesperetin, eriodictyol, luteolin, quercetin, luteolin-7-O-glucoside, hyperoside, vitexin, vitexin-4'-O-rhamnoside were isolated from *Crataegus microphylla* C. Koch. by similar procedure (Melikoğlu et al., 2004).

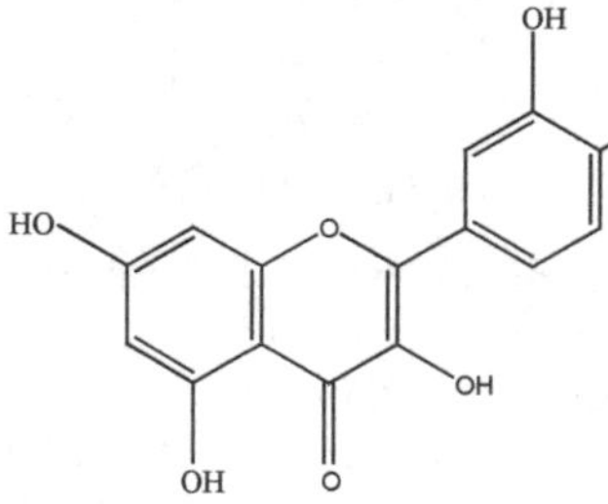

quercetin

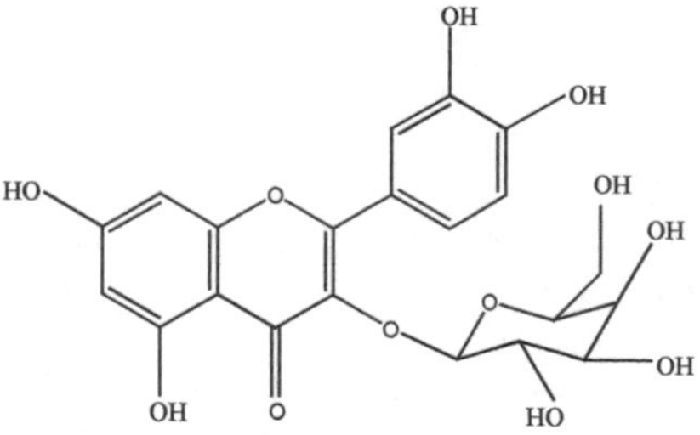

hyperoside

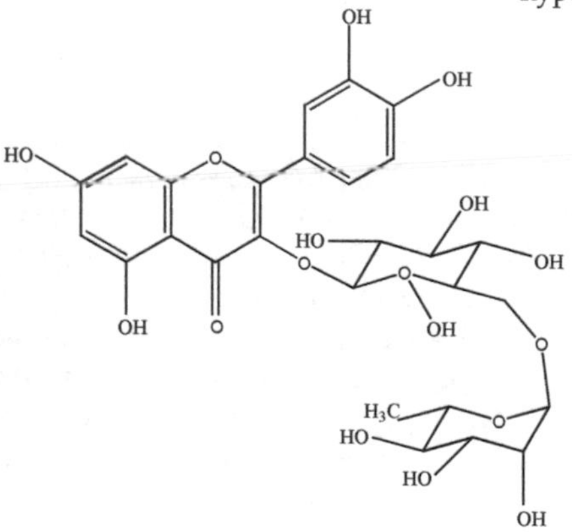

rutin

Cranberry (*Vaccinium macrocarpon*) fruits are excellent raw materials for juice production, as they contain numerous antioxidants including phenolic compounds, vitamin C, minerals and many others. Compounds present in the fruits of the *Vaccinium* species are reported to play several roles in human health maintenance. Consumption of cranberries is found to have protective effects against urinary tract infections. Health benefits such as reduced risks of cancer and cardiovascular disease, are believed to be due to the presence of various polyphenolic compounds, including anthocyanins, flavonols, and procyanidins. The potent antioxidant properties of *Vaccinium* fruits have been well documented. Biological properties of the fruit extract, rich in anthocyanins, include antioxidant capacity, astringent and antiseptic properties, ability to decrease the permeability and fragility of capillaries, inhibition of platelet aggregation, inhibition of urinary tract infection and strengthening of collagen matrices via cross linkages. Cranberry extracts also exhibited a selective tumor cell growth inhibition in prostate, lung, cervical, colon, and leukemia cell lines (Caillet et al., 2011). To isolate cranberry phenolics, cranberry fruits were crushed, macerated with aqueous acetone (80:20 acetone-$H_2O$) and extracted at room temperature for with agitation. The resulting extract was filtered, extraction was repeated on the remaining solids, and the two aqueous acetone extracts were combined. The acetone was removed by rotary evaporation at 35 °C under high vacuum and frozen at −20 °C. The extracts were pre-purified using a method described below After removal of acetone, the aqueous layer was partitioned into hexane to remove carotenoids, fats, and waxes, followed by additional partitioning into ethyl acetate to selectively extract proanthocyanidins with anthocyanin glycosides and flavonols. The ethyl acetate extract was concentrated by vacuum evaporation. The ethyl acetate extract was dissolved in a small amount of methanol for transfer to column chromatography system pre-packed with sephadex LH-20 (column size 100 × 45 mm). The column was subsequently eluted with water, 20% methanol in water, 60% methanol in water, 100% methanol and 80% acetone in water. The water (first liter) eluate yielded organic acids, with subsequent 20% methanol in water elution of anthocyanins, which were pooled together on the basis of color and TLC. The fraction eluted with 60% methanol in water, methanol, and 80% acetone in water yielded the flavonol glycoside fraction, proanthocyanidin fraction. Further purification was performed using preparative HPLC to obtain myricetin-3-β-galactoside, myricetin-3-α-arabinofuranoside, quercetin-3-β-galactoside, quercetin-3-β-glucoside, quercetin-3-rhamnopyranoside, quercetin-3-*O*-(6"-*p*-benzoyl)-β-galactoside and epicatechins (monomer, dimer and trimer) (Singh et al., 2009) .

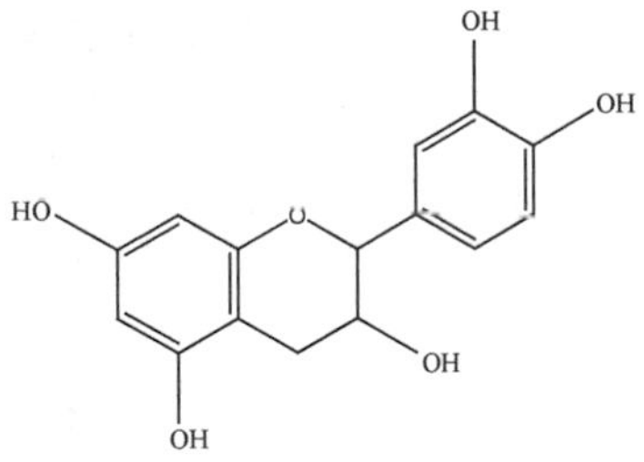

epicatechin (monomer)

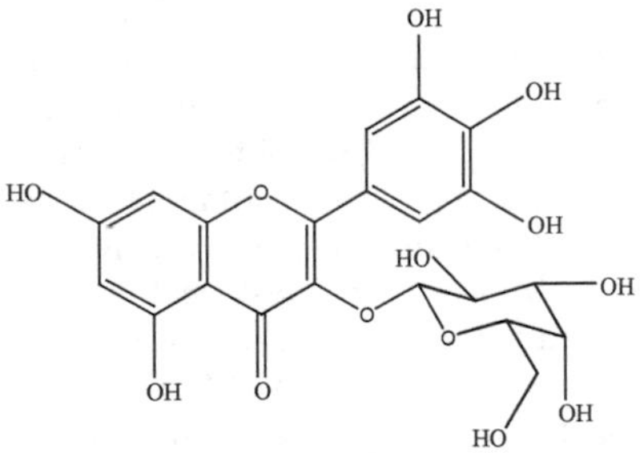

myricetin-3-galactoside

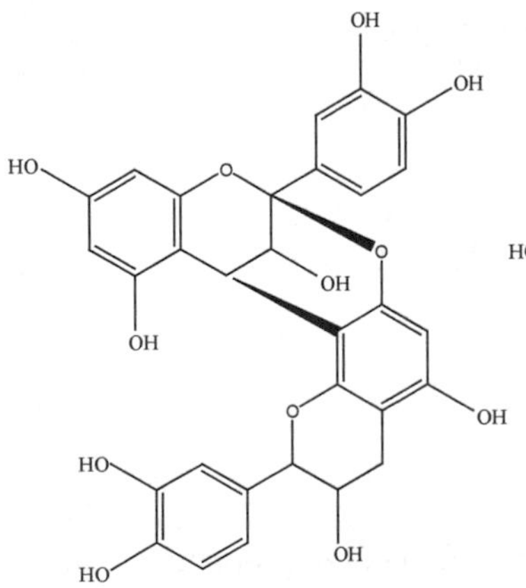

epicatechin (dimer)

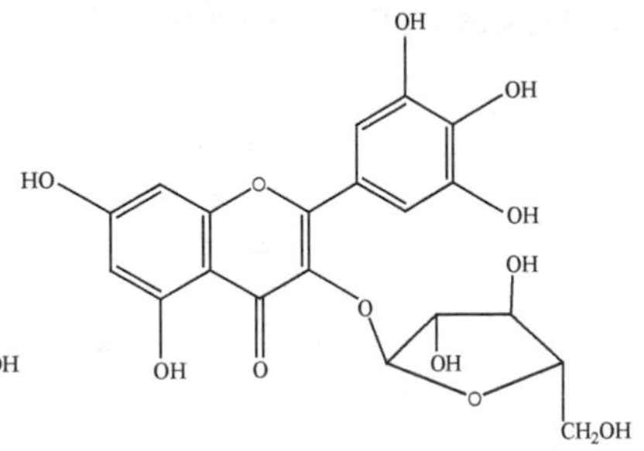

myricetin-3-arabinofuranoside

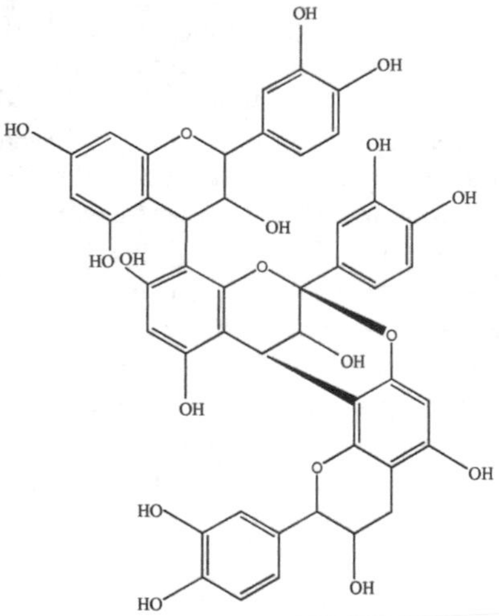

epicatechin (trimer)

Resveratrol is known as 3, 5, 4'-trihydroxystilbene, is widely used in medicine and health products due to their important biological activities such as anti-inflammatory, anticancer, and cardioprotective. Resveratrol is not widely distributed in plants and has been reported that few fruits and vegatables contain. *Polygonum cuspidatum* (Polygonaceae) is one of the richest sources of resveratrol (Soleas et al., 1997; Mantegna et al., 2012). Resveratrol was isolated from *Pleuropterus ciliinervis* belonging to Polygonaceae (J.P.Lee et al., 2003). Dried roots of the plant were extracted with MeOH. MeOH extract was evaporated to dryness and suspended in $H_2O$ and then partitioned with hexane, EtOAc, BuOH, respectively. The EtOAc soluble fraction was subsequently fractionated on silica gel column chromatography and eluted with gradient of hexane-EtOAc (2-50%) and EtOAc-MeOH (5:1) to yield six fractions. Fraction 4 rechromatographed on silica gel column using hexane-EtOAc gradient mixture to afford resveratrol (J.P.Lee et al., 2003). *Gnetum gnemon* Linn. (Gnetaceae) from *Gnetum* genus is known to contain abundant stilbene derivatives also led to the isolation of resveratrol. Successive extraction of acetone, MeOH and 70%MeOH respectively was performed. The acetone extract was subjected to silica gel column chromatography using $CHCl_3$-MeOH in increasing polarity as eluents to give eleven fractions. Fraction 1 gives resveratrol together with other stilbenes after column chromatography of Fraction 1 on Sephadex LH-20 (MeOH) (Iliya et al., 2003).

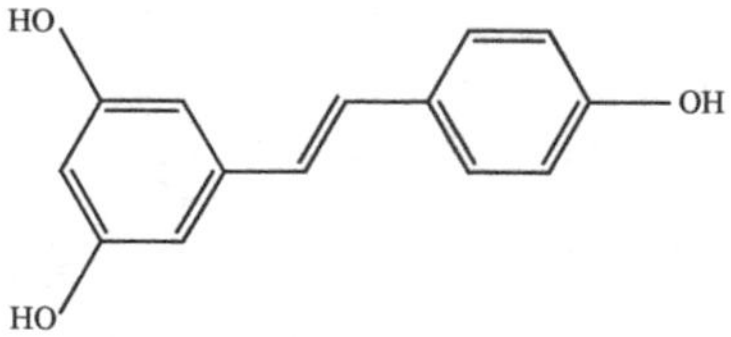

Resveratrol

## 3. Conclusion

Remarkable advances have been accomplished in natural products isolation since the discovery of chromatography. Natural products are present generally as mixtures which makes hard to separation of them. Different chromatographic techniques such as paper chromatography (PC), thin layer chromatography (TLC), gas liquid chromatography (GLC), high performance liquid chromatography and column chromatography employ for separation and purification of natural products. Amon them column chromatography which is the oldest chromatographic technique, is still widely used especially for large scale isolation procedure. Conventional open column chromatography is preferable because of its simlicity. Different stationary phases can be available according to the polarity of the test samples such as silica gel, bonded-phase silica gel, polyamide, sephadex, cellulose, ion-exchange resins. However separation procedure in conventional column chromatography is time consuming and this method suffers from a certain degree of irreversible adsorption of the solute on the column. Vacuum liquid chromatography (VLC) a modified method of conventional column chromatograhy provides practical usage for the rapid fractionation of extracts. Medium-pressure liquid chromatography (MPLC) a closed column which connected to a compressed air source or a reciprocating pump covers is also a simple alternative method to open-column chromatography or flash chromatography.

## 4. References

Abdelgaleil, S.A.M.; Abbassy, M.A.; Belal, A.H. & Abdel Rasoul, M.A.A. (2008). Bioactivity of Two Major Constituents Isolated from The Essential Oil of Artemisia judaica L. *Bioresource Technology*, Vol. ,pp. 5947-5950.

Ahmad, A.L.; Chan, C.Y.; Abd Shukor, S.R.; Mashitah, M.D. & Sunarti, A.R. (2009). Isolation of Carotenes from Palm Oil Mill Effluent and Its Use as a Source of Carotenes. *Desalination and Water Treatment*, Vol. 7, pp. 251-256.

Ali, S.M.; Saleem, M.; Ahmad, W.; Parvez, M.; & Yamdagni, R. (2002). A Chlorinated Monoterpene Ketone, Acylated β-sitosterol Glycosides and a Flavanone Glycoside from Mentha longifolia (Lamiaceae). *Phytochemistry*, Vol. 59, pp. 889-895.

Ali, Z.; Khan, S.I.; Fronczek, F.R. & Khan, I.A. (2007). 9,10-seco-9,19-Cyclolanostane Arabinosides from The Roots of Actaea podocarpa. *Phytochemistry*, Vol. 68, pp. 373-382.

Balunas, M.J. & Kinghorn, A.D. (2005). Drug Discovery from Medicinal Plants. *Life Sciences*, Vol. 78, pp. 431-441.

Beek, T.A. & Lelyveld, G.P. (1997). Preparative Isolation and Separation Procedure for Ginkgolides A, B, C, and J and Bilobalide. *Journal of Natural Products*, Vol. 60, pp. 735-738.

Bhat, S.V., Nagasampagi, B.A., & Sivakumar, M. (2005). *Chemistry of Natural Products*. Narosa Publishing House, India.

Borelli, F. & Ernst, E. (2008). Black cohosh (Cimicifuga racemosa) for Menopausal Symptoms: A Systematic Review of Its Efficacy. *Pharmacological Research*, Vol. 58, pp. 8-14.

Bos, R.; Hendriks, H.; Bruins, A.P.; Kloosterman, J. & Sipma, G. (1986) Isolation and Identification of Valerenane Sesquiterpenoids from Valeriana officinalis. *Phytochemistry*, Vol. 25, pp. 133-135.

Briskin, D.P. (2000). Medicinal Plants and Phytomedicines. Linking Plant Biochemistry and Physiology to Human Health. *Plant Physiology*, Vol. 124, pp. 507-514.

Brown, G.D.; Liang, G.Y. & Sy, L.K. (2003). Terpenoids from the Seeds of Artemisia annua. *Phytochemistry*, Vol. 64, pp. 303-323.

Burns, J.; Fraser, P.D. & Bramley, P.M. (2003). Identification and Quantification of Carotenoids, Tocopherols and Chlorophylls in Commonly Consumed Fruits and Vegetables. *Phytochemistry*, Vol. 62, pp. 939-947.

Caillet, S.; Cote, J.; Doyon, G.; Sylvain, J.F. & Lacroix, M. (2011). Effect of Juice Processing on the Cancer Chemopreventive Effect of Cranberry. *Food Research International*, Vol. 44, pp. 902-910.

Chen, S.N.; Li, W.; Fabricant, D.S.; Santarsiero, B.D.; Mesecar, A.; Fitzloff, J.F.; Fong, H.H.S.; & Farnsworth, N.R. (2002). Isolation, Structure Elucidation, and Absolute Configuration of 26-Deoxyactein from Cimicifuga racemosa and Clarification of Nomenclature Associated with 27-Deoxyactein. *Journal of Natural Products*, 65, 601-605.

Cheng, C.R.; Yue, Q.X.; Wu, Z.Y.; Song, X.Y.; Tao, S.J.; Wu, X.H.; Xu, P.P.; Liu, X.; Guan, S.H. & Guo, D.A. (2010). Cytotoxic Triterpenoids from Ganoderma lucidum. *Phytochemistry*, Vol. 71, pp. 1579-1585.

Chin, Y.; Balunas, M.J.; Chai, H.B. & Kinghorn, A.D. (2006). Drug Discovery from Natural Sources. *The AAPS Journal*, Vol. 8, pp. 239-253.

Cook, N.C. & Samman, S. (1996). Flavonoids-Chemistry, metabolism, cardioprotective effects, and dietary sources. *Nutritional Biochemistry*, Vol. 7, pp. 66-76.

Dewick, P.M. (2009). *Medicinal Natural Products* (Third Edition), John Wiley & Sons Ltd, England.

Fischer, N.H.; Isman, M.B. & Stafford, H.A. (1991). *Recent Advances in Phytochemistry* Vol. 25, *Modern Phytochemical Methods*, Plenum Press, New York, United States of America.

Fu, H.; Zhang, L.; Yi, T.; Feng, Y. & Tian, J. (2010). Two New Guaiane-Type Sesquiterpenoids from The Fruits of Daucus carota L. *Fitoterapia*, Vol. 81, pp. 443-446.

Fukuda, N.; Tanaka, H. & Shoyama, Y. (2000). Isolation of The Pharmacologically Active Saponin Ginsenoside Rb1 from Ginseng by Immunoaffinity Column Chromatography. *Journal of Natural Products*, Vol. 63, pp. 283-285.

Garg, S.N. & Mehta, V.K. (1998). Acyclic Monoterpenes from The Essential Oil of *Tagetes minuta* Flowers. *Phytochemistry*, Vol. 48, pp. 395-396.

Garg, S.N.; Charles, R. & Kumar, S. (1999). A New Acyclic Monoterpene Glucoside from the Capitula of Tagetes patula. *Fitoterapia*, Vol. 70, 472-474.

Gill, N.S. & Bali, M. (2011). Isolation of Anti Ulcer Cucurbitane Type Triterpenoid from the Seeds of Cucurbito pepo. *Research Journal of Phytochemistry*, Vol. 5, pp. 70-79.

Giroud, C.; Felber, F.; Augsburger, M.; Horisberger, B.; Rivier, L. & Mangin, P. (2000). Salvia divinorum: An Hallucinpgenic Mint Which Might Become a New Recreational Drug in Switzerland. *Forensic Science International*, Vol. 112, pp. 143-150.

Gould, M.N. (1997). Cancer Chemoprevention and Theraphy by Monoterpenes. *Environmental Health Perspectives*, Vol. 105, pp. 977-979.

Gunawardena, K.; Rivera, S.B. & Epstein, W.W. (2002). The Monoterpenes of Artemisia tridentata ssp. vaseyana, Artemisia cana ssp. viscidula, Artemisia tridentata ssp.spiciformis. *Phytochemistry*, Vol. 59, pp. 197-203.

Hadju, Z.; Hohman, J.; Forgo, P.; Martinek, T.; Dervarics, M.; Zupko, I.; Falkay, G.; Cossuta, D. & Mathe I. (2007). Diterpenoids and Flavonoids from the Fruits of *Vitex agnus-castus* and Antioxidant Activity of the Fruit Extracts and Their Constituents. *Phytotheraphy Research*, Vol. 21, pp. 391-394.

Hamburger, M.; Adler, S.; Baumann, D.; Förg, A. & Weinreich, B. (2003). Preparative Purification of the Major Anti-inflammatory Triterpenoid Esters from Marigold (Calendula officinalis). *Fitoterapia*, Vol. 74, pp. 328-338.

Hanson, J.R. (2003). *Natural Products the Secondary Metabolites*. The Royal Society of Chemistry, 1-27, Cambridge, UK.

Harborne, J.B.; Mabry, T.J. & Mabry, H. (1975). *The Flavonoids*. Chapman and Hall Ltd., London, Great Britain.

Harborne, J.B. (1998). Phytochemical *Methods; A guide to Modern Techniques of Plant Analysis*,Chapman and Hall Ltd., London, Great Britain.

Heim, K.E.; Tagliaferro, A.R. & Bobilya, D.J. (2002). Flavonoid Antioxidants: Chemistry, Metabolism and Structure-Activity Relationships. *Journal of Nutritional Biochemistry*, Vol. 13. pp. 572-584.

Heinrich, M.; Barnes, J.; Gibbons, S. & Williamson, E.M. (2004). *Fundamentals of Pharmacognosy and Phytotherapy*. Churchill Livingstone, Spain.

Hirobe, C.; Qiao, Z.S.; Takeya, K. & Itokawa, H. (1997). Cytotoxic Flavonoids from Vitex agnus-castus. *Phytochemistry*, Vol. 46, pp. 521-524.

Hossain, M.A. & Ismail, Z. (2010). Isolation and Characterization of Triterpenes from the Leaves of Orthosiphon stamineus. *Arabian Journal of Chemistry*, 2010, in press.

Hong, C.H.; Kim, Y. & Lee, S.K. (2001). Sesquiterpenoids from the Rhizome of Curcuma zedoaria. *Archives of Pharmacal Research*, Vol. 24, pp. 424-426.

Iliya, I.; Ali, Z.; Tanaka, T.; Linuma, M.; Furusawa, M.; Nakaya, K.; Murata, J.; Darnaedi, D.; Matsuura, N. & Ubukata, M. (2003). Stilbene Derivatives from Gnetum gnemon Linn. *Phytochemistry*, Vol. 62, pp. 601-606.

Ikan, R. (1991). *Natural Produtcs A Laboratory Guide* (Second Edition), Academic Press, California, United States of America, 1991.

Jacobs, B.P. & Browner, W.S. (2000). Ginkgo biloba: A Living Fossil. *American Journal of Medicine*, Vol. 108, 341-342.

Jaracz, S.; Malik, S. & Nakanishi, K. (2004). Isolation of Ginkgolides A, B, C, J and Bilobalide from G. Biloba Extracts. *Phytochemistry*, Vol. 65, pp. 2897-2902.

Kimura, M. & Rodriguez-Amaya, D.B. (2002). A Scheme for Obtaining Standards and HPLC Quantification of Leafy Vegetable Carotenoids. *Food Chemistry*, Vol. 78, pp. 389-398.

Klayman, D.L.; Lin, A.; Acton, N.; Scovill, J.P.; Hoch, J.M.; Milhous, W.K. & Theoharides, A.D. (1984). Isolation of Artemisinin (Qinghaosu) from Artemisia annua Growing in the United States. *Journal of Natural Produtcs*, Vol. 47, pp. 715-717.

Kuruüzüm-Uz, A.; Güvenalp, Z.; Ströch, K.; Demirezer, Ö. & Zeeck, A. (2008). Antioxidant Potency of Flavonoids from Vitex agnus-castus L. growing in Turkey. *FABAD Journal of Pharmaceutical Sciences*, Vol. 33, pp. 11-16.

Lee, D.Y.W.; Ma, Z.; Liu-Chen, L.Y.; Wang, Y.; Chen, Y.; Carlezon, W.A. & Cohen, B. (2005). New Neoclerodane Diterpenoids Isolated from the Leaves of Salvia divinorum and their Binding Affinities for Human κ Opioid Receptor. *Bioorganic & Medicinal Chemistry*, Vol.13, pp. 5635-5639.

Lee, J.P.; Min, B.S.; An, R.B.; Na, M.K.; Lee, S.M.; Lee, H.K.; Kim, J.G.; Bae, K.H. & Kang, S.S. (2003). Stilbenes from the Roots of Pleuropterus ciliinervis and their Antioxidant Activities. *Phytochemistry*, Vol. 64, pp. 759-763.

Li, J.X.; Liu, J.; He, C.C.; Yu, Z.Y.; Du, Y.; Kadota, S. & Seto, H. (2007). Triterpenoids from Cimicifuga Rhizome, a Novel Class of Inhibitors on Bone Resorption and Ovariectomy-Induced Bone Loss. *Maturitas*, Vol. 58, pp. 59-69.

Li, Y.; Husbands, S.M.; Mahon, M.F.; Traynor, J.R. & Rowan, M.G. (2007). Isolation and Chemical Modification of Clerodane Diterpenoids from Salvia species as Potential Agonists at the kappa-opioid Receptor. *Chemistry & Biodiversity*, Vol. 4, pp. 1586-93.

Liu, J. (1995). Pharmacology of Oleanolic Acid and Ursolic Acid. *Journal of Ethnopharmacology*, Vol. 49, pp. 57-68.

Liu, J. (2005). Oleanolic acid and Ursolic Acid: Research Perspectives. *Journal of Ethnopharmacology*, Vol. 100, pp. 92-94.

Lobo, R.; Prabhu, K.S. & Shirwaikar, A. (2009). Curcuma zedoaria Rosc. (white turmeric): A Review of its Chemical, Pharmacological and Ethnomedicinal Properties. *Journal of Pharmacy and Pharmacology*, Vol. 61, pp. 13-21.

Mahmoud, A.A. & Ahmed, A.A. (2006). A-Pinene-type Monoterpenes and Other Constituents from Artemisisa suksdorfi. *Phytochemistry*, Vol. 67, pp. 2103-2109.

Mantegna, S.; Binello, A.; Boffa, L.; Giorgis, M.; Cena, C. & Cravotto, G. (2012). A One-pot Ultrasound-Assisted Water Extraction/Cyclodextrin Encapsulation of Resveratrol from Polygonum cuspidatum. *Food Chemistry*, Vol. 130, pp. 746-750.

Maoka, T.; Mochida, K.; Kozuka, M.; Ito, Y. Fujiwara Y, Hashimoto K, Enjo F, Ogata M, Nobukuni Y, Tokuda H, Nishino H. Cancer Chemopreventive Activity of Carotenoids in the Fruits of Red Paprika Capsicum annuum L. *Cancer Letters*, 2001, 172, 103-109.

Matsumura, T.; Ishikawa, T.; & Kitajima, J. (2001). New p-menthanetriols and their Glucosides from the Fruit of Caraway. *Tetrahedron*, Vol. 57, pp. 8067-8074.

McCurdy, C.R. & Scully, S.S. (2005). Analgesic substances derived from natural products (natureceuticals). *Lice Sciences*, Vol. 78, pp. 476-484.

McMurry, J. (2010). *Organic Chemistry with Biological Applications*. Brooks/Cole Cengage Learning,1015-1046. Canada.

Melikoğlu, G.; Bitiş, L. & Meriçli, A.H. (2004). Flavonoids of Crataegus microphylla. *Natural Product Research*, Vol. 18, pp. 211-213.

Nagy, V.; Agocs, A.; Turcsi, E.; Molnar, P.; Szabo, Z. & Deli, J. (2007). Latoxanthin, a Minor Carotenoid Isolated from the Fruits of Yellow Paprika (Capsicum annuum var. lycopersicum flavum). *Tetrahedron Letters*, Vol. 48, pp. 9012-9014.

Nhiem, N.X.; Tai, B.H.; Quang, T.H.; Kiem, P.V.; Minh, C.V.; Nam, N.H.; Kim, J.H.; Im, L.R.; Lee, Y.M. & Kim, Y.H. (2011). A New Ursane-Type Triterpenoid Glycoside from Centella asiatica Leaves Modulates the Production of Nitric Oxide and Secretion of TNF-α in Activated RAW 264.7 Cells. *Bioorganic & Medicinal Chemistry*, Vol. 21, pp. 1777-1781.

Ngo, K.S. & Brown, G.D. (1998). Stilbenes, Monoterpenes, Diarylheptanoids, Labdanes and Chalcones from Alpinia katsumadai. *Phytochemistry*, Vol. 47, pp. 1117-1123.

Nian, Y.; Zhang, X.M.; Li, Y.; Wang, Y.Y.; Chen, J.C.; Lu, L. & Zhou, L. (2011). Cycloartane Triterpenoids from the Aerial Parts of Cimicifuga foetida Linnaeus. *Phytochemistry*, Vol. 72, pp. 1473-1481.

Ninomiya, K.; Matsuda, H.; Shimoda, H.; Nishida, N.; Kasajima, N.; Yoshino, T.; Morikawa, T. & Yoshikawa, M. (2004). Carnosic Acid, a New Class of Lipid Absorbtion Inhibitor from Sage. *Bioorganic & Medicinal Chemistry Letters*, Vol. 14, pp. 1943-1946.

Ohshimo, Y.; Okuyama, T.; Takahashi, K.; Takizawa, T. & Shibata, S. (1988). Isolation and High Performance Liquid Chromatography (HPLC) of Isoflavonoids from the Pueraria Root. *Planta Medica*, Vol. 54, pp. 250-254.

Osorio, C.; Duque, C.; & Fujimoto, Y. (2000). Oxygenated Monoterpenoids from Badea (Passiflora quadrangularis) Fruit Pulp. *Phytochemistry*, 53, 97-101.

Park, I.H.; Han, S.B.; Kim, J.M.; Piao, L.; Kwon, S.W.; Kim, N.Y.; Kang, T.L.; Park, M.K. & Park, J.H. (2002). Four New Acetylated Ginsenosides from Processed Ginseng (Sun Ginseng). *Archives of Pharmacological Research*, Vol. 25, pp. 837-841.

Peterson, J. & Dwyer, J. (1998). Flavonoids: Dietary Occurrence and Biochemical Activity. *Nutrition Research*, Vol. 18, pp 1995-2018.

Raaman, N. (2006). *Phytochemical Techniques*. New India Publishing Agency, New Delhi, India.

Rijke, E.; Out, P.; Niessen, W.M.A.; Ariese, F.; Gooijer, C. & Brinkman, U.A.T. (2006). Analytical Separation and Detection Methods for Flavonoids. *Journal of Chromatography A*, Vol. 1112, pp. 31-63.

Rivero-Cruz, J.F.; Zhu, M.; Kinghorn, A.D. & Wu, C.D. (2008). Antimicrobial Constituents of Thompson Seedless Raisins (Vitis vinifera) Against Selected Oral Pathogens. *Phytochemistry Letters*, Vol. 1, pp. 151-154.

Robbers, J.E.; Speedie, M.K. & Tyler, V.E. (1996). *Pharmacognosy and Pharmacobiotechnology*. Williams & Wilkins, Pennsylvania, United States of America.

Ruan, C.C.; Liu, Z.; Li, X.; Liu, X.; Wang, L.J.; Pan, H.Y.; Zheng, Y.N.; Sun, G.Z.; Zhang, Y.S. & Zhang, L.X. (2010). Isolation and Characterization of a New Ginsenoside from the Fresh Root of Panax Ginseng. *Molecules*, Vol. 15, pp. 2319-2325, ISSN 1420-3049.

Sameeno, B. (2007). Chemistry of Natural Compounds. 10.09.2011. Available from: http://nsdl.niscair.res.in/bitstream/123456789/700/1/revised+terpenoids.pdf.

Sarker,SD.; Latif, Z. & Gray AI. (2006). *Natural Products Isolation* (Second Edition), Humana Press, New Jersey, United States of America.

Senilh, V.; Bleckert, S.; Colin, M.; Guenard, D.; Picot, F.; Potier, P. & Varenne, P. (1984) Journal of Natural Products, Vol. 47, pp. 131-137.

Schwaiger, S.; Zeller, I.; Pölzelbauer, P.; Frotschnig, S.; Laufer, G.; Messner, B.; Pieri, V.; Stuppner, H. & Bernhard D. (2011). Identification and Pharmacological Characterization of the Anti-inflammatory Principal of the Leaves of Dwarf Elder (Sambucus ebulus L.) *Journal of Ethnopharmacology*, Vol. 133, pp. 704-709.

Shibata, S. (2001). Chemistry and Cancer Preventing Activities of Ginseng Saponins and Some Related Triterpenoid Compounds. *Journal of Korean Medical Sciences*, Vol. 16, pp. 28-37, ISSN 1011-8934.

Shiobara, Y.; Asakawa, Y.; Kodama, M. & Takemoto, T. (1986). Zedoarol, 13-Hydroxygermacrone and Curzeone, Three Sesquiterpenoids from Curcuma zedoaria. *Phytochemistry*, Vol. 25, pp. 1351-1353.

Singh, A.P.; Wilson, T.; Kalk, A.J.; Cheong, J. & Vorsa, N. (2009). Isolation of Specific Cranberry Flavonoids for Biological Activity Assesment. *Food Chemistry*, Vol. 116, pp. 963-968.

Soleas, G.J.; Diamandis, E.P. & Goldberg, D.M. (1997). Resveratrol: A molecule Whose Time Has Come? And Gone?. *Clinical Biochemistry*, Vol. 30, pp. 91-113.

Sözer, U.; Dönmez, A.A. & Meriçli, A.H. (2006). Constituents from the Leaves of Crataegus davisii Browicz. *Scientia Pharmaceutica*, Vol. 74, pp. 203-208.

Subbiah, V. (1999). *Method Of Isolating Cucurbitacin.*United States Patent. 1999, patent number: 5,925,356.

Sur, S.V. (1991). Isolation of Oxygen-Containing Monoterpenoids of Essential Oils by Preparative Adsorbtion Chromatography with Gradient Elution. *Khimiya Prirodnykh Soedinenii*, Vol. 5, pp. 634-637.

Svedström, U.; Vuorela, H.; Kostiainen, R.; Tuominen, J.; Kokkonen, J.; Rauha, J.P.; Laakso, I. & Hiltunen, R. (2002). Isolation and Identification of Oligomeric Procyanidins from Crataegus Leaves and Flowers. *Phytochemistry*, Vol. 60, pp. 821-825.

Syu, W.J.; Shen, C.C.; Don, M.J.; Ou, J.C.; Lee, G.H. & Sun, C.M. (1998). Cytotoxicity of Curcuminoids and Some Novel Compounds from Curcuma zedoaria. *Journal of Natural Products*, Vol. 61, pp. 1531-1534.

Tiuman, T.S.; Ueda-Nakamura, T.; Garcia Cortez, D.A.; Dias Filho, B.D.; Morgado-Diaz, J.A.; Souza W. & Nakamura, C.V. (2005). Antileishmanial Activity of Parthenolide, a Sesquiterpene Lactone Isolated from Tanacetum parthenium. *Antimicrobial Agents and Chemotherapy*, Vol. 49, pp. 176-182.

Umlauf, D.; Zapp, J.; Becker, H. & Adam, K.P. (2004). Biosynthesis of the Irregular Monoterpene Artemisia Ketone, the Sesquiterpene Germacrene D and Other Isoprenoids in Tanacetum vulgare L. (Asteraceae). *Phytochemistry*, Vol. 65, pp. 2463-2470.

Vortherms, T.A. & Roth BL. Salvinorin A: From natural Product to Human Therapeutics. *Molecular Interventions* 2006 Oct;6(5):257-65.

Wall, M.E. & Wani, M.C. (1996). Camptothecin and Taxol: From Discovery to Clinic. *Journal of Ethnopharmacology*, Vol. 51, pp. 239-254.

Wang, G., Tang, W. & Bidigare, R.R. In Ed: Zhang L., & Demain, A.L., (2005). *Natural Products Drug Discovery and Therapeutic Medicine Terpenoids As Therapeutic Drugs As Pharmaceutical Agents.* Humana Press, New Jersey, United States of Amerika.

Watanabe, K.; Mimaki, Y.; Sakagami, H.; & Sashida, Y. (2002). Cycloartane Glycosides from the Rhizomes of Cimicifuga racemosa and Their Cytotoxic Activities. *Chemical Pharmaceutical Bulletin*, Vol. 50, pp. 121-125.

Yamada, K.; Murata, T.; Kobayashi, K.; Miyase, T.; & Yoshizaki, F. (2010). A Lipase Inhibitor Monoterpene and Monoterpene Glycosides from Monarda punctata. *Phytochemistry*, Vol. 71, pp. 1884-1891.

# 4

# Purification of Marine Bacterial Sialyltransferases and Sialyloligosaccharides

Toshiki Mine and Takeshi Yamamoto
*Glycotechnology Business Unit, Japan Tobacco Inc.,*
*Japan*

## 1. Introduction

Sialic acids are important components of carbohydrate chains and are usually found at the terminal position of the carbohydrate moiety of glycoconjugates (Angata & Varki, 2002; Schauer, 2004). Sialyloligosaccharides of glycoconjugates play important roles in many biological processes (Gagneux & Varki, 1999; Varki, 1993). The transfer of sialic acids to carbohydrate chains is performed by specific sialyltransferases in the cell (Angata & Varki, 2002; Vimr et al., 2004). Thus, sialyltransferases are considered to be key enzymes in the biosynthesis of sialylated glycoconjugates. Detailed investigations of the biological functions of sialylated glycoconjugates require an abundant supply of the target compounds. To date, many sialyltransferases, and the genes encoding them, have been isolated from various sources including mammalian, bacterial, and viral sources (Schauer, 2004; Sujino et al., 2000; Yamamoto et al., 2006). During our research, we have isolated over 20 bacteria that produce sialyltransferase and have revealed the characteristics of these enzymes (Kajiwara et al., 2009; Yamamoto, 2010). In this chapter, we will introduce our research activities focusing on methods for (1) screening bacteria for glycosyltransferase activity; (2) purifying native sialyltransferases from marine bacteria; and (3) synthesizing and purifying sialyloligosaccharides produced by marine bacterial sialyltransferases.

Sialic acid is a family of acidic monosaccharides comprising over 50 naturally occurring derivatives of neuraminic acid (5-amino-3,5-dideoxy-D-*glycero*-D-*galacto*-2-nonulosonic acid or Neu) (Angata & Varki, 2002; Vimr et al., 2004). Structurally, sialic acid is one of the more complicated naturally occurring monosaccharides and is based on a skeleton of nine carbons (Schauer, 2004). N-acetylneuraminic acid (Neu5Ac), N-glycolylneuraminic acid (Neu5Gc), and 2-keto-3-deoxy-D-*glycero*-D-*galacto*-nonulosonic acid (deaminoneuraminic acid, KDN), are the three most common members of this family (Angata & Varki, 2002; Schauer, 2004). The structure of Neu, Neu5Ac, Neu5Gc and KDN are shown in Figure 1. Although sialic acid is widely distributed in higher animals and some classes of microorganisms, only Neu5Ac is ubiquitous (Angata & Varki, 2004). Usually, sialic acid exists in the carbohydrate moiety of glycoconjugates, including glycoproteins and glycolipids, and is linked to the terminal positions of the carbohydrate chains of the glycoconjugates. Many studies have been carried out to clarify the structure-function relationship of carbohydrate chains containing sialic acid. These studies have revealed that Neu5Ac is the most common sialic acid component of carbohydrate chains and sialylated carbohydrate chains of glycoconjugates play significant roles in many biological processes including inflammation, glycoprotein clearance from circulation, cell-cell recognition, cancer metastasis, and virus infection (Kannagi, 2002; Paulson, 1989). Sialyltransferases commonly transfer Neu5Ac from cytidine 5'-monophospho-$N$-acetylneuraminic acid (CMP-Neu5Ac) to various acceptor substrates (Angata & Varki, 2002). Thus, sialyltransferases are thought to be one of the important enzymes in the biosynthesis of sialylated glycoconjugates.

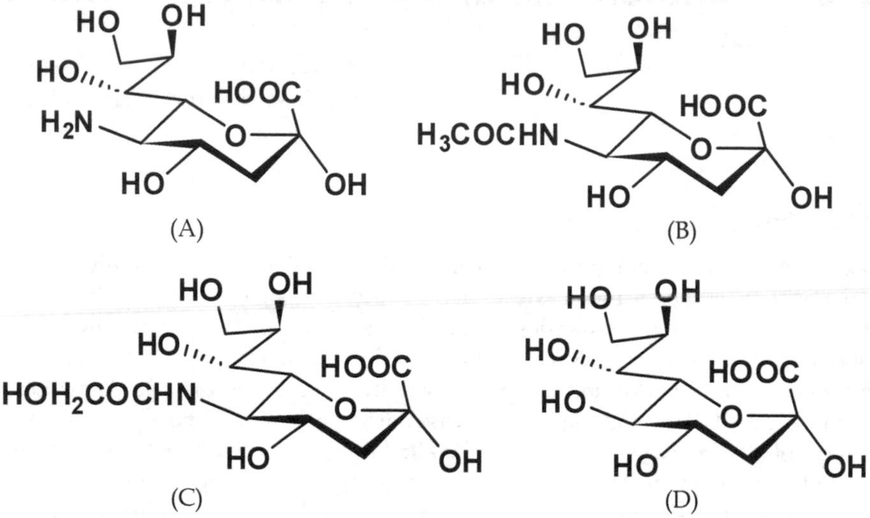

(A) Neuraminic acid (Neu), (B) $N$-acetylneuraminic acid (Neu5Ac), (C) $N$-glycolylneuraminic acid (Neu5Gc), (D) deaminoneuraminic acid (KDN).

Fig. 1. Structures of sialic acids.

Among the biological phenomena described above, the relationship between the carbohydrate chain structure of the host cell and host cell recognition by influenza virus is one of the best investigated (Suzuki, 2005; Weis et al., 1988). Many reports have shown that influenza A and B viruses bind via viral hemagglutinin to host cell surface receptors that are Neu5Ac- or Neu5Gc-linked glycoproteins or glycolipids (Suzuki, 2005). Furthermore, these influenza viruses also recognize the carbohydrate chain structure of the host cell (Connor et al., 1994). Confirming evidence has shown that avian influenza viruses recognize Neu5Acα2-3Galβ1-3/4GlcNAc structures, and that human influenza viruses recognize Neu5Acα2-6Galβ1-3/4GlcNAc structures (Connor et al., 1994; Suzuki, 2005). The host cell specificities of the influenza A and B viruses are determined mainly by the linkage of Neu5Ac or Neu5Gc to the penultimate galactose residues and core structure of the host glycoproteins or glycolipids. For this reason, the distribution of Neu5Ac and Neu5Gc and their linkage patterns on the host cell surface are important determinants of host tropism.

A large variety of oligosaccharides exist in nature. For example, many kinds of sialyloligosaccharides, such as 3'-sialyllactose, 6'-sialyllactose, and sialyllacto-$N$-neotetraose, are contained in milk of various animals (Kunz et al., 2000); however, the purification and isolation of sialyloligosaccharides from natural sources is very difficult due to their structural complexity. Therefore, the research use and development of drugs that depend on sialyloligosaccharides relies on sialyloligosaccharide synthesis by chemical or enzymatic methods. Although many methods for the synthesis of sialyloligosaccharides including chemical- and enzyme-based methods using glycosyltransferases have been developed, it is still difficult to synthesize large amounts of sialyloligosaccharides. Therefore, only a limited amount and only a few kinds of sialyloligosaccharides are currently available as research reagents. In this chapter, we introduce our results with regard to methods for screening for bacterial glycosyltransferases, purification of native sialyltransferases from bacteria, and the synthesis and purification of sialyloligosaccharides produced by marine bacterial sialyltransferases. The methodologies introduced in this chapter might to be useful for screening of bacteria that produce other types of glycosyltransferases, purification of membrane-binding proteins, and purification of oligosaccharides.

## 2. Screening bacteria for sialyltransferase activity

### 2.1 Basic screening method

Samples of seawater, sea-sand, mud, seaweed, and small animals including various kinds of fishes and shells, were collected from various coastal locations in Japan. Bacteria that grew on marine agar 2216 or nutrient agar (Becton-Dickinson, Franklin Lakes, NJ, USA) that was supplemented with 2% NaCl at 15°C, 25°C, or 30°C were isolated from the samples. Aliquots of the bacteria were suspended in 10% glycerol and stored at −80°C. For each bacterial isolate, 6 mL of marine broth 2216 (Becton-Dickinson) in a 15-mL test tube was inoculated with bacteria and cultivated at 15°C, 25°C, or 30°C for 18 h on a rotary shaker (180 rpm). After the cultivation, bacteria were harvested from 2 to 4 mL of the culture broth by centrifugation and then suspended in 200 µL of 20 mM sodium cacodylate buffer (pH 6.0) that contained 0.2% Triton X-100, lysed by sonication on ice, and measured immediately for sialyltransferase activity. Sialyltransferase activity was confirmed as follows: the reaction mixture (30 µL) consisted of the bacterial lysate as the sample of enzyme, 120 mM lactose, 2.3 mM CMP-Neu5Ac (Nakarai Tesque, Kyoto, Japan), 4620 Bq CMP-[4,5,6,7,8,9-$^{14}$C]-Neu5Ac (Amersham Biosciences, Little Chalfont, UK), 100 mM Bis–Tris buffer (pH 6.0), 0.5 M NaCl, and 0.03% Triton X-100. The reaction was carried out at 25°C for 2 h. The reaction mixture was then diluted with 5 mM sodium phosphate buffer (pH 6.8) to a final volume of 2 mL, and applied to a column (0.5 × 2 cm) of Dowex-1 × 8 (phosphate form, Bio-Rad Laboratories, Hercules, CA, USA). The eluate (2 mL) was collected directly into a scintillation vial for counting. The radioactivity of [4, 5, 6, 7, 8, 9-$^{14}$C]-Neu5Ac that had transferred to the acceptor substrate in the eluate was measured by using a liquid scintillation counter, and the amount of Neu5Ac transferred was calculated. Unreacted CMP-Neu5Ac was not eluted in this buffer concentration. Using this procedure, we have isolated many bacteria that possess sialyltransferase activity. Many of the marine bacteria that produced sialyltransferases were classified in genus *Photobacterium* or the closely related genus *Vibrio*. For instance, *Photobacterium phosphoreum* JT-ISH-467 that showed α2,3-sialyltransferase activity was isolated from the outer skin of Japanese common squid, *Todarodes pacificus* (Tsukamoto et al. 2007); *Photobacterium damselae* JT0160 that expressed α2,6-sialyltransferase activity was isolated from seawater (Yamamoto et al., 1998); *Photobacterium* sp. JT-ISH-224 that contained both α2,3- and α2,6-sialyltransferase activities was isolated from the gut of Japanese barracuda, *Sphyraena pinguis* (Tsukamoto et al., 2008); and *Photobacterium leiognathi* JT-SHIZ-145 that expressed α2,6-sialyltransferase activity was isolated from the outer skin of Japanese squid, *Loliolus japonica* (Yamamoto et al., 2007).

## 2.2 Simultaneous measurement of several glycosyltransferases activities

To assess the activities of various glycosyltransferases, not only sialyltransferases, we performed the enzyme assay using a mixture of the donor substrates of glycosyltransferases (GDP-fucose, the common donor substrate of fucosyltransferase; UDP-galactose, the common donor substrate of galactosyltransferase; and UDP-GlcNAc, the common donor substrate of $N$-acetyl-glucosaminyltransferase), and a mixture of the acceptor substrates of glycosyltransferases (4-Nitrophenyl α-D-galactopyranoside {Gal-α-pNp}; 4-Nitrophenyl β-D-galactopyranoside {Gal-β-pNp}; 4-Nitrophenyl $N$-acetyl-α-D-galactosaminide {GalNAc-α-pNp}; 4-Nitrophenyl $N$-acetyl-β-D-galactosaminide {GalNAc-β-pNp}; 4-Nitrophenyl $N$-acetyl-α-D-glucosaminide {GlcNAc-α-pNp}; 4-Nitrophenyl $N$-acetyl-β-D-glucosaminide {GlcNAc-β-pNp}; 4-Nitrophenyl α-D-glucopyranoside {Glc-α-pNp}; 4-Nitrophenyl β-D-glucopyranoside {Glc-β-pNp}; 4-Nitrophenyl α-L-fucopyranoside {Fuc-α-pNp}; 4-Nitrophenyl β-L-fucopyranoside {Fuc-β-pNp}; 4-Nitrophenyl α-D-mannopyranoside {Man-α-pNp}; and 4-Nitrophenyl β-D-mannopyranoside {Man-β-pNp}), and bacterial lysate described in 2.1 as the enzyme sample. An example of results obtained by using this method is shown in Figure 2.

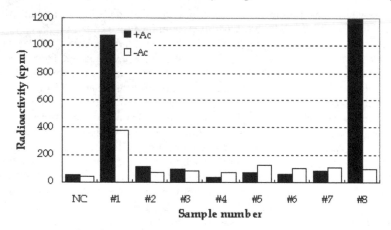

Fig. 2. Screening assay for glycosyltransferase activities.

High levels of radioactivity were observed in the eluates from the reaction mixture of samples #1 and #8, respectively, when the reaction was performed in the presence of acceptor substrate. From this result, it was strongly expected that lysates prepared from bacteria number #1 and #8 contained fucosyltransferase, galactosyltransferase and/or $N$-acetylglucosaminyltransferase. NC; negative control, +AC; containing acceptor substrate mixture in the reaction mixture, -AC; no acceptor substrate mixture in the reaction mixture.

The reaction mixture (50 µL) consisted of the following: a sample of enzyme, a mixture of 0.5 mM acceptor substrates consisting of 4-nitrophenyl compounds, as described above, a mixture of 0.5 mM donor substrates consisting of sugar-nucleotides as described above, 4620 Bq UDP-[U-14C]-galactose, 4620 Bq UDP-$N$-acetyl-D-[U-14C]-glucosamine, 4620 Bq GDP-[U-14C]-fucose (Amersham Biosciences, Little Chalfont, UK), 100 mM bis-Tris buffer (pH 6.0), 10 mM MnCl$_2$, and 3 mM ATP. The reaction was carried at 25°C for 16 to 18 h. After the reaction, 100 µL of water was added to the reaction mixture, and the mixture was applied to a Sep-Pac Vac 50cc column (Waters, Milford, MA, USA) that was conditioned with ethanol and equilibrated with water. The column was washed twice with 1 mL of water and the reaction product was eluted with 1 mL of 70% ethanol. One millilitre of scintillation cocktail was added to the eluate, and the radioactivity of the mixture was measured by using a liquid scintillation counter. In this way, we could detect glycosyltransferase activities, comprising fucosyltransferase, galactosyltransferase and/or N-acetylglucosaminyltransferase activity, simultaneously in marine bacteria. To clarify which of the glycosyltransferase activities the bacteria displayed, the enzymatic reaction was performed independently with each of the donor substrates in turn. The two bacteria that showed glycosyltransferase activity in Figure 2 were shown to specifically produce fucosyltransferase.

## 2.3 Screening by lectin staining

Lectins are sugar-binding proteins that are highly specific for their sugar moieties. The lectin *Sambucus sieboldiana* agglutinin (SSA) recognizes the Neu5Acα2-6Gal or Neu5Acα2-6GalNAc structure of sialyloligosaccharides in glycoconjugates (Shibuya et al., 1989). Kajiwara et al. carried out lectin staining of the cells of *P. damselae* JT0160, *P. leiognathi* JT-SHIZ-145, *P. phosphoreum* JT-ISH-467, and *Photobacterium* sp. JT-ISH-224, by using biotin-labeled SSA, and then examined the cells by using differential interference contrast (DIC) and fluorescence microscopy (Kajiwara et al., 2010). Lectin staining was carried out as follows: the 4 bacterial species described above were cultivated in nutrient broth supplemented with 2% (w/v) NaCl at 25°C for 18 h on a rotary shaker. The bacterial cells were collected by centrifugation (8,000g, 15 min, 4°C) and suspended in 25 mM Tris-HCl buffer 8 (pH 7.5). The suspensions were spotted onto glass slides, fixed with a 4% (w/v) paraformaldehyde solution at room temperature for 15 min, and blocked with a 5% (w/v) bovine serum albumin (BSA) phosphate-buffered saline (PBS) solution. After the glass slides were washed, biotin-labeled SSA (5 mg mL$^{-1}$) was added and the cells were incubated at room temperature for 2 h. After the cells were washed 4 times with PBS, Alexa 594-labeled streptavidin (Invitrogen, Carlsbad, CA, USA) solution (5 mg mL$^{-1}$) was added and the incubation was continued at room temperature for 1 h. After 5 washes with PBS, the cells were mounted using Prolong Gold antifade reagent (Invitrogen) and observed by using DIC and fluorescence microscopy. The SSA bound to *Photobacterium* sp. JT-ISH-224, *P. damselae* JT0160, and *P. leiognathi* JT-SHIZ-145. These *Photobacterium* strains produce α2,6-sialyltransferases, so the lectin staining indirectly detected α2,6-sialyltransferase-producing bacteria (Fig. 3). SSA did not bind to *P. phosphoreum* JT-ISH-467, which produces only α2,3-sialyltransferase. Therefore, the SSA lectin might be useful to screen for not only Neu5Acα2-6Gal and/or Neu5Acα2-6GalNAc structures on the bacterial cell surface but also to screen for the production of α2,6-sialyltransferase. We consider that this method would be applicable to the screening of other glycosyltransferases by changing the type of lectin used. We have confirmed that one of the two bacteria that showed fucosyltransferase activity, described in section 2.2, was detected by biotin-labeled *Aleuria aurantia* lectin (AAL, from Seikagaku Kogyo), which recognizes the fucose residue in carbohydrate chains (Kochibe & Furukawa, 1980).

## 3. Purification of sialyltransferase from the native bacterium

For the purification of a protein, it is necessary and important to find the appropriate conditions for enzyme solubilization, including solubilization efficiency, and the most efficient combination of chromatography processes. Each process has a different separation mode, and it is crucial to conduct a detailed study of the conditions required for each process. Crude extracts are commonly used in such studies, but care must be taken to

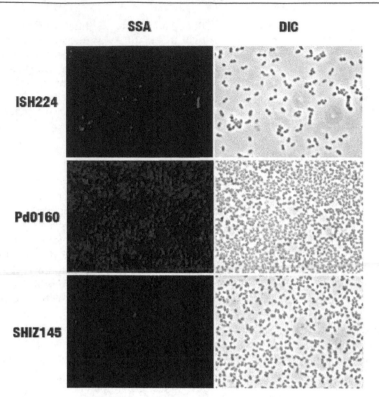

ISH224, *Photobacterium* sp. JT-ISH-224; Pd0160, *P. damselae* JT0160; SHIZ145, *P. phosphoreum* JT-SHIZ-145; SSA, fluorescence microscopy of cells stained with *Sambucus sieboldiana* agglutinin (SSA); DIC, differential interference contrast microscopy of the cells shown in the SSA panels.

Fig. 3. Lectin staining of *Photobacterium* strains by *Sambucus sieboldiana* agglutinin.

minimize protease activity, which may decompose the target enzyme. Furthermore, it is necessary to consider the pH of the buffers used in the purification steps as well as the temperature employed during the preparation of the extracts and the purification process. For details of general procedures and methods for protein purification, we recommend that you refer to other textbooks (e.g., Deutscher, 1990; Scopes, 1982). Here, we describe examples of the purification of sialyltransferase from marine bacteria.

### 3.1 Purification of α2,6-sialyltransferase from *P. damselae* JT0160

### 3.1.1 Preparation of the crude extract from *P. damselae* cells

The first step in the purification of a protein is the preparation of an extract containing the protein in a soluble form. During the purification of sialyltransferase from *P. damselae*, we examined in detail the conditions for preparing a crude extract containing the target enzyme in a soluble form (Yamamoto et al., 1998). The method that we established was deemed appropriate for the preparation of a crude extract containing sialyltransferase because no decrease in sialyltransferase activity was detected during the procedure. We determined that the most important factor in the preparation of the crude extract was the timing of the cell lysis after cultivation. For instance, almost no sialyltransferase activity was detected in crude extract prepared from cryopreserved cells of *P. damselae*. The procedure for crude extract preparation was as described below.

1. After cultivation, *P. damselae* JT0160 cells were harvested from the culture by centrifugation (6,000 $g$, 20 min).
2. The harvested cells were suspended in 20 mM sodium cacodylate buffer (pH 6.0) containing 0.2% Triton X-100 and 1 M NaCl, and were sonicated immediately (<4°C) until the absorbance at 660 nm reached 30% or less of that of the original cell suspension.
3. The sonicated solution was centrifuged (100,500$g$, 60 min) and the supernatant was dialyzed, using cellulose tubing, against 20 mM sodium cacodylate buffer (pH 6.0) containing 0.2% Triton X-100.
4. After dialysis, the precipitate was removed by centrifugation (100,500$g$, 60 min) to obtain the clarified extract.

### 3.1.2 Purification of sialyltransferase by using column chromatography

Sialyltransferase produced by *P. damselae* was then purified from the crude extract by a combination of 4 steps of column chromatography. The conditions and method used for each chromatography step are described below.

1. Q-Sepharose column chromatography. A column of Hi-Load 26/10 Q Sepharose HP (ø 2.6 × 10 cm; GE Healthcare Science, Buckinghamshire, UK) was equilibrated with 20 mM sodium cacodylate buffer (pH 6.0) containing 0.2% Triton X-100. Clarified extract was applied to the column, and the column was washed with 150 mL of the same buffer. Enzyme fractions were eluted with a linear gradient of 0 to 1 M NaCl in the buffer. The fractions exhibiting sialyltransferase activity ("active" fractions) were collected and pooled. Desalting of the "active" fractions was performed by dialysis, using cellulose tubing, against 20 mM sodium cacodylate buffer (pH 6.0) containing 0.2% Triton X-100.
2. Hydroxyapatite column chromatography. A column of hydroxyapatite (ø 2 × 10 cm; Bio-Rad Laboratories) was equilibrated with 20 mM sodium cacodylate buffer (pH 6.0) containing 0.2% Triton X-100. After the application of the enzyme solution obtained in step 1, the column was washed with the same buffer. The enzyme fraction was eluted with a gradient of 0 to 0.35 M potassium phosphate. The "active" fractions were collected and pooled, and then concentrated by ultrafiltration using Molecut L (exclusion molecular mass, 10 kDa ; Millipore, Billerica, MA, USA).
3. Gel–filtration column chromatography. A column of Hi-Load 26/60 Sephacryl S-200 HE (ø 2.6 × 60 cm; GE Healthcare Science, Buckinghamshire, UK) was equilibrated with 20 mM sodium cacodylate buffer (pH 6.0) containing 0.2% Triton X-100 and 0.1 M NaCl. The enzyme solution obtained in step 2 was applied to the column and eluted with the same buffer. The "active" fractions were collected and pooled. Desalting of these fractions was performed by dialysis, using cellulose tubing, against 20 mM sodium cacodylate buffer (pH 6.0) containing 0.2% Triton X-100.
4. CDP–hexanolamine-agarose column chromatography. A column of CDP–hexanolamine-agarose (ø 1 × 3 cm) was equilibrated with 20 mM sodium cacodylate buffer (pH 6.0) containing 0.2% Triton X-100. The enzyme solution (4 mL) obtained in step 3 was applied to the column. The column was washed with 8 mL of the same buffer. The enzyme was eluted with 6 mL of 2 M NaCl, and "active" fractions were collected and pooled.

The purity and yield of the enzyme at each step is summarized in Table 1.

| Purification step | Volume (mL) | Total protein (mg) | Total activity (U) | Specific activity (U/mg) | Yield (%) | Purification (fold) |
|---|---|---|---|---|---|---|
| Crude extract | 760 | 2584 | 21.1 | 0.008 | 100 | 1 |
| Q Sepharose | 240 | 552 | 12.4 | 0.022 | 59 | 2.8 |
| Hydroxyapatite | 120 | 85 | 8 | 0.094 | 38 | 11.8 |
| Sephacryl S-200 | 30 | 20.1 | 6.7 | 0.3 | 32 | 37.5 |
| CDP-hexanolamine-agarose | 15 | 0.75 | 4.1 | 5.5 | 19 | 687.5 |

Table 1. Purification of sialyltransferase from cell lysate of *Photobacterium damselae.*

The enzyme was purified 688-fold, with a yield of 19%. The purified enzyme migrated as a single polypeptide with a molecular mass of 61 kDa by SDS-polyacrylamide gel electrophoresis under denaturing conditions.

## 3.2 Purification of α2,3-sialyltransferase from *P. phosphoreum* JT-ISH-467

### 3.2.1 Preparation of the crude extract from *P. phosphoreum* cells

The crude extract containing sialyltransferase from *P. phosphoreum* cells was prepared by the method described in section 3.1.1, with slight modifications (Tsukamoto et al., 2007), and then crude extract containing the soluble form of the enzyme was prepared.

### 3.2.2 Purification of sialyltransferase from *P. phosphoreum* by using column chromatography

Sialyltransferase produced by *P. phosphoreum* was purified from the crude extract by a combination of 5 steps of column chromatography. The conditions and method used for each of the column chromatography steps are described below.

1. DEAE column chromatography. The clarified crude extract was applied to a Hi-Prep 16/10 DEAE FF column (ø 1.6 × 10 cm; GE Healthcare Science) equilibrated with 20 mM bis-Tris buffer (pH 6.0) containing 0.3% Triton X-100. After the column was washed with the same buffer, the enzyme was eluted with a linear gradient of 0 to 1 M NaCl in the same buffer. The fractions with sialyltransferase activity were pooled and then diluted to three times the original volume with 20 mM potassium phosphate buffer (pH 6.0) containing 0.3% Triton X-100.

2. Hydroxyapatite column chromatography. The enzyme solution obtained in step 1 was applied to a hydroxyapatite column (ø 1.5 × 11.3 cm; Bio-scale CHT20-I; Bio-Rad Laboratories) that was equilibrated with 20 mM potassium phosphate buffer (pH 6.0) containing 0.3% Triton X-100. After the column was washed with the same buffer, the enzyme was eluted with a linear gradient of 20 to 500 mM potassium phosphate. The "active" fractions were pooled, and then diluted to two times the original volume with 20 mM potassium phosphate buffer (pH 6.0) that contained 0.3% Triton X-100.

3. Mono Q column chromatography (pH 6.0). The enzyme solution obtained in step 2 was loaded onto a column of Mono Q 10/100 GL (ø 1 × 10 cm; GE Healthcare Science) that was equilibrated with 20 mM potassium phosphate buffer (pH 6.0) containing 0.3% Triton X-100. After the column was washed with the same buffer, the enzyme was

eluted with a linear gradient of 0 to 1 M NaCl in the same buffer. The "active" fractions were pooled, and then diluted to three times the original volume with 20 mM bis-Tris buffer (pH 7.0) containing 0.3% Triton X-100.

4.  Mono Q column chromatography (pH 7.0). The enzyme solution obtained in step 3 was applied to a column of Mono Q 10/100 GL equilibrated with 20 mM bis-Tris buffer (pH 7.0) containing 0.3% Triton X-100. After the column was washed with 20 mM bis-Tris buffer (pH 7.0) containing 0.3% Triton X-100, the enzyme was eluted with a linear gradient of 0 to 1 M NaCl in the same buffer. The "active" fractions were pooled.

5.  Superdex 200 column chromatography. The enzyme solution obtained in step 4 was loaded onto Hi-Load 16/60 Superdex 200 pg (ø 1.6 × 60 cm; GE Healthcare Science) that was equilibrated with 20 mM bis-Tris buffer (pH 7.0) containing 0.3% Triton X-100 and 0.2 M NaCl and eluted with the same buffer. The "active" fractions were collected and pooled.

The results for the purification of the enzyme are summarized in Table 2.

| Purification step | Volume (mL) | Total protein (mg) | Total activity (U) | Specific activity (mU/mg) | Yield (%) | Purification (fold) |
|---|---|---|---|---|---|---|
| Crude extract | 3155 | 6159 | 8.40 | 1.4 | 100 | 1 |
| DEAE | 410 | 932 | 3.10 | 3.4 | 37 | 3 |
| Hydroxyapatite | 264 | 153 | 1.30 | 8.2 | 15 | 6 |
| Mono Q (pH 6.0) | 12 | 24 | 0.96 | 39 | 11 | 29 |
| Mono Q (pH 7.0) | 1.5 | 1.7 | 0.52 | 315 | 6.2 | 29 |
| Superdex 200 | 1.5 | 0.2 | 0.10 | 457 | 1.2 | 333 |

Table 2. Purification of sialyltransferase from cell lysate of P. phosphoreum.

The enzyme was purified 333-fold, with a yield of 1.2%. Because no affinity chromatography step was used, the yield of the protein purification in this case was very low. Therefore, preparing affinity gels with the appropriate ligand for the target enzyme is very important in the purification process.

## 4. Enzymatic synthesis and purification of sialyloligosaccharides

### 4.1 Synthesis of sialyloligosaccharides by recombinant sialyltransferases from marine bacteria

Chemoenzymatic synthesis of various sialyloligosaccharides by mammalian sialyltransferases, and the purification of the product, has been reported (Sabesan & Paulson, 1986). However, mass-production of sialyloligosaccharides by using mammalian-derived sialyltransferases remains problematic because the enzymes are unstable and difficult to produce as recombinant proteins in *Escherichia coli*. In comparison to mammalian sialyltransferases, bacterial sialyltransferases are generally more stable and productive in *E. coli* protein-expression systems (Tsukamoto et al., 2007, 2008; Yamamoto et al., 2006), and they show a broader acceptor substrate specificity (Izumi & Wong, 2001; Yu et al., 2005). Here, we report the methods that we developed to use recombinant sialyltransferases from marine bacteria to successfully produce large quantities of 6'-sialyllactose and synthesize various sialyloligosaccharides.

### 4.1.1 Synthesis of 6'-sialyllactose from lactose and CMP-Neu5Ac by using purified recombinant α2,6-sialyltransferase from *P. damselae* JT0160

Purified recombinant α2,6-sialyltransferase from *P. damselae* JT0160 shows broader acceptor substrate specificity than that of the mammalian enzymes. For example, it could transfer Neu5Ac to not only disaccharides but also mono- and tri-saccharides efficiently, and provided the corresponding sialosides (Fig. 4; Kajihara et al., 1996; Yamamoto et al., 1998).

Below, we describe an example of the enzymatic synthesis of 6'-sialyllactose (sialoside 1).

1.  The reaction mixture was composed of 20 mg (55 μmol) of lactose (Galβ1-4Glc), 79 mg (110 μmol) of CMP-Neu5Ac, and 0.6 U of the purified enzyme in 0.5 mL of 100 mM bis-Tris buffer (pH 6.0).
2.  The reaction mixture was incubated at 30°C for 2 h.
3.  The product formed by the enzymatic reaction was analyzed by using thin layer chromatography (TLC) as follows: a small amount of the enzymatic reaction mixture was applied to a pre-coated silica gel plate (60 F254, Merck, Darmstadt, Germany), which was then developed with 2-propanol/acetic acid/water (3:2:1 v/v); for visualization of the organic compounds, the plate was dipped into a solution of 5% v/v sulfuric acid in ethanol and then heated.

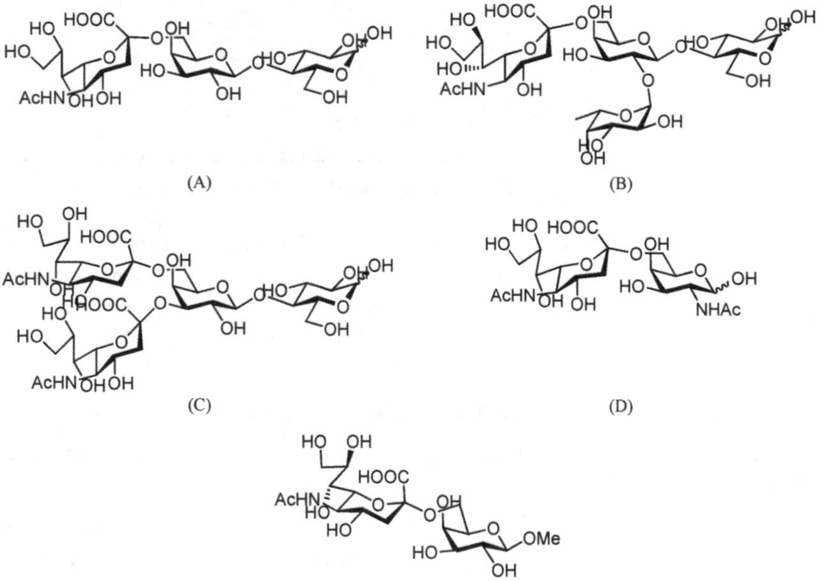

(A) 6'-sialyllactose (sialoside 1), (B) 2'-fucosyl-6'-sialyllactose (sialoside 2), (C) 3', 6'-disialyllactose (sialoside 3), (D) 6-sialyl-*N*-acetylgalactosamine (sialoside 4), (E) 6-sialyl-methyl-β-D-galactopyranoside (sialoside 5).

Fig. 4. Structures of sialosides 1–5.

### 4.1.2 *In vivo* synthesis of 6'-sialyllactose by using genetically engineered *E. coli*

Recently, we succeeded in mass-producing 6'-sialyllactose by using a genetically engineered *E. coli* strain expressing the *Photobacterium* sp. JT-ISH-224 gene for α2,6-sialyltransferase

(Drouillard et al., 2010). Our method was developed from a microbiological system for the large-scale production of 3'-sialyllactose that used high cell-density cultures of a genetically engineered *E. coli* strain expressing the *Neisseria meningitidis* gene for α2,3-sialyltransferase (Fierfort & Samain, 2008). To date, we have achieved the production of 6'-sialyllactose with a final concentration greater than 30 g L$^{-1}$ of culture medium, by continuously feeding the culture with an excess of lactose. A detailed report of the production conditions is provided in Drouillard et al. (2010).

### 4.1.3 Synthesis of various sialyloligosaccharides by using purified recombinant α2,3-sialyltransferase from *Photobacterium* sp. JT-ISH-224

Using the procedure described in section 4.1.1, we could enzymatically produce 3'-sialyllactose (sialoside 6) by using α2,3-sialyltransferase instead of α2,6-sialyltransferase. While using a recombinant α2,3-sialyltransferase derived from *Photobacterium* sp. JT-ISH-224 to produce 3'-sialyllactose, we detected a by-product in the enzymatic reaction mixture and determined its structure to be 2,3'-disialyllactose (sialoside 7; Mine et al., 2010a). This recombinant α2,3-sialyltransferase can also transfer Neu5Ac from CMP-Neu5Ac to the β-anomeric hydroxyl groups of mannose and 6-mannobiose to produce sialosides 8 & 9, respectively (Mine et al., 2010b), and transfer Neu5Ac to inositols to produce sialosides 10 & 11 (Mine et al., 2010c). The structures of sialosides 6–11 are shown in Figure 5.

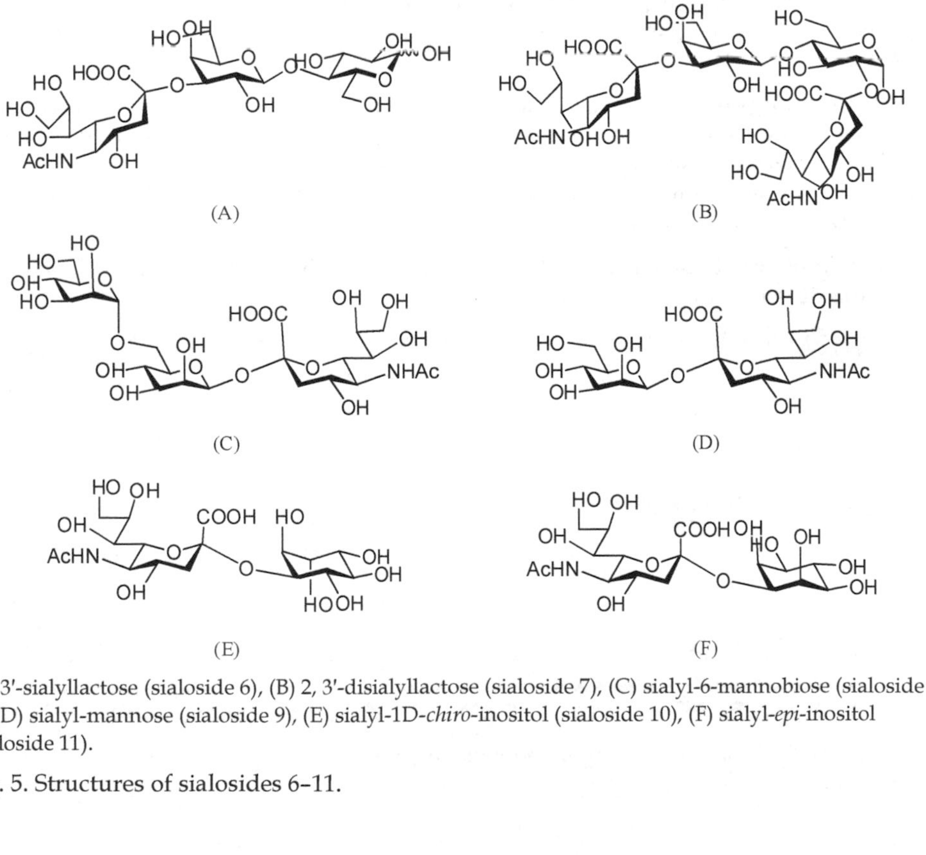

(A) 3'-sialyllactose (sialoside 6), (B) 2, 3'-disialyllactose (sialoside 7), (C) sialyl-6-mannobiose (sialoside 8), (D) sialyl-mannose (sialoside 9), (E) sialyl-1D-*chiro*-inositol (sialoside 10), (F) sialyl-*epi*-inositol (sialoside 11).

Fig. 5. Structures of sialosides 6–11.

## 4.2 Purification of sialyloligosaccharides by use of column chromatography

In general, for the separation of oligosaccharides, it is convenient to utilize high-performance liquid chromatography (HPLC), and various types of columns, such as reverse-phase columns, ion-exchange columns, and gel-filtration columns, that are commercially available. Because Neu5Ac is negatively charged, it is comparatively easy to separate sialyloligosachharide(s) from other neutral oligosaccharides by using anion–exchange column chromatography (Sabesan & Paulson, 1986). For further purification of the compound, gel–filtration column chromatography is effective.

The conditions and method used for each column chromatography step are described below.

### 4.2.1 Separation of the sialyloligosaccharide and unreacted substrates from the enzymatic reaction mixture

The basic procedure for anion–exchange column chromatography is as follows:

1.  The reaction mixture was diluted with 10 mL of deionized water and introduced onto an Econo column (ø1.0 cm × 10 cm; Bio-Rad Laboratories) containing AG1-X2 ion–exchange resin (phosphate form; 200–400 mesh).
2.  The column was washed with 3 column volumes (~ 30 mL) of deionized water.
3.  Elution of the sialyloligosaccharide was performed twice with 10 mL each of 5, 10, 50, 100, 500, or 1000 mM potassium phosphate buffer (pH 6.8).
4.  An aliquot of each eluted fraction was analyzed by using TLC, as described in section 4.1.1.

The column volume required for separation is dictated by the scale of the synthetic reaction. For 10 mg or less of acceptor substrate, all of the reaction product will bind to the resin described above. For more than 100 mg of acceptor substrate, it is desirable to either perform the chromatography process at least twice, or to increase the amount of resin by using a larger column (e.g., ø 2.5 cm × 10 cm).

An example of results obtained for the separation of sialyloligosaccharide by using the above procedure is shown in Figure 6.

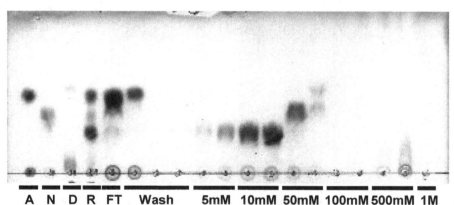

Fig. 6. TLC analysis of fractions separated by using anion–exchange column chromatography.

The reaction solution after enzymatic reaction of substrates with recombinant α2,6-sialyltransferase from *P. damselae* JT0160 strain contained unreacted lactose and CMP-Neu5Ac, free Neu5Ac as result of hydrolysis of CMP-Neu5Ac, and the product. The contents of the fractions eluted with 5, 10, 50, 100, 500, and 1000 mM potassium phosphate buffer (pH 6.8) are shown. A, lactose; N, Neu5Ac; D, CMP-Neu5Ac; R, reaction solution after enzymatic reaction; FT, flow-through.

Many mono-sialyloligosaccharides composed of di-, tri- or tetra-saccharide eluted with 5 to 10 mM potassium phosphate buffer. We also demonstrated that disialyloligosaccharides, such as sialosides 3 (Fig. 4) and 7 (Fig. 5), eluted with 100 mM potassium phosphate buffer. In contrast, many of the unreacted acceptor substrates passed through the column because of their electrically neutral property. Unreacted CMP-Neu5Ac and free Neu5Ac, resulting from the hydrolysis of CMP-Neu5Ac during the reaction, were eluted with 500 and 50 mM potassium phosphate buffer, respectively (Fig. 7). Therefore, it is easy to separate these compounds in the enzymatic reaction mixture with this column chromatography process.

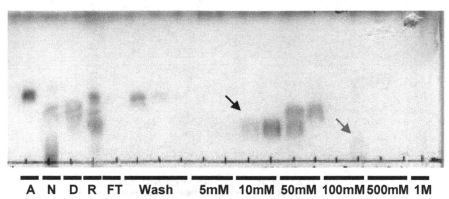

Fig. 7. Separation of mono-sialyloligosaccharide and di-sialyloligosaccharide from the reaction solution by using anion–exchange column chromatography.

The reaction solution after enzymatic reaction with recombinant α2,3-sialyltransferase from *Photobacterium*.sp. JT-ISH-224 strain contained unreacted lactose, CMP-Neu5Ac, free Neu5Ac, and both the mono-sialyloligosaccharide (sialoside 6) as main product (black arrow) and the di-sialyloligosaccharide (sialoside 7) as by-product (red arrow). The contents of the fractions eluted with 5, 10, 50, 100, 500, and 1000 mM potassium phosphate buffer (pH 6.8) are shown. A, lactose; N, Neu5Ac; D, CMP-Neu5Ac; R; reaction solution after enzymatic reaction, FT, flow-through.

During the stepwise elution described above, we sometimes observed that both the reaction product and free Neu5Ac were present in the same fraction. In this case, the separation of these compounds can be improved by increasing the volume of 10 mM potassium phosphate buffer (e.g., using 3–5 column volumes of the buffer).

This basic procedure for the separation of sialyloligosaccharide in the enzymatic reaction mixture is more effective when the enzyme reaction produces a single mono-sialyloligosaccharide. If the reaction mixture contains a variety of mono-sialyloligosaccharides, it is preferable to perform the preparative chromatography using a different column, such as TSKgel Amide-80 (Tosoh Bioscience, Tokyo, Japan) (Endo et al., 2009).

For further purification of the sialyloligosaccharide, we performed gel–filtration column chromatography. The procedure is as follows:

1. The fractions containing glycosidic Neu5Ac were evaporated to dryness.
2. The dried residue was dissolved in 2.5 mL of deionized water and then loaded onto a Sephadex G-15 column (ø 1.6 × 70 cm) and eluted with deionized water under a 2.5 mL/min flow rate and collected in increments of 1 mL.
3. The fractions containing glycosidic Neu5Ac were pooled and evaporated to dryness.

The purpose of this process is to remove salt carried from the former chromatography process. The product was eluted in the 30th to 50th fractions (Fig. 8). When Neu5Ac was mixed with the product, it could be separated from the product under a lower flow rate (e.g., 1.0 mL/min). The purity of sialyloligosaccharides obtained by using a combination of the two chromatography processes described above is usually more than 95% (data not shown).

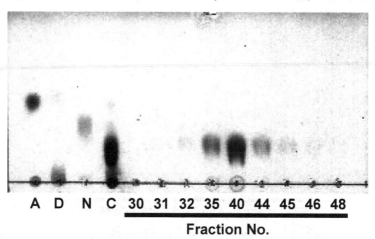

Fig. 8. TLC analysis of the fractions separated by use of gel–filtration column chromatography.

The product is usually contained in the 30th to 50th fraction eluted during the gel–filtration column chromatography; a typical example is shown. A; lactose, D; CMP-Neu5Ac, N; Neu5Ac, C; 6'-sialyllactose standard.

## 4.2.2 Alternative anion-exchange column chromatography method for large-scale purification of 6'-sialyllactose

As mentioned in section 4.1.2, a large volume of solution containing 6'-sialyllactose could be prepared by using high cell-density cultures of a genetically engineered *E. coli* strain. In such cases, we performed an alternative anion–exchange column chromatography process. At the end of the fermentation, the whole culture was permeabilized by autoclaving at 100°C for 50 min. The mixture was centrifuged at 7,000g for 30 min and the supernatant containing the oligosaccharides was removed. The pH of the extracellular fraction was lowered to 3.0 by the addition of a strong cation-exchange resin (Amberlite IR120 H+ form, Sigma-Aldrich Japan, Tokyo), and the proteins that were precipitated by this process were removed by centrifugation. The pH of the clear supernatant was then adjusted to 6.0 by the addition of a weak anion exchanger (Dowex 66 free base form; Sigma-Aldrich Japan) and, after decanting, the supernatant was loaded onto a Dowex 1 (HCO$_3$ form, Sigma-Aldrich Japan) column (ø 5 x 20 cm). After the column was washed with distilled water, the acidic oligosaccharides retained on the Dowex 1 resin were eluted with 100 mM NaHCO$_3$. The eluted fractions containing acidic oligosaccharides were pooled and the NaHCO$_3$ was removed by treatment with Amberlite IR120 (H$^+$ form) until the pH reached 3. The pH was then adjusted to 6.0 with NaOH and the acidic oligosaccharide fraction was freeze-dried.

### 4.2.3 Separation of various sialyl-compounds by using HPLC

Anion–exchange column chromatography is a powerful method for separating a single mono-sialyloligosaccharide from the other components in the sample solution; however, it is impossible to separate one mono-sialyloligosaccharide from another mono-sialyloligosaccharide, such as 6'-sialyllactose or 3'-sialyllactose, by using this method. In such cases, HPLC can be used successfully for the separation, as described by Endo et al. (Endo et al., 2009). In this protocol, the HPLC system is equipped with a TSKgel Amide-80 column (particle size 5 mm, ø 4.6 × 250 mm; Tosoh Bioscience), and is run at 40°C with a flow rate of 1mLl/min, and the eluates are monitored with a UV (195 nm) detector. The elution conditions are as follows: 0 to 15 min isocratic elution with 75% acetonitrile in 15 mM potassium phosphate buffer (pH 5.2); and 15 to 45 min linear gradient elution with a gradient of 75% to 50% acetonitrile in 15 mM potassium phosphate buffer (pH 5.2).

Some examples of HPLC chromatograms produced for sialyl-compounds synthesized as described in section 4.1 are shown in Figure 9.

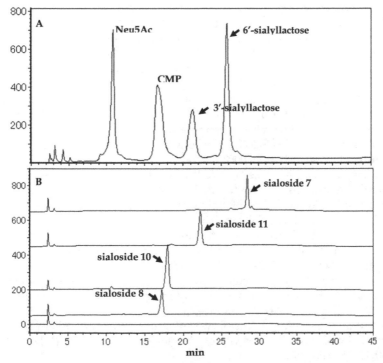

Fig. 9. Chromatograms of some sialyl-compounds.

The sample solution (150 mL) containing sialyl-compound(s) was separated as described in section 4.2.3. A; an example of the separation of two mono-sialyloligosaccharides, B; examples of the chromatogram of various mono-sialyloligosaccharides.

## 5. Conclusion

It is now possible to produce large amounts of sialyloligosaccharides by using newly developed methods, including chemoenzymatic methods and fermentation methods. It is also possible to produce huge quantities of sialyltransferase enzymes. However, large-scale production of other glycosyltransferases, such as $N$-acetylglucosaminyltransferase or fucosyltransferase, is still difficult. For this reason, it is of great importance to identify enzymes that could be used in the production of other glycosyltransferases and to establish mass-production methods for these enzymes.

## 6. Acknowledgment

The authors would like to thank Ms. Hitomi Kajiwara for her valuable comments and all of their collaborators.

## 7. References

Angata, T., & Varki, A. (2002) Chemical diversity in the sialic acids and related α-keto acids: an evolutionary perspective. Chemical Reviews, Vol.102, Issue 2, (February, 2002), pp.439-469, ISSN: 0009-2665

Connor, R. J.; Kawaoka, Y.; Webster, R. G. & Paulson, J. C. (1994) Receptor specificity in human, avian, and equine H2 and H3 influenza virus isolates. Virology, Vol.205, Issue 1, (Nobember 15, 1994), pp.17–23, ISSN 0042-6822

Deutscher, M.P. (Ed.)., (1990), Guide to Protein Purification, ACADEMIC PRESS, INC., ISBN 0-12-213585-7, San Diego

Drouillard, S.; Mine, T.; Kajiwara, H.; Yamamoto, T. & Samain, E. (2010) Efficient synthesis of 6′-sialyllactose, 6′,6-disialyllactose and 6′-KDO-lactose by metabolically engineered *Escherihiae coli* expressing a multifunctional sialyltransferase from the *Photobacterium* sp. JT-ISH-224. Carbohydrate Research, Vol.345, Issue 10, (July 2,2010), pp.1394-1399, ISSN 0008-6215

Endo, S.; Morita, M.; Ueno, M.; Maeda, T. & Terabayashi, T. (2009) Fluorescent labeling of a carboxyl group of sialic acid for MALDI-MS analysis of sialyloligosaccharides and ganglioside. Biochemical and Biophysical Research Communications, Vol.378, Issue 4, (January 23, 2009), pp.890-894, ISSN 0006-291X

Fierfort, N. & Samain, E. (2008) Genetic engineering of Escherichia coli for the economical production of sialylated oligosaccharides. Journal of Biotechnology, Vol.134, No.3, (April 30, 2008), pp.261-265, ISSN 0168-1656

Gagneux, P. & Varki, A. (1999) Evolutionary considerations in relating oligosaccharide diversity to biological function. Glycobiology, Vol. 9, No. 9, (August 1, 1999), pp.747-755, ISSN 0959-6658

Izumi, M. & Wong, C-H. (2001) Microbial sialyltransferases for carbohydrate synthesis. Trends in Glycoscience and Glycotechnology, Vol.13, No.72, (July, 2001), pp.345-360, ISSN 0915-7352

Kajihara, Y.; Yamamoto, T.; Nagae, H.; Nakashizuka, M.; Sakakibara, T. & Terada, I. (1996) A novel α-2,6-sialyltransferase: transfer of sialic acid to fucosyl and sialyl trisaccharides. The journal of Organic Chemistry, Vol.61, No.24, (November 29, 1996), pp.8632-8635, ISSN 0022-3263

Kajiwara, H.; Mine, T. & Yamamoto, T. (2009) Sialyltransferases Obtained from Marine Bacteria. Journal of Applied Glycoscience, Vol.56, No.2, (October 15, 2009), pp.77-82, ISSN 1344-7882

Kajiwara, H.; Toda, M.; Mine, T.; Nakada, H.; Wariishi, H. & Yamamoto, T. (2010) Visualization of Sialic Acid Produced on Bacterial Cell Surfaces by Lectin Staining. Microbes Environ. Vol. 25, No. 3, (September, 2010) pp.152–155, ISSN 1342-6311

Kannagi, R. (2002) Regulatory roles of carbohydrate ligands for selectins in homing of lymphocytes. Current Opinion in Structural Biology, Vol.12, Issue5, (October 1, 2002), pp.599-608, ISSN 0959-440X

Kochibe, N. & Furukawa, K. (1980) Purification and properties of a novel fucose-specific hemagglutinin of *Aleuria aurantia*. Biochemistry, Vol.19, No.13, (June, 1980), pp.2841–2846, ISSN 0006-2960

Kunz, C.; Rudloff, S.; Baier, W.; Klein, N. & Strobel, S. (2000) Oligosaccharides in human milk: structural, functional and metabolic aspects. Annual Review of Nutrition, Vol.20, (July, 2000), pp.699-722, ISSN 0199-9885

Mine, T.; Kajiwara, H.; Murase, T.; Kajihara, Y. & Yamamoto, T. (2010a) An α2,3-sialyltransferase cloned from *Photobacterium* sp. JT-ISH-224 transfers N-acetylneuraminic acid to both O-2 and O-3' hydroxyl groups of lactose. Journal of Carbohydrate Chemistry, Vol.29, Issue 2, ( March, 2010), pp.51-60, ISSN 0732-8303

Mine, T.; Miyazaki, T.; Kajiwara, H.; Naito, K.; Ajisaka, K. & Yamamoto, T. (2010b) Enzymatic synthesis of unique sialyloligosaccharides using marine bacterial α-(2→3)- and α-(2→6)-sialyltransferases. Carbohydrate Research, Vol.345, Issue 10, (July 2,2010), pp.1417-1421, ISSN 0008-6215

Mine, T.; Kajiwara, H.; Tateda, N.; Miyazaki, T.: Ajisaka, K. & Yamamoto, T. (2010c) Recombinant α-(2→3)-sialyltransferase with extremely broad acceptor substrate specificity from *Photobacterium* sp. JT-ISH-224 could transfer N-acetylneuraminic acid to inositols. Carbohydrate Research, Vol.345, Issue 10, (July 2,2010), pp.1394-1399, ISSN 0008-6215

Paulson, J. C. (1989) Glycoproteins: What are the sugar chains for? Trends in Biochemical Sciences, Vol.14, Issue 7, (July, 1989), pp.272-276, ISSN 0968-0004

Sabesan, S. & Paulson, J.C. (1986) Combined chemical and enzymatic synthesis of sialyloligosaccharides and characterization by 500 MHz [1]H and [13]C NMR spectroscopy. Journal of the American Chemical Society, Vol.108, Issue 8, (April, 1986), pp.2068-2080, ISSN 0002-7863

Schauer, R. (2004) Sialic acid: fascinating sugars in higher animals and man. *Zoology*, Vol.107, Issue 1, (March 16, 2004), pp.49-64, ISSN 0944-2006

Scopes, R.K. (1982), *Protein Purification –Principles and Practice*, Springer-Verlag, ISBN 0-387-90726-2, New York

Shibuya, N.;Tazaki, K.; Song, Z.;Tarr, G. E.;Goldstein,I. J. & Peumans, W. J. A. (1989) Comparative study of bark lectins from three elderberry (*Sambucus*) species. Journal of Biochemistry, Vol.106, No.6, (June, 1989), pp.1098–1103, ISSN 0021-924X

Sujino, K.; Jackson, R.J.; Chan, N.W.C.; Tsuji, S. & Palcic, M.M. (2000) A novel viral α2,3-sialyltransferase (v-ST3Gal I): transfer of sialic acid to fucosylated acceptors. Glycobiology, Vol.10, Issue 3, (March 1, 2000), pp.313-320, ISSN 0959-6658

Suzuki, Y. (2005) Sialobiology of influenza: molecular mechanism of host range variation of influenza virus. Biological & Pharmaceutical Bulletin, Vol.28, No.3, (March, 2005), pp.399-408. ISSN 0918-6158

Tsukamoto, H.; Takakura, Y. & Yamamoto, T. (2007) Purification, cloning and expression of an α-/β-galactoside α2,3-sialyltransferase from a luminous marine bacterium, *Photobacterium phosphoreum*. Journal of Biological Chemistry, Vol.282, Issue 41, (October 12, 2007), pp.29794–29802, ISSN 0021-9258

Tsukamoto, H.; Takakura, Y.; Mine, T. & Yamamoto, T. (2008) *Photobacterium* sp. JT-ISH-224 Produces Two Sialyltransferases, α-/β-Galactoside α2,3-Sialyltransferase and β-Galactoside α2,6-Sialyltransferase. Journal of Biochemistry, Vol. 143, No.2, (February, 2008), pp.187-197, ISSN 0021-924X

Varki, A. (1993) Biological roles of oligosaccharides: all of the theories are correct. Glycobiology, Vol.3, Issue 2, (April 3, 1993), pp.97-130, ISSN 0959-6658

Vimr, E. R.; Kalivoda, K. A.; Denzo, E. L. & Steenbergen, S. M. (2004) Diversity of microbial sialic acid metabolism. Microbiology and Molecular Biology Reviews, Vol.68, No.1, (March, 2004), pp.132-153, ISSN 1092-2172

Weis, W.; Brown, J. H.; Cusack, S.; Paulson, J. C.; Skehel, J. J. & Wiley, D. C. (1988) Structure of influenza virus haemagglutinin complexed with its receptor, sialic acid. Nature, Vol.333, (June 2, 1988), pp.426-431, ISSN 0028-0836

Yamamoto, T. (2006) Bacterial Sialyltransferases. Trends in Glycoscience and Glycotechnology, Vol.18, N0.102, (July, 2006), pp.253-265, ISSN 0915-7352

Yamamoto, T. (2010) Marine Bacterial Sialyltransferases. Marine Drugs, Vol.8, Issue 18, (November 10, 2010), pp.2781-2794, ISSN 1660-3397

Yamamoto, T.; Nakashizuka, M. & Terada, I. (1998) Cloning and expression of a marine bacterial β-galactoside α2,6-sialyltransferase gene from *Photobacterium damsela* JT0160. Journal of Biochemistry, Vol.123, No.1, (January, 1998), pp.94–100, ISSN 0021-924X

Yamamoto, T.; Hamada, Y.; Ichikawa, M.; Kajiwara, H.; Mine, T.; Tsukamoto, H. & Takakura, Y. (2007) A β-galactoside α2,6-sialyltransferase produced by a marine bacterium, *Pho&bacterium leiognathi* JTSHIZ-145, is active at pH 8. Glycobiology, Vol.17, No.11, (Nobember, 2007), pp.1167–1174, ISSN 0959-6658

Yamamoto, T.; Nagae, H.; Kajihara, Y. & Terada, I. (1998) Mass production of bacterial α2,6-sialyltransferase and enzymatic syntheses of sialyloligosaccharides. Bioscience, Biotechnology, and Biochemistry, Vol.62, No.2 , (Febraury, 1998), pp.210-214, ISSN 0916-8451

Yu, H.; Chokhawala, H.; Karpel, R.; Yu, H.; Wu, B.; Zhang, J.; Zhang, Y.; Jia, Q. & Chen, X. (2005) A multifunctional *Pasteurella multocida* sialyltransferase: a powerful tool for the synthesis of sialoside libraries. Journal of the American Chemical Society, Vol.127, Issue 50, (November, 2005), pp.17618-17619, ISSN 0002-7863

# 5

# Wound Healing and Antibacterial Properties of Leaf Essential Oil of *Vitex simplicifolia* Oliv. from Burkina Faso

Magid Abdel Ouoba[1], Jean Koudou[2*] , Noya Some[3],
Sylvin Ouedraogo[2,3] and Innocent Pierre Guissou[1,2]
[1]*Unit of Teaching and Research (UFR) of Health Sciences, University of Ouagadougou,*
[2]*Laboratory of Chemistry of Natural Products, Faculty of Sciences,*
*University of Bangui, Bangui,*
[3]*Institute of Research in Health Sciences, CNRST, Ouagadougou,*
[1,3]*Burkina Faso*
[2]*Central African Republic*

## 1. Introduction

*Vitex simplicifolia* Oliv. (Verbenaceae) is a perennial shrub or small tree which grows to a height of aproximatively 8 m and is widely distributed from Egypt to Guinea. In Burkina Faso, the plant is used for internal or external use to treat various diseases like skin diseases, dermatitis, bilharzia, migraines, fever, aches, amoebiasis, sore teeth, colic, infant tetanus(Nacoulma,1996). Our ethnobotanical investigations have revealed that this plant is also used in the treatment of skin infections and wounds healing. In Burkina Faso, infectious diseases are the leading cause of infant mortality (2.37%) and maternal (14.6%), therefore they constitute public health problems. The treatment of skin diseases dates back to ancient times, and many treatments were using medicinal plants. About 30% of traditional remedies are used to treat wounds and skin lesions, compared to only 1-3% of modern drugs (Mantle et al., 2001). The healing process is an immune response that begins after injury and takes place in three stages: vascular and inflammatory stage, phase of tissue repair and phase of maturation. A drug having simultaneously the potential antioxidant and antimicrobial activities may be a good therapeutic agent to accelerate cicatrization and wound healing [Houghton et al., 2005; Phillips et al., 1991; Heike et al., 1999]. Aromatherapy is now considered to be another alternative way in healing people, and therapeutic values of aromatic plants lie in their volatile constituents such as monoterpenoids, sesquiterpenoids and phenolic compounds that produce a definite physiological action on the human body [Bruneton, 1993].To the best of our knowledge, there is no report on pharmacological studies of this plant. The present work reported results of a detailed investigation of

---

* Corresponding Author

cicatrization and antibacterial activities of the leaf essential oil with the aim to contributing to the search for beneficial uses of this plant.

## 2. Materials and methods

### 2.1 Animals

The experiments were conducted using six male and female rabbits aged 7 months and weighing between $1\pm0.05$ and $2\pm0.06$ kg, housed alone under 12h light-dark cycle, controlled humidity (75%), and temperature (20-25°C) conditions with free access to food and water [Draize et al., 1944]. In addition, three other rabbits were taken as controls. The experiments were carried out according to the method as described previously [Draize et al., 1944] in accordance with the guidelines for the care of laboratory animals and ethical guidelines for the investigation of experimental pain in conscious animals and revised by Official Journal of France (1971/04/21).

### 2.2 Plant material and extraction

The leaves of *V. simplicifolia* Oliv were collected in January 2007 from the Kadiogo region, village of Balgui, 10 km near Ouagadougou, Burkina Faso. A voucher specimen was identified by Pr. Jeanne Millogo, botanist (University of Ouagadougou) and deposited at the herbarium of IRSS of Ouagadougou.

Dried and powdered leaves (500g) were subjected to hydrodistillation for 4h with a clavenger-type apparatus. The essential oil was collected and dried, after decantation, over anhydrous sodium sulfate and stored in refrigerator at 4°C for further use [Ouoba et al., 2009].

### 2.3 Reference bacteria strains

Microorganisms used in this study were:

*Bacillus cereus* LMG13569BHI, *Listeria innocua* LMG13568BHI, *Staphylococcus aureus* ATCC 25293BHI, *Staphylococcus camorum* LMG 13567BHI, *Staphylococcus aureus* ATCC9144BHI, *Enterococcus faecalis* CIP103907BHI, *Proteus mirabilis* CIP104588, *Shigella dysenteria* CIP5451, *Salmonella enterica* CIP105150, *Escherichia coli* CIP105182. These strains were identified by the conventional methods and tested. Bacteria were obtained from stock cultures of the laboratory of pharmacology and clinic biochemistry of CRSBAN, University of Ouagadougou. The bacteria stock cultures were maintained on Müller-Hinton agar and which were stored at 4°C.

### 2.4 Wound healing activity

Assessment of the healing power of the oil was performed using the method [Draize et al., 1944], on 6 male and female rabbits housed in individual cages. Both flanks of each rabbit were shaved, deeply incised prior to application of the essential oil. Rabbits were fixed horizontally from their ears and legs. One flank was covered with a compress soaked with 0.44 mg (0.50 ml) of the pure oil and held by a sticking-plaster, the other untreated flank serving as control. The same operation was repeated with Cicatryl as a reference standard, with a dose of 1 g per flank. The rabbits were returned to their cages after treatment.

Observation of the evolution of wound healing versus time was carried out at 48h and 96h after treatment. All of the tests were made in duplicate.

## 2.5 Antibacterial activity

### Determination of the strain sensitivity

The test was performed using Müller-Hinton medium for bacteria using disk diffusion method following the National Committee for Clinical Laboratory Standards methods [Kiehlbauch et al., 2000]. Overnight broth cultures of each strain were prepared in nutriment Broth (Diagnostic Pasteur, France). The final concentration of each inoculums was got making dilution of each strain in 9 % NaCl solution. The turbidity of each inoculum was compared with McFarland 0.5 solution. The final concentration of each inoculum (approximatively $5.10^5$ CFU / ml) was confirmed by viable count on Plate Count Agar (Merck, Germany). 3µl of essential oil was put on every disk (8 mm diameter).

Positive and negative growth controls were performed for every test. The plates were incubated aerobically at 30°C or 37 °C for 24 hours. The bacterial sensitivity to the essential oil was assessed by measuring the diameter of inhibition zone and recorded if the zone of inhibition is greater than 9 mm. The inhibition zones were compared with that of ampicilline (BIO-RAD Marne- la- coquette, France) and tetracycline (BIO-RAD Marne- la-coquette, France). All of the tests were made in triplicate.

### Determination of MIC and MBC values

A broth microdilution method was used to determine the minimum inhibitory concentration (MIC) and the minimum bactericidal concentration (MBC) [Bassolé et al., 2003] All tests were performed in Mueller-Hinton Broth (Becton Dickinson, USA). A serial of double of each essential oil was prepared in 96 well-plates over the range 0.03-8% (v/v). The broth was supplemented with tween 80 at a concentration of 0.1% in order to enhance essential oil solubility. The tween 80 was at the final concentration of 0.001% (v/v). Overnight broth cultures of each strain were prepared in Nutrient Broth (Diagnostic Pasteur, France) and the final concentration in each well was adjusted to $5 \times 10^5$ CFU/ml following inoculation. The concentration of each inoculum was confirmed by viable count on Plate Count Agar (Merck, Germany).

Positive and negative growth controls were included in every test. The tray was incubated aerobically at 30 °C (Reference Gram-negative strain) or 37 °C (Reference and isolated Gram-positive) and MICs were determined. The MIC defined as the lowest concentration of the essential oil at which the microorganism tested does not demonstrate visible growth. To determine MBCs, 10µl suspension were taken from each well and inoculated in Mueller-Hinton Agar (Becton Dickinson, USA) for 24 h at 30 or 37 °C. The MBC is defined as the lowest concentration of the essential oil killing 99.9% of bacteria inocula [Michel Briand, 1986]. All tests were performed in triplicate.

### Statistical analysis

Data were expressed as mean±SEM. A one way variance was use to analyse data. $p < 0.01$ represented significant difference between means (Duncans multiple range test).

## 3. Results

### 3.1 Analyses

GC and GC/MS analyses of the essential oil composition of *Vitex simplicifolia* were as previously described [Ouoba et al., 2009] The oil contained monoterpenoids as predominant (71.02%). Among monoterpene hydrocarbon, myrcene (53.50%) had been found as the major component and four components were detected as predominant: α-pinene (5.13%), β-pinene (2.48%) and β-phellandrene (1.38%). In the oxygenated fraction, 10 monoterpenes (6.32%) and 12 sesquiterpenes (5.58%) were present with linalool (4.70%) and humulen-1,2- epoxyde (1.15%) as the major constituents. Among mono and sesquiterpenes three ketones are detected as minor compounds piperitone (0.05%) cis-jasmone (0.11%) and salvia-4(14)en-1-one (0.07%). No phenolic compound has been detected in the oil.

### 3.2 Wound healing activity

Experiments on rabbits showed that the oil has a healing effect. As shown in Table1, the different stages of evolution of healing capacity of essential oil in comparison with those of cicatryl and natural immunity of rabbit. The different stages of evolution of healing were characterized by changes in the color of the wound over time, the closure of lacerations and the absence of erythema and edema of the wounds.

| rabbits treated | wounds at 48h | wounds at 96h | wounds from 6 to 10 days |
|---|---|---|---|
| Essential oil | vascular and inflammatory stage tissue répair stage maturation stage (start) | maturation stage (end of cicatrization) **complete healing** | |
| Cicatryl | vascular and inflammatory stage tissue répair stage maturation stage (start) | tissue repair stage (end) maturation stage (start) | maturation stage (end of cicatrization) **complete healing** |
| Rabbits untreated | vascular and inflammatory stage tissue répair stage maturation stage (start) | tissue repair stage (end) maturation stage (start) | maturation stage (end of cicatrization) **complete healing** |

Table 1. Different stages of evolution of wounds healing.

### 3.3 Determination of the strain sensitivity

The results showed that almost of the bacterial strains were sensitive to *Vitex simplicifolia* essential oil (Table2). *Staphylococcus aureus* ATCC9144 BHI (zone of inhibition 34.5mm) was the most sensitive bacteria tested. Only *Bacillus cereus* LMG 13569 BHI was not sensible to *Vitex simplicifolia* (zone of inhibition 8.5mm) (Fig1).

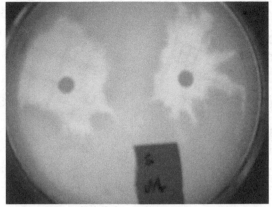

(1a) *Salmonella enterica*

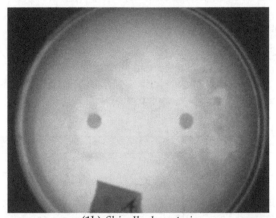

(1b) *Shigella dysenteria*

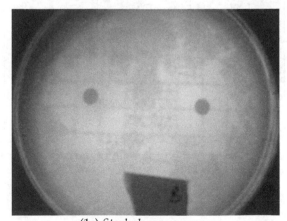

(1c) *Staphylococcus aureus*

Fig. 1. Inhibition zones for some bacterial strains.

| Reference strains | Origin | V. s | Amp | Te |
|---|---|---|---|---|
| *Enterococcus faecalis* CIP 103907 BHI | CIP | 21 | 35 | 20 |
| *Bacillus cereus* LMG 13569 BHI | LMG | 8.5 | 31 | 22 |
| *Listeria innocua* LMG 13568 BHI | LMG | 16.5 | 45 | 12 |
| *Staphylococcus aureus* ATCC 25293 BHI | ATCC | 24.5 | 56 | 27 |
| *Staphylococcus camorum* LMG 13567 BHI | LMG | 19.5 | 54 | 21 |
| *Staphylococcus aureus* ATCC 9144 BHI | ATCC | 34.5 | 40 | 32 |
| *Escherichia coli* CIP 105182 | CIP | 10.5 | 43 | 21 |
| *Proteus mirabilis* CIP 104588 | CIP | 11.5 | 21 | 18 |
| *Shigella dysenteria* CIP 5451 | CIP | 24.5 | 11 | 22 |
| Salmonella enterica *CIP 105150* | CIP | 28 | 36 | 24 |

V.s.: *Vitex simplicifolia* Amp: ampicilline Te: tetracycline

Table 2. Diameter of inhibition zone (mm) of bacteria growth.

### 3.4 Determination of antibacterial activity

The MICs and MBCs of *Vitex simplicifolia* essential oil were consigned in Table3. Five bacterial strains were selected and tested because of their highest sensitivity to essential oil. The oil inhibited the growth of these bacteria with MIC of 0.50% except for *Staphylococcus aureus* ATCC 9144BHI that was more sensitive with MIC of 0.25%. The results of MBC demonstrated a bacteriostatic effect.

| Reference strains | Origin | MIC | MBC | MBC/MIC |
|---|---|---|---|---|
| *Enterococcus faecalis* CIP 103907 | CIP | 0,5 | 8 | 16 |
| *Escherichia coli* CIP 105182 | CIP | 0,5 | 4 | 8 |
| *Listeria innocua* LMG 135668 | LMG | 0,5 | 4 | 8 |
| *Staphylococcus aureus* ATCC 25293 | ATCC | 0,5 | 8 | 16 |
| *Staphylococcus camorum* LMG 13567 | LMG | 0,5 | 8 | 16 |
| *Staphylococcus aureus* ATCC 9144 BHI | ATCC | 0.25 | 1 | 4 |

Table 3. Minimum inhibitory concentration, minimum bactericidal concentration data (%v/v) obtained by microdilution method.

## 4. Discussion

Wound healing is very complex, it involves a sequence of multifactorial events including several cellular and biochemical processes. These processes aim to ensure the regeneration and reconstruction of anatomical and functional disturbances of the skin[Chattopadhyay et al., 2002]. The repair of damaged tissues occurs as a sequence of events that included inflammation, proliferation and migration of different cell types [Sidhu et al., 1999]. At the dose of 0.44 mg used only once the essential oil healed wounds for 96h. Essential oil of *Vitex simplicifolia* accelerated the three stages of cicatrization process: vascular and inflammatory, tissue repair and maturation. While at the dose of 1000 mg used only once, cicatryl exhibited a complete healing with for 7 days against 10 days for the effect of natural immunity of rabbit. In the phase of maturation a renewal of the skin was seen, the old skin started to fall and made way for the new. The essential oil of *Vitex simplicifolia* exhibited a stronger healing effect than cicatryl and natural immunity. This effect could be due to the presence of ketones in the oil that activated the healing process with stimulating of new cell growth, reducing old scare tissue in wound and were highly immunostimulatory [Willem, 2004]. The presence of minor compounds as aldehydes and sesquiterpenes activated anti inflammatory, calming and sedative effects. Thus, their low proportion allowed to consider possible synergistic effects of these compounds in the oil. The significant presence of monoterpenoids in the oil might cause analgesic, antioxidative, antiseptic effects and stimulating the immune system [Mertz et al., 1993].

In other hand, the skin infections are in most cases due to staphylococci with the pathogenic species is Staphylococcus aureus. It is responsible for suppurative infections, widespread and food poisoning. Thus, wound infections are most common in developing countries because of poor sanitation. *Staphylococcus aureus, Shigella dysenteria, Salmonella enterica, Escherichia coli* are important microorganisms causing an infection of the wound [Mansouri et al., 2011]. The best sensitivity to *Vitex simplicifolia* essential oil was, respectively, obtained on *Staphylococcus aureus* ATCC 9144 BHI (34.5mm), *Salmonella enterica* CIP 105150 (28mm), *Shigella dysenteria* CIP 5451(24.5mm), *Staphylococcus aureus* ATCC 25293 BHI (24.5mm), *Enterococcus faecalis* CIP 103907 BHI(21mm). Following the results in Table 2 the different strains were less sensitive to V.s than ampicilline, while *shigella dysenteria* CIP 5451 was sensitive to V.s. The most important information was that essential oil exhibited more activity on *Staphylococcus aureus* ATCC 9144 BHI (34.5mm), *Salmonella enterica* CIP 105150 (28mm) *Shigella dysenteria* CIP5451(24.5mm) than tetracycline and ampicilline (*Shigella dysenteria* CIP 5451, 11mm). The essential oil failed to inhibit *Staphylococcus aureus* ATCC 9144 BHI at the lowest MIC 0.25%. The essential oil was bacteriostatic for *Staphylococcus aureus* ATCC 9144 BHI, *Escherichia coli* CIP 105182 and *Listeria innocua* LMG 135668. The most resistant strains with highest MBC (8%) were *Enterococcus faecalis* CIP 103907, *Staphylococcus aureus* ATCC 25293 and *Staphylococcus camorum* LMG 13567. Considering MICs and MBCs no significant difference could be seen between Gram-positive and Gram-negative bacteria. The chemical composition of the oil consisted of various constituents. Therefore, the determination of the component responsible for activity was very difficult. Furthermore the essential oil consists of complex mixture of numerous constituents. Major or minor compounds might cause the bacteriostatic and cicatrization activities exhibited, terpinene-4-ol and other monoterpenes in essential oil may act as antiseptic, anti-inflammatory and antimicrobial: myrcene, sabinene, terpinene, cadinene and limonene [Sinan Dayisoylu et al. 2009]. In addition, the presence of α-pinene, β-pinene [Houghton Peter, 2004] terpinen-4-ol [Lee et al., 2001] and γ-terpinene[Sonboli et al., 2005] were responsible of antioxidative and antiseptic activities of essential oils studied. However, caryophyllene oxide, E-nerolidol humulene epoxide-1,2 possessed antiinflammatory activity [Chavan et al., 2010; Yu-Tang et al., 2008; Wanjohi Mwangi et al., 2009] Possible synergistic and antagonistic effects compounds in *V. simplicifolia* essential oil should also be taken into consideration. These reports are compatible with our results in the present study.

## 5. Conclusion

This study shows in *vivo* wound healing activity and in *vitro* bacteriostatic effect of *Vitex simplicifolia* essential oil. The oil demonstrates the strongest wound cicatrization activity than cicatryl and natural immunity. In addition the oil may help to prevent wound infections and others such diarrhoea, dysentery and skin diseases. These results indicate that the plant could be use as a natural potential remedy for healing wounds and antiseptic agent. Further investigations will be performed by determination of analgesic, antioxidant and anti inflammatory activities of the essential oil and to expand to other *Vitex* species.

## 6. Remarks

1. Choice of rabbits: we have chosen rabbits because of they were available in the laboratory and cheaper. They were also very easy to be used in the cicatrization effect than mice and rats
2. The resolution of photographs depend of the quality of the apparatus, we deleted them because we have not a best quality. We are sorry for the bad quality of photos. Thank you for your understanding.

## 7. References

Bassole, I.H.N., Ouattara, A.S., Nebié, R., Ouattara, C.A.T., Kaboré, Z.I., Traoré, S.A., 2003. Chemical composition and antibacterial activities of essential oils of *Lippia chevalieri* and *Lippia multoflora*. Phytochemistry 62, 209-212

Bruneton, J., 1993. Pharmacognosy, phytochemistry, medicinal plants. 2nd Tech and Doc. Lavoisier Paris p915

Chattopadhyay, D., Arunachalam, G., Mandal, A.B., Sur, T.K., Mandal ,S.C., Bahattacharya, S.K., 2002. Antimicrobial and anti inflammatory activity of folklore: Mellotus peltatus leaf extract. Journal of Ethnopharmacognosy 82, 229-237

Chavan, M.J., Wakte, P.S., Shinde, D.B., 2010. Analgesic and anti-inflammatory activity of Caryophyllene oxide from Annona squamosa L. bark. Phytomedicine international journal of phytotherapy and phytopharmacology 17(2), 149-151

Draize, J.H., Woodward, G., Calvery, H.O., 1944. Methods for study of irritation and toxicity of substances applied topically to the skin and mucous membranes. Journal of Pharmacology and Experimental Therapeutics 2, 377-390

Heike, S., Munz, B., Werner, S., Brauchle, M., 1999. Different types of ROS-scavenging enzymes are expressed during cutaneous wound repair. Experimental Cell Research 247, 484-494

Houghton Peter, J., 2004. Activity and constituents of sage relevant to the potential treatment of symptoms of Alzheimer's disease. Herbal Gram 61, 38-54

Houghton Hiylands, P.J.., Mensahb, A.Y., Hensel, A., Deters, A.M., 2005.In vitro tests and ethnopharmacological investigations: wound healing as an example. Journal of Ethnopharmacology 100:107-100

Kiehlbauch, Julia A., Hannett, G.E, Salfinger, M., Archinal, W., Monserra, C., Carlin, C., 2000.Use of the National Committee for Clinical Laboratory Standards Guidelines for Disk Diffusion Susceptibility Testing in New York State Laboratories. J. Clin. Microbiol. 38(9), 3341-3348

Lee, K.G., Shibamoto, T.J., 2001.Antioxidant activities of volatile components isolated from Eucalyptus species. J. Sci. Food. Agric. 81, 1573-1579

Mansouri, S.B., Ghanmi, M., El Ghadraoui, L., Guedira, A, AAFI, A., 2011.Composition chimique, activité antimicrobienne et antioxydante de l'huile essentielle de juniperus communis du maroc. Bulletin de la Société Royale des Sciences de Liège 80, 791- 805

Mantle, D., Gok ,M.A., Lennard T.W.J., 2001. Adverse and beneficial effects of plant extracts on skin and skin disorders. Adverse Drug Reactions and Toxicological Reviews 20(2): 103-89

Mertz ,P., Ovington L., 1993.Wound healing microbiogy. Dermatologic Clinics 11(7), 739

Michel Briand, Y., 1986. Mécanismes moléculaires de l'action des antibotiques. Collections de Biologie moléculaire. Edition Masson p370

Ouoba, A.M., Koudou, J., Somé, N., Guissou, I.P., Figueredo, G., Chalchat ,J.C., 2009. Asian Journal of Chemistry 21(4), 3304-3306

Phillips, G.D., Whitehe, R.A., Kinghton, D.R. 1991. Initiation and pattern of angiogenesis in wound healing in the rat. American Journal of Anatomy 192, 257-262

Sidhu, G.S., Mani, H., Gaddipati, Singh, J.P., Seth, P., Banaudha, K.K., Patnaik, G.K., Maheshwari, R.K., 1999.Curcumin enhances wound healing in streptozotocin induced diabetic rats and genetically diabetic mice. Wound Repair and Regeneration 7, 362-374

Sinan Dayisoylu, K., Duman, A.D., Hakki Alma, M., Digrak, M. 2009. Antimicrobial activity of the essential oils of rosin from cones of Abies cilicica subsp. Cilicica. African Journal of Biotechnology 8(19), 5021-5024

Sonboli, A., Saleli, P., Kanani, M.R., Ebrahimi, S.N., 2005. γ-terpinene, pcymene, antibacterial and antioxidant activities. Z. Naturforsch 60c, 534-538

Wanjohi Mwangi Julius, Njeri Thoithi Grace, Ongubo Kibwage Isaac, 2009.Essential Oil Bearing Plants from Kenya: Chemistry, Biological Activity and Applications. Rodolfo Juliani, H., Simon James, E., Ho Chi-Tang. African natural plant products: new discoveries and challenges in chemistry and quality. Ed. Washington, DC, American Chemical Society, New York, distributed by Oxford University Press chap27, 495-525

Yu-Tang Tung, Meng-Thong Chua, Sheng-Yang Wang, Shang-Tzen Chang, 2008. Anti-inflammation activities of essential oil and its constituents from indigenous *Cinnamon* (*Cinnamomum osmophloeum*) twig. Bioresource Technology 99, 3908-3913

# 6

# Simple Preparation of New Potential Bioactive Nitrogen-Containing Molecules and Their Spectroscopy Analysis

Vladimir V. Kouznetsov, Carlos E. Puerto Galvis,
Leonor Y. Vargas Méndez and Carlos M. Meléndez Gómez
*School of Chemistry, Industrial University of Santander, Bucaramanga,*
*Colombia*

## 1. Introduction

The impact of research on the small molecules chemistry is difficult to quantify and currently, it is still one of the most active areas of organic chemistry, medicinal chemistry and lately chemical biology. In recent years, a lot of interest has been shown in the preparation of nitrogen-containing compounds due to their numerous biologically significant activities. But it is the separation and purification process of the new synthetized organic molecules, the ones that take a key role in drug design and development.

Many texts about the simple and optimal preparation of bioactive compounds have been published, and in this chapter the multicomponent reactions and efficient linear process, which allow the synthesis of this kind of structures, will be discussed. However, the purpose of this chapter is to reveal those important aspects that finally determined why a molecule can be used and distributed as a drug: their preparation, purification, and characterization.

In almost all organic synthetic methodologies the purification process use simple column chromatography techniques (gravity or external pressure) using different support materials (solid adsorbents) as the stationary phase. Column chromatography is advantageous over most of the other chromatographic techniques because it can be used in both analytical and preparative applications. After the preparation and purification of a new compound has been realized, it becomes the characterization step. New purified molecules must be strongly characterized to determine its structural configuration. Among different analysis techniques, NMR experiments and X-Ray crystallography are the most efficient ways to determine the relative stereochemistry and, in suitable cases, also the absolute configuration of the obtained products.

In the development of our medicinal program directed to small molecules for drug delivery, the strategies for the preparation of nitrogen-containing molecules such as substituted indoles, tetrahydroquinolines, and N-substituted amides of carboxylic acids are illustrated in this chapter as well as their synthetic applications and analytic characterization. The discussion is complemented with a deep explanation of the analytical techniques employed for their isolation and purification including the spectroscopic and spectrometric techniques using for the elucidation structure for every new compound.

## 2. Simple preparation of new N-aryl-N-(3-indolmethyl) acetamides and their spectroscopic analysis

The vital importance of derivatives like indole-3-acetic acid, as a hormone responsible for the plant growth, and tryptophan, as a constituent of proteins and indispensible precursor of indole alkaloids, enhances the current interest for the design of simple and efficient synthetic routes for the preparation of molecules with the indole skeleton.

The research of the indol chemistry has been and it is still one of the most active areas of heterocyclic chemistry. In the past decades, a lot of interest has been shown in the preparation of substituted indoles due to their numerous biologically significant activities (Gribble, 2003). The derivatives of 3-indolylmethanamine 1 are the important intermediates of natural and natural-like products, such as hydro-γ-carboline and pyrido[4,3-b]indole derivatives (Wynne & Stalick, 2002; Molina et al., 1996). This 3-indolyl methanamine motif is also embedded in numerous indole alkaloids from the simple alkaloid gramine 2 to complex aspidospermine alkaloid 3 (Baxter et al., 1999; Saxton, 1998) (Fig. 1).

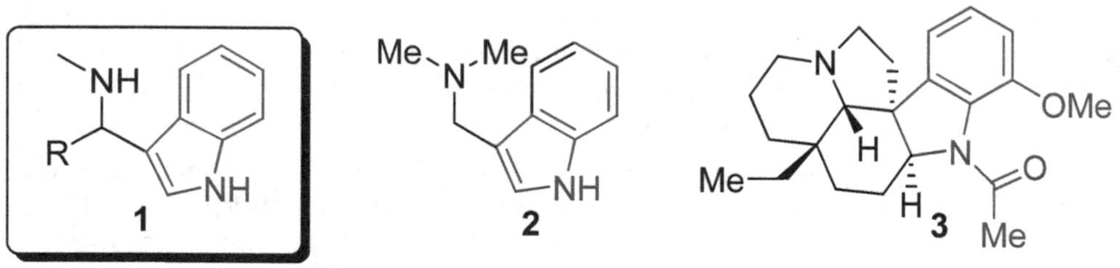

Fig. 1. Relevant natural alkaloids derived from the 3-indolylmethanamine system.

As a result of their biological and synthetic importance, a variety of methods have been reported for the preparation of 3-substituted indoles, using indol or 3-indolcarboxyaldehyde as starting materials. Generally, the Mannich reaction (Dai et al., 2006) and the catalyzed Friedel-Crafts alkylation reactions of indoles (Ke et al., 2005; Zhao et al., 2006; Jiang et al., 2005; Shirakawa & Kobayashi, 2006) are considered as a powerful carbon-carbon bond process to afford the 3-indolylmethanamine derivatives 1. However, another synthetic route to access to these compounds by using 3-indolcarboxyaldehyde, via its imino derivatives, is valid. This route has been used by our laboratory, which recently started an own medicinal program directed to small molecules for drug delivery. The particular interest in 3-indolylmethanamine derivatives molecules, that could serve as useful precursors to many drug-like indolic or quinolinic compounds, is based on the evaluated antiparasitic properties of some analogues (Kouznetsov et al., 2004a, Kouznetsov et al., 2004b; Vargas et al., 2003). In this novel direction, the simple preparation of new (3-indolmethyl)acetamide and (1-acetylindolmethyl-3)acetamide, regulating only the solvent nature, is the relevant fact that has not been described and it gives the opportunity to prepare more of this kind of compounds.

## 2.1 Synthesis and purification of the new the (3-indolmethyl)acetamide and (1-acetylindolmethyl-3)acetamide

Aldimines are valuable starting materials not only for different N-containing heterocycles but also to diverse secondary heteroaromatic amines (Hutchins & Hutchins, 1991), which represent good candidates for bio-screening with diverse types of activities (Kleemann & Engel, 2005; Evers et al., 2005). Thus, the N-aryl imine **6**, the main and value starting, can be prepared from commercially available 3-indolaldehyde **4** and 2-cyanoaniline **5**, according to published methods (Colyer et al., 2006). This aldimine is obtained as a white and stable solid after purification by recrystallization in 95 % yield (Fig. 2). The method employed to obtain the precursor **7**, is one of the most known procedures to reduce an aldimines with an excess of NaBH$_4$ in methanol, protocol that is still the reaction of choice to produce secondary amines in reasonably good yield. Thus, N-(2-cyanophenyl)-N-(3-indolylmethyl)amine **7** is prepared as was described and was obtained as white solid in 70 % yield after purification through recrystallization (Bello et al., 2010). This example, in which the main precursor is prepared in a linear process and purified by recrystallization, represents an excellent technique that avoids the loss of value material and it will be increase the global yields of the final product.

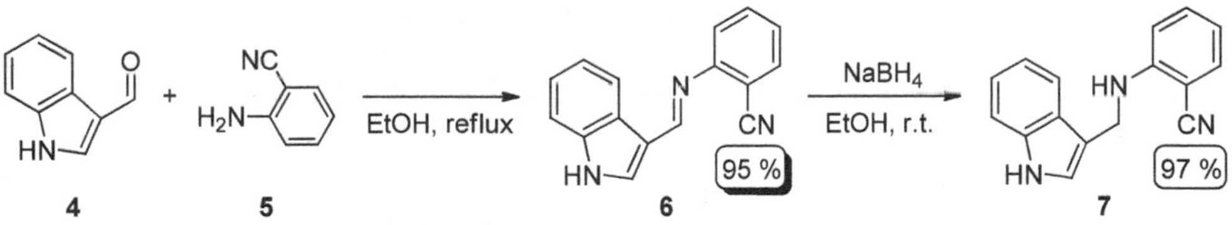

Fig. 2. Synthesis and reduction of the aldimine **6** to give the desired secondary amine **7** in excellent overall yields.

Besides the efficient preparation and easily purification of compound **7**, this amine has interesting structural elements to use in the synthesis of different indolic heterocycles. In this case, the study of its acetylation by acetic anhydride is showed.

First, to a stirred solution of amine **7**, using in toluene as solvent due to the insolubility of compound **7** in polar common solvents (CH$_3$CN, CH$_2$Cl$_2$, AcOEt and DMF), it is added Et$_3$N and acetic anhydride, the mixture is refluxed for appropriate time to obtain the N-(2-cyanophenyl)-N-(3-indolmethyl)acetamide **8** in acceptable yield (50 %) after purification using silica gel 60 Mesh and using a mixture of hexane: ethyl acetate (2:1) as an eluent.

Then, the acetylating reaction described above was performed between the amine and an excess acetic anhydride in the presence of Et$_3$N at 100 °C, without the organic solvent (toluene). After the usual workup, the diacetylated indole **9** is obtained in good yield (85 %) using the same parameters employed to the purification of compound **8**. This simple change in the reaction conditions could afford different acetamides based on the 3-indolyl methanamine motif (Fig. 3). This finding reveals a selective process to protect different amino groups and represents a good protocol to the synthetic organic chemistry, especially within those processes that require a particular position protection.

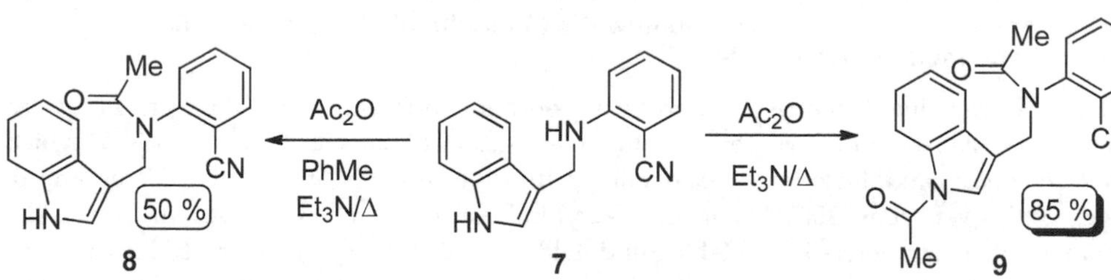

Fig. 3. Synthesis and purification of the N-aryl-N-(3-indolmethyl)acetamides **8** and **9**.

## 2.2 Characterization of the *N*-aryl-*N*-(3-indolmethyl)acetamides 8 and 9 by spectroscopy and spectrometric techniques

The structures of the C-3 substituted indoles **7-9** were confirmed on the basis of recorded analytical and spectral data and are supported by inverse-detected 2D NMR experiments. The IR spectrum of compound **7** illustrates the characteristic absorption bands at 3402 and 3352 $cm^{-1}$ assignable to tension vibrations of $CH_2$-N-H and N-$H_{indol}$, respectively. Its [1]H NMR spectrum displays a duplet at δ 4.56 ppm ($J$ = 4.9 Hz) ppm corresponding to two protons coupling with the neighbor N-H proton (br. s, 4.84 ppm), which suggest the presence of the methylenic unit linked to the N-H function. The peaks find at δ 7.17-7.3, 7.36-7.40 and 7.63 ppm reveals the presence of aromatic protons of the indole moiety. The [13]C NMR spectra, also showed all expected characteristic peaks at δ 39.4 ($CH_2$), 117 (CN), and 95.6-150.2 (aromatic carbons).

The mass spectrometric analysis of compound **8** gives a molecular ion peak M$^+$·, at $m/z$ 289, suggesting the molecular formula $C_{18}H_{15}N_3O$, and indicating that the acetyl group is coupling with **7**. The acetamide **8** displayed characteristic infrared absorption bands with a single amine absorption band at 3342 $cm^{-1}$ and with a carbonyl sign at 1701 $cm^{-1}$ suggesting the acetylation reaction involvement of the $CH_2$-N; this is the band appearing at high wave number of the corresponding N-$H_{indol}$ vibration tension in the IR spectrum. Its [1]H NMR spectra analysis showed a singlet at 2.58 ppm corresponding to three protons which belong to the acetyl group and another singlet at 4.55 ppm due to the presence of the methylenic 3-$CH_2$-N indolic protons. This signal's multiplicity is explained by assuming the proton N-H next to it, substituted now for the acetyl group, which leaves no possibility to H,H coupling, while it does happen with the amine **7**.

The [13]C NMR spectrum of the acetamide **8** displayed characteristic carbonyl signal at 168 ppm, this is strong evidence to the new acetyl group bonded to the molecule; in addition to a signal at 39.3 and 23.9 ppm, showing the presence of $CH_2$ and $CH_3$ in the molecule. Introduction of an acetyl group into the molecule affects the chemical shift of the hydrogen bonded to the aromatic carbon close to the acetylatin nitrogen; this appears at 116 ppm for the compound **7** and at 123 ppm for the acetamide **8**. The signal at 39.3 ppm for $CH_2$-N has been distinguished on the basis of the DEPT-135 experiment. On the basis of these spectral studies, compound **8** was characterized as the N-(2-cyanophenyl)-N-(1H-indol-3-ylmethyl)acetamide.

The new compound **9** gave a molecular ion peak at $m/z$ 331 in the mass spectrometric analysis, corresponding to the molecular formula $C_{20}H_{17}N_3O_2$ as indicated by its EI-MS. The loss of 43 units (one acetyl group) generates the same mass spectrum as the acetamide **8**. The IR spectrum of this molecule shows bands at 1704 and 1654 cm$^{-1}$, assignable to two carbonyl groups while the N-H absorption bands are not observed in the region of 3300-3400 cm$^{-1}$. The $^1$H NMR spectrum showed, as expected, two singlets at δ 22.4 and 23.9 ppm, which integrated for three protons each. In the case of the methylenic protons, they appeared to be diasterotopic resonating at the high field frequencies δ 4.75 and 5.46 ppm with a coupling constant $J$ = 15 Hz, usual constant value to a germinal coupling. Of course, the aromatic protons were also assigned.

The $^{13}$C NMR spectrum showed all expected characteristic peaks at δ 169.4 (ArN-CO-), 168.5 (Ar$_{indol}$N-CO-) ppm, in addition to a signal at δ 117.3 ppm showing the presence of C≡N in the molecule. Besides, methyl carbons at 23.9 (Ar$_{indol}$NCO-CH$_3$) and 22.4 (ArNCO-CH$_3$) ppm and the methylene carbon at δ 42.8 ppm were also displayed in the $^{13}$C NMR.

With respect to the characterization of the diacetamide **9**, through X-ray diffraction, the monoclinic system was determined with the compound crystallized at 25°C from heptane-ethyl acetate (2:1) (Fig. 4).

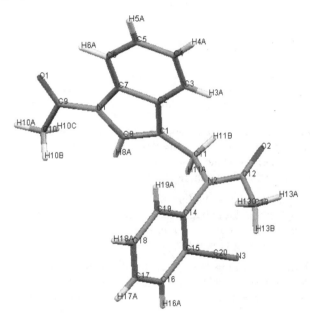

Fig. 4. X-Ray structure of the diacetamide **9**.

The crystallized material has the following cell constants: a = 11.1184(19) Å, b = 8.0048(13) Å, c = 20.534(4) Å and space group P 2$_1$/n (Table 1), possessing the different bond lengths of the molecule constituent atoms was also extracted with this technique (Table 1).

From this data, the different bond lengths of the two amide bonds present within the structure were as expected. Even knowing the double bond character of the amide bonds, in this case, the amide bond distance between the aliphatic nitrogen N2 and C12 is 1.364 Å, while the distance between the aromatic nitrogen N1 and C9 is 1.388 Å.

| Crystal morphology | White parallelepipede |
|---|---|
| Chemical formula | $C_{20}H_{17}N_3O_2$ |
| Molecular weight | 331.13 |
| Crystal system | Monoclinic |
| Space group | $P2_1/n$ |
| Cells constants | $a$ = 11.1184(19) Å, $b$ = 8.0048(13) Å and $c$ = 20.534(4) Å, $\alpha$ = 90°, $\beta$ = 94.281(4), $\gamma$ = 90° |
| Volume | 1822.4(5) Å³ |
| Absorption coefficient | 0.82 mm⁻¹ |
| Temperature | 293(2) K |
| Range for data collection | 1.99-28.09 |
| Index range | $h$ = -13 → 28, $k$ = -9 → 7, $l$ = -23 → 23 |
| $R$ | 0.0567 |
| $R_w$ | 0.0582 |
| Threshold expression | >2sigma(I) |
| Diffraction radiation | $M_0K\alpha$ |
| $\lambda$ | 0.71070 Å |

Table 1. Crystal data and structure refinement parameters of diacetamide 9.

From this data, the different bond lengths of the two amide bonds present within the structure were as expected. Even knowing the double bond character of the amide bonds, in this case, the amide bond distance between the aliphatic nitrogen N2 and C12 is 1.364 Å, while the distance between the aromatic nitrogen N1 and C9 is 1.388 Å.

These data correspond with the thought that the amide bond N2-C12 is shorter because of the electron withdrawing inductive effect from the $\alpha$-cyanophenyl substituent and the possibility of the nitrogen non-shared electrons to be delocalized on the amide bond through a mesomeric effect giving this bond a stronger double bond character.

On the other hand, the amide bond N1-C9 is longer because the nitrogen non-shared electrons are compromised with the aromatic system and they are not as available to be delocalized on the amide bond giving it less double bond character (Table 2).

## 2.3 Conclusions

An efficient, economic, and fast synthetic route was designed and its illustrated in this section showing the possible construction of the N-aryl-N-(3-indolmethyl)acetamides with the incorporation of the indolic core as a structural analogues of some alkaloids.

The acylation method is worth as a regioselective process because the conditions variations lead to the mono- or di-acetamide. The characterization of the obtained compounds through different techniques gives evidence enough and strong support with regard to the success of the proposed scheme.

| Number | Atom 1 | Atom 2 | Length (Å) |
|--------|--------|--------|------------|
| 1 | O1 | C9 | 1.220 |
| 2 | O2 | C12 | 1.221 |
| 3 | N1 | C7 | 1.415 |
| 4 | N1 | C8 | 1.405 |
| 5 | N1 | C9 | 1.388 |
| 6 | N2 | C11 | 1.473 |
| 7 | N2 | C12 | 1.364 |
| 8 | N2 | C14 | 1.430 |
| 9 | N3 | C20 | 1.143 |
| 10 | C1 | C2 | 1.449 |
| 11 | C1 | C8 | 1.338 |
| 12 | C1 | C11 | 1.489 |
| 13 | C2 | C3 | 1.385 |
| 14 | C2 | C7 | 1.404 |
| 15 | C3 | H3A | 0.931 |
| 16 | C3 | C4 | 1.371 |
| 17 | C4 | H4A | 0.929 |
| 18 | C4 | C5 | 1.378 |
| 19 | C5 | H5A | 0.931 |
| 20 | C5 | C6 | 1.377 |
| 21 | C6 | H6A | 0.930 |
| 22 | C6 | C7 | 1.384 |
| 23 | C8 | H8A | 0.930 |
| 24 | C9 | C10 | 1.481 |
| 25 | C10 | H10A | 0.960 |
| 26 | C10 | H10B | 0.961 |
| 27 | C10 | H10C | 0.961 |
| 28 | C11 | H11A | 0.971 |
| 29 | C11 | H11B | 0.971 |
| 30 | C12 | C13 | 1.497 |
| 31 | C13 | H13A | 0.961 |
| 32 | C13 | H13B | 0.960 |
| 33 | C13 | H13C | 0.960 |
| 34 | C14 | C15 | 1.387 |
| 35 | C14 | C19 | 1.364 |
| 36 | C15 | C16 | 1.388 |
| 37 | C15 | C20 | 1.427 |
| 38 | C16 | H16A | 0.930 |
| 39 | C16 | C17 | 1.363 |
| 40 | C17 | H17A | 0.929 |
| 41 | C17 | C18 | 1.360 |
| 42 | C18 | C18A | 0.929 |
| 43 | C18 | C19 | 1.397 |
| 44 | C19 | H19A | 0.931 |

Table 2. Bond lengths between the molecule atoms of diamide 9.

## 3. A convenient procedure for the synthesis of new quinoline derivatives and their spectroscopic analysis

Quinoline and tetrahydroquinoline structures are essential feature of many natural products. These heterocycles play a key role in heterocyclic and medicinal chemistry. Their syntheses by various methodologies have been published extensively (Kouznetsov et al, 2005; Kouznetsov et al., 1998; Katrizky et al., 1996; Jones, 1984).

Polyfunctionalized tetrahydroquinolines (THQs) are molecules of great interest in organic synthesis due to the fact that many natural products present this system in their structure, and these compounds exhibit diverse biological activities (Glushenko et al., 2008; Broch et al., 2008; Ichikawa et al., 2004; Morsali, et al., 2004; Chen et al., 2000).

Apart from their marked bioactivities, THQs are also important and reliable precursors in quinoline preparation, another group of heterocyclic molecules that has a great number of pharmacological properties (Trpkovska et al., 2003). An efficient route for the preparation of THQs is the acid-catalyzed Povarov reaction that is classified as imino Diels Alder cycloaddition (Kouznetsov, 2009; Paazderski et al., 2007; Kouznetsov & Mora, 2006; Youssed et al., 2003) that permits the condensation of anilines, aldehydes, and electron-rich alkenes using acidic catalysts under mild conditions to afford new substituted tetrahydroquinolines.

For any Direct Oriented Synthesis (DOS) methodology towards the synthesis of bioactive substituted tetrahydroquinolines and quinolones, this route represents an easily and scalable approach for the investigations on the synthesis of small drug-like (tetrahydro)quinoline molecules containing C-2 aryl fragment, those synthesis could be accomplished via cycloaddition reactions. In this order, this section explain the simple preparation of new *N-(2-nitrophenyl-1,2,3,4-tetrahydroquinolin-4-yl)* pyrrolidin-2-ones using BiCl$_3$-catalyzed three component Povarov reaction between nitrobenzaldehydes, toluidine and N-vinylpyrrolidin-2-one, and their transformations into potentially bioactive 2-aryl-tetrahydroquinoline derivatives, *N-amidyl* substituted at the C-4 position.

### 3.1 Synthesis of new *N-(2-nitrophenyl-1,2,3,4-tetrahydroquinolin-4-yl)* pyrrolidin-2-ones

Having an experience in the construction of diverse heterocycles containing nitrogen via multi-component Povarov reaction, (Kouznetsov et al., 2010; Kouznetsov et al., 2007; Kouznetsov et al., 2006). The preparation of the selected tetrahydroquinoline compounds **14** and **15** was achieved using BiCl$_3$ as a catalyst for the three-component imino Diels-Alder cycloaddition between toluidine **10**, nitrobenzaldehydes **12** and **13** and N-vinylpyrrolidin-2-one **11** (NVP) (Bermudez et al., 2011) (Fig. 5).

These reactions proceeded smoothly in MeCN and at room temperature giving final products, substances easy to purify by column chromatography using silica gel as a support and a mixture of petroleum ether and ethyl acetate (2:1) as an isocratic eluent. The *N-(2-nitrophenyl-1,2,3,4-tetrahydroquinolin-4-yl)* pyrrolidin-2-ones **14** and **15** can be obtained with good respective yields 95% and 70% (Table 3) as a easily handle solids.

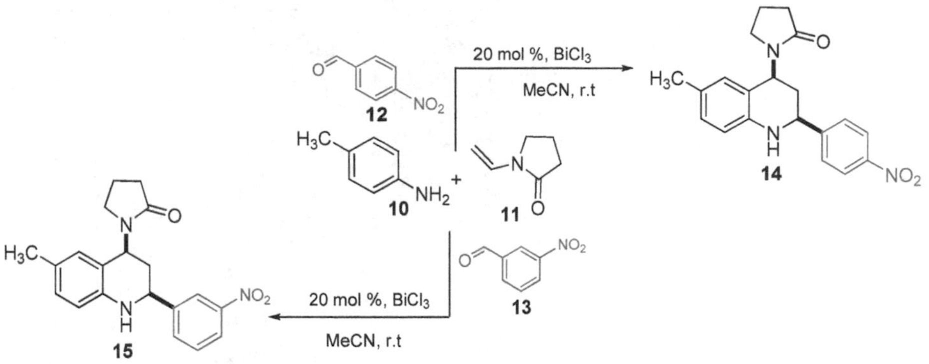

Fig. 5. Synthesis of nitrophenyl-tetrahydroquinolines using the multi-component imino Diels-Alder reaction.

| Comp. | Molecular Formula | Molecular Weight | IR (KBr), v, cm$^{-1}$ | mp, $^{\circ}$C | Yield (%) |
|---|---|---|---|---|---|
| 14 | $C_{20}H_{21}N_3O_3$ | 351.40 | 3394, 2947, 2916, 1666, 1620 | 222-223 | 95 |
| 15 | $C_{20}H_{21}N_3O_3$ | 351.40 | 3271, 2972, 2916, 2854, 1666 | 242-243 | 70 |

Table 3. Physical description, IR data and yields of the 2-nitrophenyl tetrahydroquinolines **14,15**.

### 3.2 Characterization of the *N-[6-methyl-2-(4´-nitrophenyl)-1,2,3,4-tetrahydroquinolin-4-yl]* pyrrolidin-2-one 14 by spectroscopy and spectrometric techniques

The structures of the C-2 substituted tetrahydroquinolines **14** and **15** were confirmed on the basis of analytical and spectral data and were supported by inverse-detected 2D NMR experiments. The IR spectrum show the characteristic absorption bands of the compound **14** at 3394 and 1666 cm$^{-1}$, assignable to the amine and amide group, respectively, and the nitro group signals at 1512 and 1342 cm$^{-1}$. Their mass spectrum showed a molecular ion m/z: 351, it coincided with the molecular weight (351 g/mol). The $^1$H NMR spectrum of this compound presented the 4-H proton signal at 5.69 ppm, observed as a double doublet with the coupling constants 6.4 Hz and 11.1 Hz.

This fact suggested axial-axial and axial-equatorial interactions between 4-H and 3-H protons. On the other hand, the 2-H proton signal was observed at 4.65 ppm with the coupling constants 3.1 Hz and 10.7 Hz that indicated at vicinal axial-axial and axial-equatorial interactions (Fig. 6).

The high value of the coupling constant (10.7-11.1Hz) of the 4-H and 2-H protons confirmed the axial proton configurations; therefore, substituents of the C-2 and C-4 positions of tetrahydroquinoline ring have the equatorial disposition, respectively.

On the other hand, it was found by the COSY experiment that the signal at 2.13-1.99 ppm belongs to the 3-H proton, observing the 3-H (4"-H) (2.13–1.99 ppm) and 4.65 ppm (2-H) and 5.69 ppm (4-H) cross peaks interactions (Fig. 7).

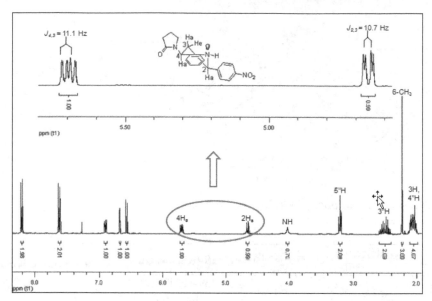

Fig. 6. ¹H NMR spectra of *N-[6-methyl-2-(4'-nitrophenyl)-1,2,3,4-tetrahydroquinolin-4-yl]* pyrrolidin-2-one **14**.

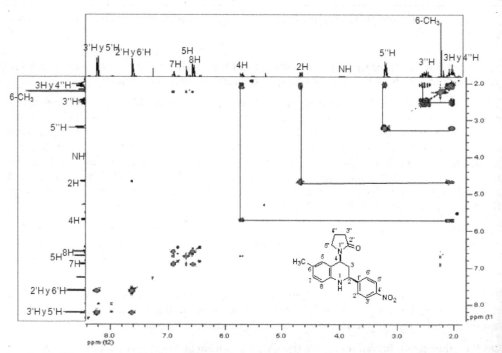

Fig. 7. COSY spectrum of *N-[6-methyl-2-(4'-nitrophenyl)-1,2,3,4-tetrahydroquinolin-4-yl]* pyrrolidin-2-one **14**.

The nitro-isomer **15** has similar chemical behavior in the spectra data. The chemical structures of the obtained *N-(1,2,3,4-tetrahydroquinolin-4-yl)* pyrrolodin-2-one molecules were strongly confirmed through IR, $^1$H and $^{13}$C NMR analyses.

However, having a possible mechanism of realized multi-component condensation, we could anticipate the various diastereomers the *cis* or *trans* configuration. For these reasons, further structural studies were realized.

### 3.3 X-Ray diffraction single crystal study

Samples crystals of both compounds of interest were growth by slow evaporation in ethanol. However, the difficulty to obtain suitable crystals from compound **15** only allows performing the study for the compound **14**. The diffraction data of the compound **14** were collected at 273K using a CCD area detector with graphite-monochromatic Mo $K_\alpha$ radiation ($\lambda = 0.71073$ Å).

This data were computed using Bruker-AXS software. For solution and refinement of the structure Shelxs-97 and Shelxl-97 (Sherldrick, 1997a; Sheldrick, 1997b) were used respectively. Molecular and crystal structures were obtained using Mercury software (Allen, 2002).

The molecular structure for the compound is presented in the Figure 8, where a *cis* conformation of the C-2 and C-4 substituents is evident, as well as a chair configuration that adopts tetrahydroquinoline system.

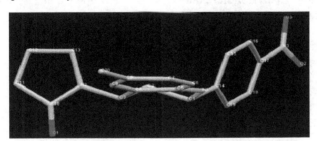

Fig. 8. Representation of the unit cell of *N-[6-methyl-2-(4´-nitrophenyl)-1,2,3,4-tetrahydroquinolin-4-yl]* pyrrolidin-2-one **14**.

The details of cell data and refinement for the compound **14** are summarized in Table 4.

| | |
|---|---|
| Unit cell parameters | $a = 9.109$ (2) Å <br> $b = 9.2812$ (5) Å <br> $c = 11.011$ (3) Å <br> $\alpha = 90.939°$ (6) <br> $\beta = 100.023°$ (6) <br> $\gamma = 93.309°$ (6) |
| Volumen | 913.998 Å$^3$ |
| System | Triclinic |
| Space Group | P-1 (No. 2) |
| Z | 2 |

Table 4. Crystallographic data obtained by four-circle diffractometry.

The structure packing is showed in the Figure 9 and finally, the powder profile simulated by the single crystal data is shown in Figure 10.

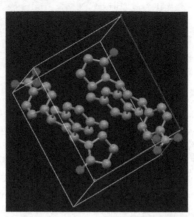

Fig. 9. Molecular packing of the unit cell of *N-[6-methyl-2-(4´-nitrophenyl)-1,2,3,4-tetrahydroquinolin-4-yl]* pyrrolidin-2-one **14**.

By means of single crystal study of the compound *N*-[6-methyl-2-(4-nitrophenyl)-1,2,3,4-tetrahydroquinolin-4-yl] pyrrolidin-2-one **15** was determined that crystals obtained from ethanol crystallizes in the triclinic system with space group P-1 (No 2).

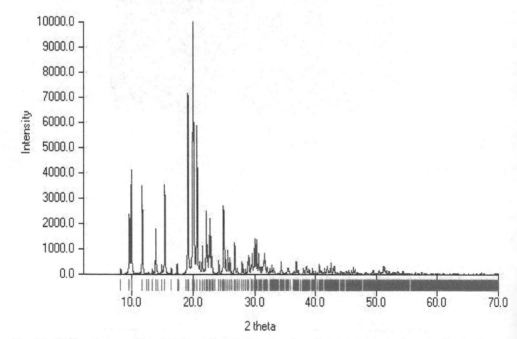

Fig. 10. Diffraction profile of *N-[6-methyl-2-(4´-nitrophenyl)-1,2,3,4-tetrahydroquinolin-4-yl]* pyrrolidin-2-one **14** simulated in Mercury software.

Table 5 shows the atomic positions. Carbon-bound H-atoms positions were idealized (C-H=0.93 Å), with H atoms riding on the atoms to which they were attached.

| Number | Label | Xfrac | Yfrac | Zfrac |
|--------|-------|-------|-------|-------|
| 1 | O1 | 0.3653 | 0.1204 | 0.8134 |
| 2 | O2 | -0.2751 | 0.1442 | -0.069 |
| 3 | O3 | -0.2013 | 0.3615 | -0.0633 |
| 4 | N1 | 0.4385 | 0.1546 | 0.3218 |
| 5 | N2 | 0.3999 | 0.2993 | 0.6826 |
| 6 | N3 | -0.1902 | 0.242 | -0.0246 |
| 7 | C1 | 0.2924 | 0.1252 | 0.3587 |
| 8 | C2 | 0.281 | 0.2232 | 0.4685 |
| 9 | C3 | 0.4102 | 0.2053 | 0.576 |
| 10 | C4 | 0.5578 | 0.2277 | 0.5329 |
| 11 | C5 | 0.5642 | 0.1906 | 0.4089 |
| 12 | C6 | 0.7043 | 0.1999 | 0.3716 |
| 13 | C7 | 0.8288 | 0.2495 | 0.4491 |
| 14 | C8 | 0.8236 | 0.2933 | 0.5723 |
| 15 | C9 | 0.6885 | 0.2777 | 0.6119 |
| 16 | C10 | 0.3777 | 0.2485 | 0.7914 |
| 17 | C11 | 0.3736 | 0.3751 | 0.8765 |
| 18 | C12 | 0.3898 | 0.5036 | 0.8068 |
| 19 | C13 | 0.4078 | 0.4576 | 0.6777 |
| 20 | C14 | 0.1688 | 0.1496 | 0.2534 |
| 21 | C15 | 0.1696 | 0.2758 | 0.1861 |
| 22 | C16 | 0.0548 | 0.3036 | 0.0958 |
| 23 | C17 | -0.0664 | 0.2057 | 0.0701 |
| 24 | C18 | -0.0721 | 0.0818 | 0.135 |
| 25 | C19 | 0.0456 | 0.0548 | 0.2264 |
| 26 | C20 | 0.9644 | 0.3544 | 0.6587 |

Table 5. Atomic positions in the unit cell of N-[6-methyl-2-(4´-nitrophenyl)-1,2,3,4-tetrahydroquinolin-4-yl] pyrrolidin-2-one (14).

## 3.4 Conclusions

The synthesis of two new nitro-isomers of N-(tetrahydroquinolinyl) pyrrolidin-2-ones using a versatile and simple methodology called the three component imino Diels-Alder cycloaddition is illustrated as an excellent route for the preparation of novel kind of structures, the spectral analysis showed the 2-$H_{axial}$, 4-$H_{axial}$ configuration; therefore the di-equatorial disposition of the C-2 and C-4 substituent that confirmed the formation of the endo-adduct during a Diels-Alder cycloaddition process.

The full characterization of N-[6-methyl-2-(4´-nitrophenyl)-1,2,3,4-tetrahydroquinoline-4-yl] pyrrolidin-2-one 14 was possible due to the single crystal X-ray diffraction studies, given the following data: the compound 14 crystallizes in the triclinic system with a = 9.109(2) Å, b = 9.281(5) Å, c = 11.011(3) Å, α = 90.939 (6)°, β = 100.023 (6)°, γ = 93.309 (6)°, Z = 2, space group P-1 [No. 2], and V = 1054.0 A³.

## 4. Convenient and scaleable three-step synthesis of new N-benzyl (pyridyl) cinnamamides from substituted (hetero)aromatic aldehydes

The relatively stable amide bond is not only common in natural-occurring materials like peptides and vitamins, it is also found in many synthetic substances. Among these important and interesting class of organic molecules: *N-benzylcinnamides* and *N-phenylcinnamides* have been always the focus of attention of organic, bioorganic and medicinal chemists due to their many useful synthetic applications (Takasu et al., 2003; Bernini et al., 2006; Nair et al., 2007) as well as their diverse (bio)chemical properties (Curtis et al., 2003; Tamiz et al., 1999; Lewis et al., 1991). Moreover, *N-amide* cinnamic acid derivatives are frequently presented in the nature (Mbaze et al., 2009; Vasques et al., 2002).

Both types of amides could be prepared generally via acylation reactions of the corresponding benzylamines or anilines and cinnamic acid derivatives (anhydrides or acyl halides) that are frequently used in the preparation of drug candidate molecules (Carey et al., 2006). However, the known acylation reactions for *N-benzylcinnamides* preparation employ hazardous and corrosive reagents (e.g., $SOCl_2$, $PCl_3$, $(COCl)_2$, $NEt_3$) besides the starting functionalized benzylamines are not commercially available products, and almost any of their functional groups needs to be protected to ensure chemoselective amide formation (Nesterenko et al., 2003).

Considering the reported biological properties of certain *N-arylcinnamides* and *N-benzylcinnamides* that have showed *in vitro* antimycotic activity (Leslie et al., 2010) and the inhibition of the transcription of carcinogenic genes in infected cells (Sienkiewicz et al., 2007). The need of an easy, rapid and "green" protocol for the synthesis of these kind of compounds is necessary to explore new pharmacological targets, and in this order the challenge of the organic chemistry is currently focused on the design of novel methodologies that suppress the use of acyl chloride (including any dangerous reagent required for their synthesis) and promoting the direct condensation between carboxylic acids and amines, coupling that currently is performed using efficient promoters such as *N,N'*-dicyclohexylcarbodiimide (DCC) (Amma & Mallouk, 2004), (benzotriazol-1-yloxy)tripyrrolidinophosphonium hexafluorophosphate (ByPOB) (McCalmont, 2004; Baures et al; 2002), triethylamine (Walpole, et al., 1993) and boron-based catalysts like $BH_3 \cdot THF$ (Huang et al., 2007).

The search of bioactive not only requires a good synthetic route, all the reagents must be stable and inexpensive to improve the structural complexity as in the biological properties. In this case, the integrity of most of the benzylamines is compromised when they are exposed to air moisture or even stored at low temperatures for a long period of time (Nazih & Heissler, 2002). A way to prevent this effects is based on choosing an appropriate strategy that enables the preparation of the desired amount benzylamines at lower cost than those acquired commercially, which could be possible by reduction of nitriles (Haddenham et al., 2009) and oximes (Gannett et al., 1988).

The "green" protocol described in this section is directed to enhance the disadvantages already described for the existing methodologies. Considering first the excellent chemical reactivity of aldehyde group against hydroxylamine's nucleophilicity to give oximes, this part was based on the existing methodologies which were improved according to the green chemistry principles with the use of less hazardous $Na_2CO_3$ as a base to release the nucleophile (Dallinger & Kappe, 2007; Roberts & Strauss, 2005; Yadav & Meshram, 2001; Varma, 1999). The posterior reduction of the prepared oximes give the functionalized benzylamines (pyridylmethylamines) in quantitative yield, this allows the coupuling of the corresponding amines, without further purification, with cinnamic acid in the presence of boric acid to afford the final products in agreement to our previous experience synthesizing this type of compounds (Hernandez et al., 2008).

## 4.1 Synthesis of new *N-benzyl* (pyridyl) cinnamamides from substituted (hetero)aromatic aldehydes

The choice of diverse (hetero)aromatic aldehydes **16a-k** was motived by the interest in quest for N-hetarylmethyl cinnamamides with antioxidant and anticancer properties. The direct condensation of **16a-k** and hydroxylamine in the presence of $Na_2CO_3$ is realize in an ethanol/water medium. Thus, a mixture of aldehydes **16a-k**, $Na_2CO_3$ and hydroxylamine hydrochloride was mixed in deionized water for 5 min. at room temperature, and then a small amount of the respective aldehyde was added for a period of another 5 minutes. The reaction mixture was stirring for 15 min. and after traditional work-up, the pure products, the substituted aldoximes **17a-k** were obtained in quantitative yield (Fig. 11).

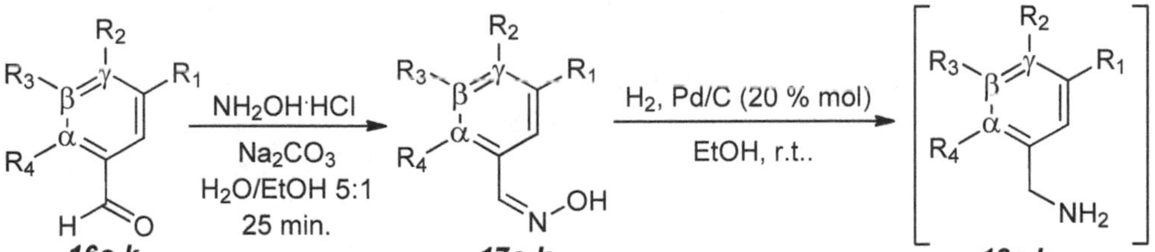

Fig. 11. Preparation of the respective (hetero)benzylamines from the aldehydes **16a-k** in a scalable methodology.

Taking into consideration that one of the most "atom-economical" procedures for the preparation of an amine is hydrogenation of an oxime in which the only by-product is water, we addressed also to this approach (Trost, 1995; Trost, 1991). Having in our hands the eleven solid and stable aldoximes **17a-k** obtained in first step, each of them was hydrogenated at room temperature overnight under $H_2$ atmosphere using 10 % palladium on charcoal in ethanol (Fig. 11).

The hydrogenation mixture obtained in the second step is filtered through celite and the filtrate was concentrated to dryness allowing the crude amine **18a-k**, which was quickly added, without any further purification, to an anhydrous toluene solution of *trans*-cinnamic acid **19** in the presence of $B(OH)_3$ (10 % mol) at 110 °C for 6-10 h (Fig. 12).

The required workup at the end of the reaction can be perform in two ways: one consist in the precipitation of the product of interest with a solution of $NaHCO_3$ and their subsequent washing with water or that can be purified with column chromatography depending on the complexity of the final crude. For the second choice, the recommended support is neutral or basic alumina ($Al_2O_3$) due to the acidity of the *trans*-cinnamic acid that will remain from the

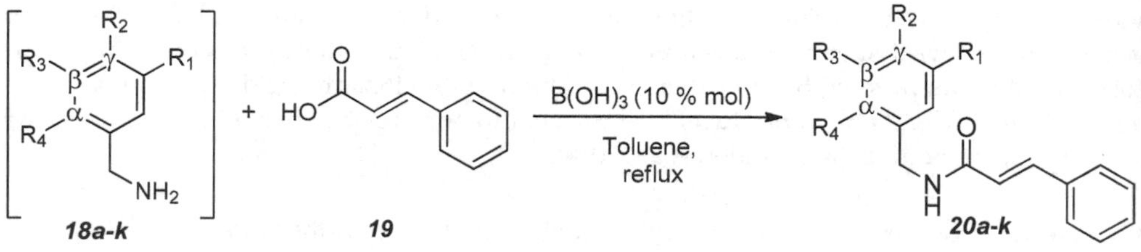

Fig. 12. Rational design of the three-step synthesis of the corresponding N-*benzyl cinnamides* from substituted (hetero) aromatic aldehydes.

reaction. An acid support like common silica gel ($SiO_2$) will retain both substances (the amide of interest and the residual cinnamic acid). After the method of preference for the purification of the amides has applied, the final products the N-benzylcinamides 20a-k were obtained in excellent yields and with a high purity level (Table 6).

| Comp. 20 | $R_1$ | $R_2$ | $R_3$ | $R_4$ | Yield (%) |
|---|---|---|---|---|---|
| a | H | Cl | H | H | 80 |
| b | H | $OCH_3$ | H | H | 87 |
| c | H | $OCH_3$ | $OCH_3$ | H | 91 |
| d | H | OH | $OCH_3$ | H | 67 |
| e | H | $OCH_3$ | OH | H | 72 |
| f | $OCH_3$ | $OCH_3$ | $OCH_3$ | H | 95 |
| g | -$OCH_2O$- | | H | H | 94 |
| h | H | H | -$C_4H_4$- | | 92 |
| i | H | H | H | $\alpha$=N | 84 |
| j | H | H | $\beta$=N | H | 89 |
| k | H | $\gamma$=N | H | H | 90 |

Table 6. Synthesized N-*benzyl cinnamides* 20a-k.

The $^1H$ NMR spectra of the compounds 20 display a general group of characteristic signals for this series. For example, in the cinnamide 20g spectrum the methylene protons at 4.47 ppm (2H, d, $J$ = 5.7 Hz, -$CH_2$) is the signal that is observed in high fields, signal that is coupling with the N-H signal, observed as a triplet at 5.99 ppm (1H, $J$ = 5.7 Hz, NH). The analysis of olefinic protons indicate the *trans* configuration of the final products when the high value of the coupling constant observed is compared with the typical value for the *cis* configuration: in the case of compound 20g, and for the entire 11 synthetized molecules, the assignment of the proton at 6.41 ppm (1H, d, $J$ = 15.7 Hz, =CHCO) and the coupling in trans form with the other olefinic proton, the one that appears at lower fields, at 7.66 ppm (1H, $J$ = 15.7 Hz, =CHPh) confirmed the configuration of all the products (Fig. 13).

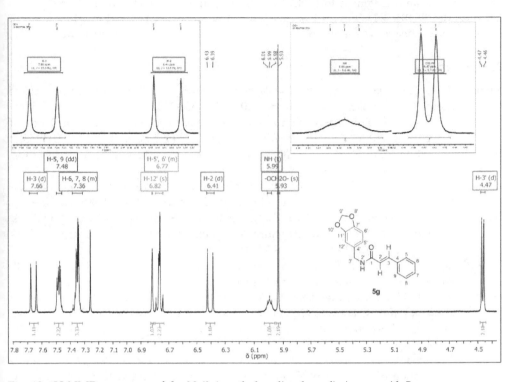

Fig. 13. $^1$H-NMR spectrum of the *N-(3,4-methylenedioxybenzyl) cinnamamide* **5g**.

## 4.2 Conclusions

The improvement of the existing methodologies for the preparation of benzylamines, is described as protocol that enhances the efficiently, easily, rapidly and safety way in which the oximes can be obtained, leading to explore their synthetic use in the preparation of more complex systems or evaluate their pharmacological properties as a potential reactivators of the acetylcholinesterase enzyme (Sinko et al., 2010) or their allergenic activity (Bergström et al., 2008).

Taking into account that boronic compounds have showed catalytic activity in peptide synthesis, it is demonstrated also that boric acid is a practical and useful catalyst for amidation between cinnamic acid and the prepared benzylamines due to its remarkable catalytic potential. The notable features of this procedure are mild and green reaction conditions, good reaction rates, cleaner reaction profiles and excellent global yields for a linear synthesis of three steps. The recollected spectral data described for *N-benzyl cinnamides* should be reliable in the structural analysis of natural cinnamides and these substances could serve as a model for small-molecule screening towards new bioactive compounds.

## 5. Acknowledgment

The authors acknowledgment to COLCIENCIAS: contract No. 432-2004, RC-3662011 and CENIVAM: CPS 015-2011, and the Universidad Industrial de Santander: throught to the VIE division, for the financial support given during the development of these researches.

## 6. References

Allen, F. H. (2002). *Cambridge Structural Database: a quarter of a million crystal structures and rising. Acta Crystallographica Section B*. 2002, B58, 380-388.

Amma, A. & Mallouk, T. E. (2004). Synthesis of an amide cyclophane building block of shape-persistent triangular molecular wedges. *Tetrahedron Lett.*, 45, pp. 1151-1153.

Baures, P. W.; Kaliyan, K. & Desper, J. (2002). N-α-Urocanylhistamine: A Natural Histamine Derivative. *Molecules*, 7, pp. 813-816.

Baxter, H. Harbone, J. B. & Pmoss, G. (1999). *Phytochemical Dictionary. A Handbook of Bioactive Compounds From Plants* (Second Edition). CRC Press, ISBN 978-0748406203, USA.

Bergström, M. A.; Andersson, S. I.; Broo, K.; Luthman, K. & Karlberg, A.-T. (2008). Oximes: Metabolic Activation and Structure-Allergenic Activity Relationships. *J. Med. Chem.*, 51, pp. 2541–2550.

Bernini, R.; Cachia, S.; De Salve, I. & Fabrizi, G. (2006). The Heck Reaction of β-Arylacrylamides: An Approach to 4-Aryl-2-quinolones *Synlett*, pp. 2947-2952.

Bermúdez, J. H.; Pinto, J. L.; Meléndez, C. M.; Henao, J. A. & Kouznetsov, V. V. (2011). Simple preparation of new N-(6-methyl-2-nitrophenyl-1,2,3,4-tetrahydroquinolin-4-yl) pyrrolidin-2-ones and their spectroscopic analysis, *Universitas Scientiarum*, 16, pp. 161-168.

Bello, J. S.; Amado, F.; Henao; A.; Atencio, R. & Kouznetsov, V. V. (2010). Simple preparation of new N-aryl-N-(3-indolmethyl) acetamides and their spectroscopic analysis *Universitas Scientiarum*, 14, pp. 216-224.

Carey, J. S.; Laffan, D.; Thomon, C. & Williams, M. K. (2006). Analysis of the reactions used for the preparation of drug candidate molecules. *Org. Biomol. Chem.*, 4, pp. 2337-2347.

Colyer, J. T.; Andersen, N. G.; Tedrow, J. S.; Soukup, T. S. & Faul, M. M. (2006). Reversal of Diastereofacial Selectivity in Hydride Reductions of N-Tert-butanesulfinyl Imines. *J. Org. Chem.*, 71, pp. 6859–6862.

Curtis, N. R.; Diggle, H. J.; Kulagowski, J. J.; London, C.; Grimwood, S.; Hutson, P. H.; Murray, F.; Richards, P.; Macaulay, A. & Wafford, K. A. (2003). Novel N1-(benzyl)cinnamamidine derived NR2B subtype-selective NMDA receptor antagonists. *Bioorg. Med. Chem. Lett.*, 13, pp. 693-696.

Dai, H. G. ; Li, J. T. ; & Li, T. S. (2006). Efficient and Practical Synthesis of Mannich Bases Related to Gramine Mediated by Zinc Chloride. *Synth. Commun.*, 36, pp. 1829-1835.

Dallinger, D. & Kappe, C. O. (2007). Microwave-Assisted Synthesis in Water as Solvent. *Chem. Rev.*, 107, pp. 2563-2591.

Evers, A.; Hessler, G.; Matter, H. & Klabunde, T. (2005). Virtual Screening of Biogenic Amine-Binding G-Protein Coupled Receptors: Comparative Evaluation of Protein- and Ligand-Based Virtual Screening Protocols. *J. Med. Chem.*, 48, pp. 5448-5465.

Gannett, P. M.; Nagel, D. L.; Reilly, P. J.; Lawson, T.; Sharpe, J. & Toth, B. (1988). The Capsaicinoids: Their Separation, Synthesis, and Mutagenicity. *J. Org. Chem.*, 53, pp. 1064-1071.

Gribble, G. W. (2003). Novel Chemistry of Indoles in the Synthesis of Heterocycles. *Pure Appl. Chem.*, 75, pp. 1417-1432.

Haddenham, D.; Pasumansky, L.; DeSoto, J.; Eagon, S. & Singaram, B. (2009). Reductions of Aliphatic and Aromatic Nitriles to Primary Amines with Diisopropylaminoborane. *J. Org. Chem.*, 74, pp. 1964-1970.

Hernandez Barajas J. G.; Vargas Méndez L. Y.; Kouznetsov V. V. & Stashenko E. E. (2008). Efficient Synthesis of New N-benzyl- or N-(2-Furylmethyl)cinnamamides Promoted by the "Green" Catalyst Boric Acid, and Their Spectral Analysis. *Synthesis*, pp. 377-382.

Huang, Z.; Reilly, J. R. & Buckle, R. N. (2007). An Efficient Synthesis of Amides and Esters via Triacyloxyboranes. *Synlett*, pp. 1026-1030.

Hutchins, R. O. & Hutchins M, K. (1991). *Comprehensive Organic Synthesis* (Vol 8), Oxford: Pergamon Press, ISBN 00800359299, pp. 25-78.

Jones G. (1984). Pyridines and their Benzoderivatives, Synthesis, in: *Comprehensive Heterocyclic Chemistry*, Katrizky AR., Vol, 8, pp. 380-412, Pergamon Press Oxford, ISBN: 978-0-08-044992-0, FL, USA.

Jiang, B. & Huang, Z.-G. (2005). Synthesis of α-(3-Indolyl)glycine Derivatives via Spontaneous Friedel-Crafts Reaction between Indoles and Glyoxylate Imines. *Synthesis*, pp. 2198-2204.

Katrizky, A. R.; Rachwal, S. & Rachwal, B. (1996). Recent progress in the synthesis of 1,2,3,4,-tetrahydroquinolines. *Tetrahedron*, 52, pp. 15031-15070.

Ke, B., Quin, Y., He, Q., Huang, Z. & Wang, F. (2005). Preparation of bisindolylalkanes from N-*tert*-butanesulfinyl aldimines. *Tetrahedron Lett.*, 46, pp. 1751-1753.

Kleemann, A. & Engel, J. (1999). *Pharmaceutical Substances* (Revised Edition), Thieme, New York, USA.

Kouznetsov, V. V. (2009). Recent synthetic developments in a powerful imino Diels-Alder reaction (Povarov reaction): application to the synthesis of N-polyheterocycles and related alkaloids. *Tetrahedron*, 65, pp. 2721-2750.

Kouznetsov, V. V.; Meléndez, C. M. & Bermúdez, J. H. (2010). Transformations of 2-aryl-4-(2-oxopyrrolidinyl-1)-1,2,3,4-tetrahydroquinolines, cycloadducts of the $BiCl_3$-catalyzed three component Povarov reaction: oxidation and reduction processes towards new potentially bioactive 2-arylquinoline derivatives. *J. Heterocycl. Chem.*, 47, pp. 1148-1152.

Kouznetsov, V. & Mora Cruz, U. (2006). Transformations of 2-(α-furyl)-4-(2-oxopyrrolidinyl-1)-1,2,3,4-tetrahydroquinolines, cycloadducts of the imino Diels-Alder reaction: A simple synthesis of new quinoline derivatives. *Lett. Org. Chem.*, 3, pp. 699-702.

Kouznetsov, V.; Mora Cruz, U.; Zubkov, F. I. & Nikitina, E. V. (2007). An efficient synthesis of isoindolo[2,1-*a*]quinoline derivatives via imino Diels-Alder and intramolecular Diels-Alder with furan methodologies. *Synthesis*, pp. 375-380.

Kouznetsov, V.; Palma, A.; Ewert, C. & Varlamov, A. (1998). Some aspects of reduced quinoline chemistry. *J. Heterocycl. Chem.*, 35, pp. 761-785.

Kouznetsov, V.; Rodríguez, W.; Stashenko, E.; Ochoa, C.; Vega, C.; Rolón, M.; Montero Pereira, D.; Escario, J. A. & Gómez-Barrio, A. (2004a). Transformation of Schiff Bases Derived from alpha-Naphthaldehyde. Synthesis, Spectral Data and Biological Activity of New-3-Aryl-2-(alpha-naphtyl)-4-thiazolidinones and N-Aryl-N-[1-(alpha-naphthyl)but-3-enyl]amines. *J. Heterocycl. Chem.*, 41, pp. 995-999.

Kouznetsov, V., Vargas Mendéz, L. Y. & Meléndez, C. M. (2005). Recent progress in the synthesis of quinolines. *Curr. Org. Chem.*, 9, pp. 141-161.

Kouznetsov, V.; Vargas Méndez, L. Y.; Tibaduiza, B.; Ochoa, C.; Montero Pereira, D.; Nogal Ruiz, J. J.; Portillo, C. F.; Gómez, A.B.; Bahsas, A. & Amaro-Luis, J. (2004b). 4-Aryl(benzyl)amino-4-heteroarylbut-1-enes as Building Blocks in Heterocyclic Synthesis. 4.[1] Synthesis of 4, 6-Dimethyl-5-nitro(amino)-2-pyridylquinolines and their Antiparasitic Activities. *Archiv Pharmazie*, 337, pp. 127-132.

Leslie, B. J.; Holaday, C. R.; Nguyen T. & Hergenrother, P. J. (2010). Phenylcinnamides as Novel Antimitotic Agents. *J. Med. Chem.*, 53, pp. 3964-3972.

Lewis, F. D.; Elbert, J. E.; Upthagrove, A. L. & Hale, P. D. (1991). Lewis-acid catalysis of photochemical reactions. 9. Structure and photoisomerization of (E)- and (Z)-cinnamamides and their Lewis acid complexes. *J. Org. Chem.*, 56, pp. 553-561.

Mbaze, L. M.; Lado, J, A.; Wansi, J. D.; Shiao, T. C.; Chiozem, D. D.; Mesaik, M. A.; Choudhary, M. I.; Lacaille-Dubois, M-A.; Wandji, J.; Roy, R. & Sewald, N. (2009). Oxidative burst inhibitory and cytotoxic amides and lignans from the stem bark of *Fagara heitzii* (Rutaceae). *Phytochemistry*, 70, pp. 1442-1447.

McCalmont, W. F.; Heady, T. N.; Patterson, J. R.; Lindenmuth, M. A.; Haverstick, D. M.; Gray, L. S. & Macdonald, T. L. (1996). Design, synthesis, and biological evaluation of novel T-Type calcium channel antagonists. *Bioorg. Med. Chem. Lett.*, 14, pp. 3691-3695.

Molina, P., Alcantara, J. & Lopez-Leonardo, C. (1996). Regiospecific preparation of γ-carbolines and pyrimido[3, 4-a]indole derivatives by intramolecular ring-closure of heterocumulene-substituted indoles. *Tetrahedron*, 52, pp. 5833-5844.

Nair, V.; Mohanan, K.; Suja, T. D. & Biju, A. (2007). Stereoselective Synthesis of 3,4-*trans*-Disubstituted γ-Lactams by Cerium(IV) Ammonium Nitrate Mediated Radical Cyclization of Cinnamamides. *Synthesis*, pp. 1179-1184.

Nazih, A. & Heissler, D. (2002). One-pot Conversion of t-Butyl Carbamates to Amides with Acyl Halide-Methanol Mixtures. *Synthesis*, pp. 203-206.

Nesterenko, V.; Putt, K. S. & Hergenrother, P. J. (2003). Identification from a Combinatorial Library of a Small Molecule that Selectively Induces Apoptosis in Cancer Cells. *J. Am. Chem. Soc.*, 125, pp. 14672-14673.

Pazderski, L., Tousek, J., Sitkowski, J., Kozerski, L. & Szłyk, E. (2007). Experimental and quantum-chemical studies of $^1$H, $^{13}$C and $^{15}$N NMR coordination shifts in Pd(II) and Pt(II) chloride complexes with quinoline, isoquinoline, and 2,2'-biquinoline. *Magn. Reson. Chem.*, 45, pp. 1059-1071.

Roberts, B. A. & Strauss, C. R. (2005). Toward Rapid, "Green", Predictable Microwave-Assisted Synthesis. *Acc. Chem. Res.*, 38, pp. 653-661.

Saxton, J. E. (1998). *The Alkaloids: Chemistry and Biology* (First Edition), Vol. 51, Academic Press, ISBN-13: 978-0124695597, Sannn Diego, USA.

Sienkiewicz, P.; Ciolino, H. P.; Leslie, B, J.; Hergenrother, P. J.; Singletary, K. & Chao Yeh, G. (2007). A novel synthetic analogue of a constituent of *Isodon excisus* inhibits transcription of CYP1A1, -1A2 and -1B1 by preventing activation of the aryl hydrocarbon receptor. *Carcinogenesis*, 28, pp. 1052-1057.

Sinko, G.; Brglez, J. & Kovarik, Z. (2010). Interactions of pyridinium oximes with acetylcholinesterase. *Chem. Biol. Interact.*, 187, pp. 172-176.

Sheldrick, G. M. (1997). SHELXS97. University of Göttingen. Germany.

Sheldrick, G. M. (1997). SHELXL97. University of Göttingen. Germany.

Shirakawa, S. & Kobayashi, S. (2006). Carboxylic Acid Catalyzed Three-Component Aza-Friedel–Crafts Reactions in Water for the Synthesis of 3-Substituted Indoles. *Org. Lett.*, 8, pp. 4939-4942.

Takasu, K.; Nishida, N. & Ihara, M. (2003). A direct entry to substituted piperidinones from α,β-unsaturated amides by means of aza double Michael reaction. *Tetrahedron Lett.*, 44, pp. 7429-7432.

Tamiz, A. P.; Cai, S. X.; Zhou, Z.-L.; Yuen, P.-W.; Schelkun, R. M.; Whittemore, E. R.; Weber, E.; Woodward, R. M. & Keana, J. F. W. (1999). Structure–Activity Relationship of N-(Phenylalkyl)cinnamides as Novel NR2B Subtype-Selective NMDA Receptor Antagonists. *J. Med. Chem.*, 42, pp. 3412-3420.

Trost. B. M. (1995). Atom economy-A challenge for organic synthesis: Homogeneous catalysis leads the way. *Angew. Chem. Int. Ed. Engl.*, 34, pp. 259-281.

Trost, B. M. (1991). The atom economy-a search for synthetic efficiency. *Science*, 254, 1471-1477.

Varma, R. S. (1999). Solvent-free organic syntheses using supported reagents and microwave irradiation. *Green Chem.*, 1, pp. 43-55.

Vargas Méndez, L. Y.; Castelli, M. V.; Kouznetsov, V.; Urbina Gonzalez, J. M., López, S. N.; Sortino, M.; Enriz, R. D.; Ribas, J. C. & Zacchino, S. A. (2003). In vitro antifungal activity of new series of homoallylamines and related compounds with inhibitory properties of the synthesis of fungal cell wall polymers. *Bioorg. Med. Chem.*, 11, pp. 1531-1550.

Vasques da Silva, R.; Debonsi Navickiene, H. M.; Kato, M. J.; da S. Bolzani, V.; Méda, C. I.; Young, M. C. M. & Furlan, M. (2002). Antifungal amides from *Piper arboreum* and *Piper tuberculatum*. *Phytochemistry*, 59, pp. 521-527.

Walpole, C. S. J.; Wrigglesworth, R.; Bevan, S.; Campbell, E. A.; Dray, A.; James, I. F.; Perkins, M. N.; Reid, D. J. & Winter, J. (1993). Analogues of Capsaicin with Agonist Activity as Novel Analgesic Agents; Structure-Activity Studies. 1. The Aromatic "A-Region". *J. Med. Chem.*, 36, pp. 2362-2372.

Wynne, J. H. & Stalick, W. M. (2002). Synthesis of 3-[(1-Aryl)aminomethyl]indoles. *J. Org. Chem.*, 67, 16, pp. 5850-5853.

Yadav, J. S. & Meshram, H. M. (2001). Green twist to an old theme. An eco-friendly approach Dry media reactions. *Pure Appl. Chem.*, 73, pp. 199-203.

Youssef, A. O.; Khalil, M. H.; Ramadan, R. M. & Soliman, A. A. (2003). Molybdenum and tungsten complexes of biquinoline. Crystal structure of W(CO)₄(2,2'-biquinoline). *Tran. Met. Chem.*, 28, pp. 331-335.

Zhao, J.L.; Liu, L.; Zhang, H.B.; Wu, Y.C.; Wang, D. & Chen, Y. J. (2006). Three-Component
    Friedel-Crafts Reaction of Indoles, Glyoxylate, and Amine under Solvent-Free and
    Catalyst-Free Conditions - Synthesis of (3-Indolyl)glycine Derivatives. *Archiv
    Pharmazie*, 337, pp. 127-132.

# 7

# Use of Associated Chromatographic Techniques in Bio-Monitored Isolation of Bioactive Monoterpenoid Indole Alkaloids from *Aspidosperma ramiflorum*

Talita Perez Cantuaria Chierrito[1], Ananda de Castro Cunha[1], Luzia Koike[2],
Regina Aparecida Correia Gonçalves[1] and Arildo José Braz de Oliveira[1]
*[1]Department of Pharmacy, State University of Maringá, Maringá, Paraná State,*
*[2]State University of Campinas, Chemistry Institute, Campinas-SP,*
*Brasil*

## 1. Introduction

The genus *Aspidosperma* (Apocynaceae) have been commonly used in folk medicine as potential antimalarial agents; in the treatment of leishmaniasis; uterus and ovary inflammations; as a contraceptive; in diabetes; stomach disorders; against cancer; fever and rheumatism (Oliveira et al, 2009). It commonly grows in tropical America, extracted from trees ranging 2 to 60 m in height. It is found in a variety of habitats from the dry fields of south-central Brazil, Paraguay, and Argentina to the inundated river margins of the Amazon basin (Tanaka, 2006). The main constituents of the *Aspidosperma* genus are indole alkaloids, a class of substances with a wide range of pharmacological activities such as cholinesterase inhibitors, analgesic, anti-inflammatory, bactericidal, oestrogenic, stimulant and depressant of the central nervous system (CNS) (Zocoler et al, 2005).

### 1.1 Chromatographic systems in *Aspidosperma* genus

Several methods are available for the analysis and identification of known indole alkaloids. Analysis of complex mixtures is frequently done by means of Thyn Layer Chromatography (TLC), through comparison of Rf values obtained in different solvent systems, and also by comparison between specific color reactions of components of the mixture and reference compounds. TLC remains one of the preferred methods for qualitative analysis of known compounds since it requires neither sophisticated equipment nor extensive sample preparation. For quantitative analysis, High Performance Liquid Chromatography (HPLC) systems linked to a UV detector are commonly used. By coupling HPLC to a photodiode array UV detector makes it possible to combine the information over retention times and the UV spectrum of each compound, and in some cases, it also enables the quantification of overlapping peaks. Capillary Gas Chromatography (GC) analysis has been described for several classes of alkaloids. A major advantage of GC over the above-mentioned methods is its enhanced sensitivity and high resolution. Another advantage is its easy coupling to a mass spectrometer that allows the identification of new and minor compounds of a mixture without laborious isolation procedures, which makes it a particularly attractive method when no decomposition due to the high temperatures applied in GC occurs (Dagnino, 1991). This chapter shows different chromatographic techniques (TLC, Preparative Thyn Layer Chromatography (PTLC), Classical Liquid Column Chromatography (CLCC), GC and HPLC) to isolate and characterize indole alkaloids of *Aspidosperma ramiflorum* species.

### 1.1.2 *Aspidosperma ramiflorum* species

*Aspidosperma ramiflorum* Muell. Arg. species commonly known as "guatambu", had been studied in 1996 by Reis et al., who were able to isolate the monoterpenoid indole alkaloids, ramiflorine A (1), ramiflorine B (2), 10-methoxy-geissoschizol (3) and β-yohimbine (4). After that, Oliveira 1999 isolated beyond these compounds the 16-(E)-isositrikine (5), all them from stem barks (Figure 1).

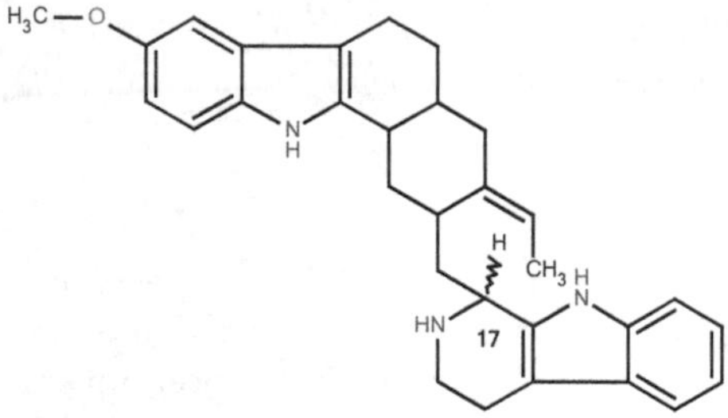

H-17-α ramiflorine A ( 1)
H-17-β ramiflorine B ( 2)

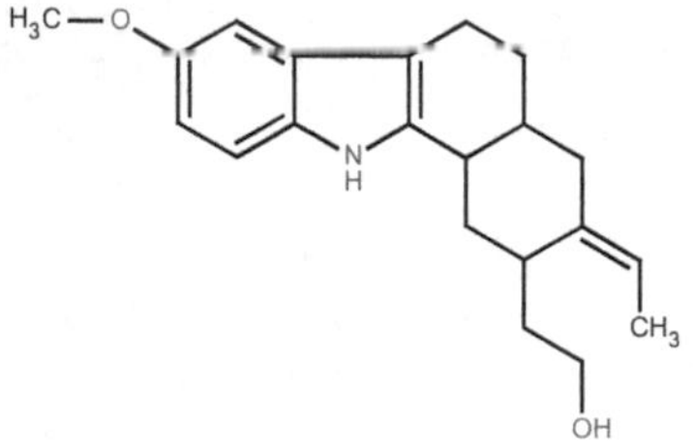

10-methoxy-geissoschizol (3)

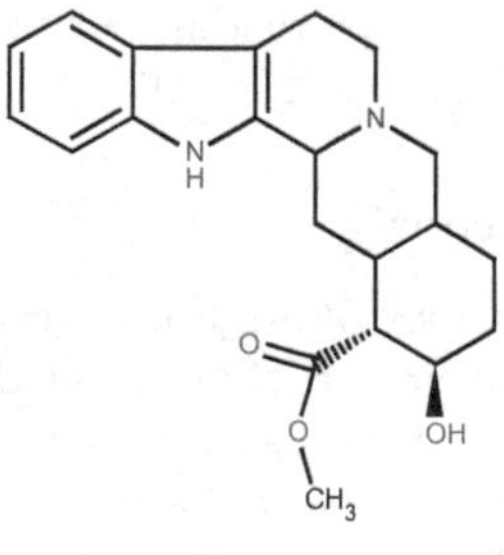

β-yohimbine ( 4)

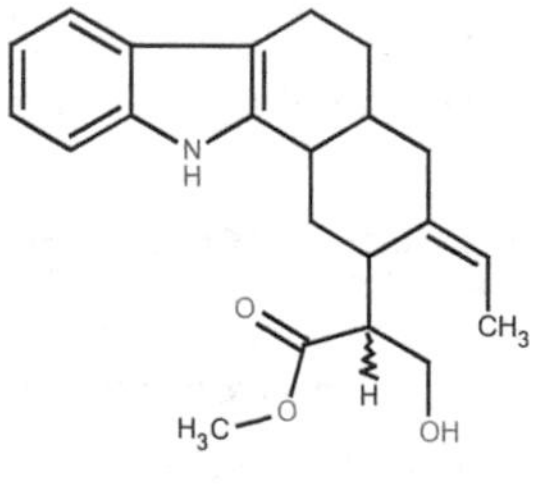

(+/-)-16-( E)-isositrikine ( 5)

Fig. 1. Major monoterpenoid indole alkaloids from *Aspidosperma ramiflorum*. Marvin was used for drawing, displaying and characterizing chemical structures, substructures and reactions, Marvin 5.4.1.1, 2011, ChemAxon available on ( http://www.chemaxon.com ).

The basic crude extract from stem barks of *A. ramiflorum* showed a good antileishmanial activity (Ferreira et al., 2004), which we attributed to the presence of indole alkaloids, and soon after, we described the fractionation, purification and isolation of alkaloids responsible

for the activity against *Leishmania (L.) amazonensis* (Tanaka et al., 2007). Our results revealed
that dimeric corynanthe alkaloids Ramiflorines A **(1)** and B **(2)** were responsible for the
activity against promastigote forms of *L. amazonensis* with significant activity ($LD_{50}$ values of
16.3 ± 1.6 μg/ml and 4.9 ± 0.9 μg/ml, respectively). Tanaka et.al. (2006) evaluated the
antibacterial activities of the crude methanol extract, fractions obtained after acid-base
extraction and pure compounds from the stem barks of *Aspidosperma ramiflorum* and both
Ramiflorines showed significant activity against *S. aureus* (MIC = 25 μg/mL) and *E. faecalis*
(MIC = 50 μg/mL), with $EC_{50}$ of 8 and 2.5 μg/mL for Ramiflorines A and B, respectively,
against *S. aureus*.

## 2. General methods of alkaloids extraction

From crude alcoholics or hidroalcoholics extracts obtained from parts of the plant material,
two methods already described in the literature may be used for preliminary extraction of
alkaloids; both methods take into account the basic nature of these compounds in order to
concentrate them in pKas nearby and produce more purified fractions, preceding the phases
of isolation and identification of substances. The first method consists of a complete acid-
base partition (Figure 2), employing solvents such as chloroform or dichloromethane,

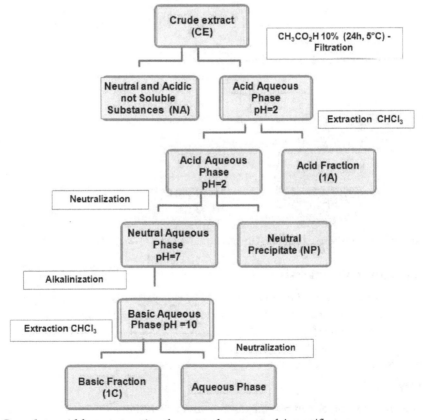

Fig. 2. Complete acid-base extraction from crude extract of *A. ramiflorum*.

extracting alkaloids in three levels of pH (acid, neutral and basic) (Marques et al., 1996). In this extraction is possible to obtain four alkaloidal rich fractions: acidic fraction (1A), neutral precipitate fraction (NP), neutral fraction (1B) and basic fraction (1C). The second consists of a simplified acid-base partition (Figure 3), employing the same solvents, but with the absence of extraction at neutral pH in order to eliminate sample neutral substances, getting only two alkaloidal rich fractions: acidic fraction (1A) and basic fraction (1C) (Oliveira et al., 1999; Tanaka et al., 2007).

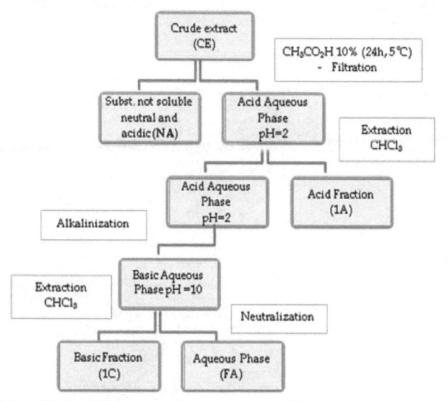

Fig. 3. Simplified acid-base extraction from crude extract of *A. ramiflorum*.

## 3. Chromatographic analysis

Chromatography is a physical-chemical method of separation and several are available for the analysis and identification of known indole alkaloids, the applicable to species *A. ramiflorum* are described below.

### 3.1 Thin Layer Chromatographic – TLC

### 3.1.1 Analytical TLC

As said before, analysis of complex mixtures is frequently done by TLC. It remains one of the preferred methods for qualitative analysis of known compounds since it requires neither

Use of Associated Chromatographic Techniques in Bio-Monitored Isolation of Bioactive Monoterpenoid Indole Alkaloids
from Aspidosperma ramiflorum

123

sophisticated equipment nor extensive sample preparation (Dagnino et al., 1991). The analytical TLC contributes greatly to a preliminary characterization of the alkaloidal extract and fractions obtained from *A. ramiflorum*. Although the substances present in various parts of this species, ramiflorine A (**1**), ramiflorine B (**2**), 10-metoxy-geissoschizol (**3**), $\beta$-yohimbine (**4**), ($\pm$)-16-(*E*)-Isositsirikine (**5**), they are already known and described (Marques, 1998; Marques et al, 1996; Oliveira, 1999; Ferreira et al., 2004), TLC plates at $R_f$ values of substances determined and revealed by UV light and specific reagents such as *p*-anisaldehyde and Dragendorff allow a prior identification of substances quickly and conclusively. Under UV light, substances with chromophore, i.e. conjugated systems absorb radiation and become fluorescent, since the stationary phase contains fluorescence indicator, the reactive *p*-anisaldehyde blush indole nucleus of purple and the reactive Dragendorff blush nitrogen compounds of orange. Figure 4 shows two TLC plates with samples of *A. ramiflorum* obtained by the team of Laboratory of Biotechnology of Synthetic and Natural Products of the State University of Maringá – PR, Brazil (LABIPROS), under the following conditions: stationary phase – commercial chromatoplate of aluminum in normal phase silica gel 60 $F_{254}$ fluorescence indicator; mobile phase (S1) – chloroform, dichloromethane, ethyl acetate, methanol in the proportions of (4:1:4.5:0.5 v/v) in environment saturated with ammonia hydroxide; samples applied in strips by mini glass capillaries revealed with *p*-anisaldehyde and Dragendorff, respectively, confirming the presence of monoterpene indole alkaloids indólicos (MIA) in samples of the species *A. ramiflorum* through stains well demarcated and staining characteristics.

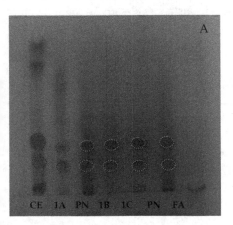 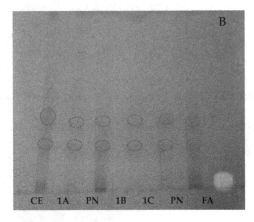

Fig. 4. A- Alkaloidal factions of *A. ramiflorum* revealed with *p*-anisaldehyde; B- Alkaloidal fractions of *A. ramiflorum* revealed with Dragendorff.

Variations on the combination of solvents and eluents generated changes in $Rf$ values for the same substance, as exemplified in Table 1, in which the same substances subject to the eluent system 1 (S1) used by the team LABIPROS have $Rf$ values distinct from the subject to the eluent system 2 (S2) used by Oliveira (1999).

| Composition of Eluent Systems (v/v) | | Stationary Phase | Substances applied on TLC plates | $R_f$ values |
|---|---|---|---|---|
| (S1) | Chloroform,dichloromethane, ethyl acetate, methanol in the proportions of 4.5:4:1.5:0.5 + saturation with ammonia hydroxide | Sílica Gel 60 $F_{254}$ | (1) | 0.45 |
| | | | (2) | 0.32 |
| | | | (3) | 0.22 |
| (S2) | Chloroform, ethyl acetate, methanol + saturation with ammonia hydroxide | Sílica Gel 60 $F_{254}$ | (1) | 0.76 |
| | | | (2) | 0.30 |
| | | | (3) | 0.36 |

Table 1. Variations of $R_f$ values of the same substances spotted on TLC plates subject different systems of eluents.

### 3.1.2 Preparative TLC – PTLC

The difference between the PTLC and the TLC is that, while the last employed to separation, identification and determination, not allows the retrieval of the sample, the preparative chromatography is a purification process and allows the isolation of pure substances contained in a mixture. Although there are many techniques developed for the isolation of natural products, PTLC is pretty used, being considered a relatively simple and inexpensive technique for separation and purification of small quantities of substances intended for chemical and physical-chemical studies. Although the migration time and the removal of the bands of substances are not easy and efficient, the LABIPROS team isolated by means of PTLC, stem barks and leaves of *A. ramiflorum* the same substances already isolated by other teams using different techniques (Marques et al, 1996; Ferreira et al, 2004). Figure 5 shows a preparative board applied, eluted and reveled in UV light 254 nm, one of alkaloidal fractions obtained from acid-base extraction of *A. ramiflorum*, composed of three main substances, of class MIA. The conditions employed in this technique were the same as described in the previous section for TLC, with the exception of the plates which were in size 20 x 20 cm with 1.0 mm thickness of silica gel 60 $F_{254}$ , and the sample was apllied in continuous line, with the help of a cotton brush. Figure 6, presents the TLC made to monitor and confirms the isolation by PTLC, demonstrated by natural spots in each application.

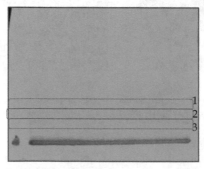

Fig. 5. PTLC alkaloidal fraction of *A. ramiflorum* composed of three majority substances of class MIA, viewed by UV irradiation at 254 nm.

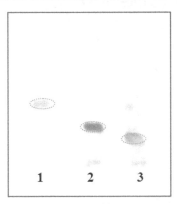

Fig. 6. TLC of the three major substances isolated by PTLC *A. ramiflorum*.

## 3.2 Classical liquid column chromatography

The majority of classical liquid column chromatography (CLCC) procedures for separating indole alkaloids from *Aspidosperma* use an adsorbent stationary phase as silicagel and the eluents employed reflect the polarity of alkaloids under investigation (Henrique et al., 2010; Barbosa et al., 2010). The majority of indole alkaloids are tertiary bases with fairly low polarity, so that mixtures containing a major part of a less polar solvent (e.g., chloroform, toluene) with a small proportion of more polar solvent (e.g., acetone, ethanol, methanol) are frequently cited (Henrique et al., 2010; Barbosa et al., 2010; Kobayashi et al., 2002). The use of technical variations of CLCC as flash, vacuum, low and medium pressure liquid chromatography has become common in the indole alkaloids isolation. *Aspidosperma ramiflorum* alkaloidal extract was also fractionated on a silica gel column, elutting with $CHCl_3$ with increasing amounts of MeOH (Marques et al., 1996).

## 3.3 Gas chromatography

Capillary gas chromatography (GC) analysis has been described for several classes of alkaloids. The major advantage of GC over the above-mentioned methods is its enhanced sensitivity and high resolution. Another advantage, it is easy coupling to a mass spectrometer (Biemann, 2002) that allows the identification of new and minor compounds of a mixture without laborious isolation procedures, which makes it a particularly attractive method when no decomposition due to the high temperatures applied in GC occurs but the widespread application of this technique is limited by volatility of more polar alkaloids.

The number of articles describing capillary GC analysis of underivatized alkaloids is continuously increasing, including indole alkaloids (Dagnino et al., 1991; Gallagher et al., 1995; Cardoso et al., 1997; Cardoso et al., 1998, Zocoler et al., 2005). Based on some of these works we developed a simple, rapid, sensitive, and reproducible technique for the qualitative and quantitative analyses of *Aspidosperma* alkaloids by capillary GC. The analysis can be done because of a slightly polar and relatively short capillary column which we have used here, with 10 m long, which enabled a rapid analysis but with a good resolution of different components (Figure 7). Thus, we separated alkaloids representing different chemical skeleton types from *A. pyricollum*, *A. olivaceum*, *A. pyrifolium*, *A. polyneuron* and *A. ramiflorum*. Using capillary GC, we were able to achieve complete separations of the

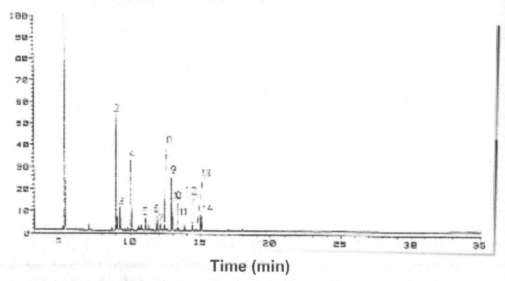

### Time (min)

Fig. 7. GC-MS chromatogram of basic fraction (1C). Peaks - 1: internal standard
(Tryptophol); 2: uleine; 3:Aspidofractine; 4: Apparicine; 5: 12-Demethoxy-aspidospermine; 6:
Aspidospermine; 7: Olivacine; 8: 15-Demethoxy-pyrifoline; 9: Pyrifolidine; 10: 15-Methoxy-
pyrifolidine; 11: Aspidofiline; 12: Pyrifolidine; 13: 15-Methoxy-pyrifolidine; 14:
Aspidoscarpine (Oliveira , 1999).

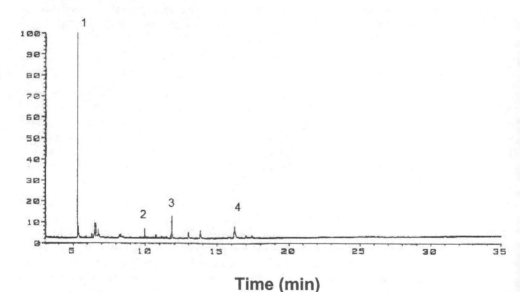

### Time (min)

Fig. 8. GC-MS chromatogram of basic fraction (1C). Peaks – 1: internal standard
(Tryptophol); 2: 1,2-Dihydro-olivacine; 3: Methoxy-1,2-dihydro-olivacine; 4: 10-Methoxy-
geissoschizol (Oliveira et al., 1999).

compounds as example in the *A. olivaceum* analysis, Figure 7. The mass spectrum of each individual peaks was determined by GC-MS under identical conditions, and the expected molecular weight and fragmentation patterns were observed for all the compounds. For *A. ramiflorum* analysis by GC five indole alkaloids were isolated and characterized as follows: 10-methoxy-geissoschizol (**3**), (*E*)-isositsirikine (**5**), ramiflorine A(**1**) e ramiflorine B (**2**) from stem barks and β-yohimbine (**4**) from seeds, but only 10-methoxy-geissoschizol, (*E*)-isositsirikine and β-yohimbine could be used as standards in GC analysis as shown in Figure 8. The compounds ramiflorine A and ramiflorine B are dimeric basic alkaloids, with high molecular weight and less volatile and they are therefore not analyzable under the used conditions. For this reason, we developed a HPLC analysis for *A. ramiflorum* alkaloids which is shown below.

### 3.3.1 Gas chromatography analysis

The stem barks of *A. olivaceum* and *A. ramiflorum* were extracted on $EtOH:H_2O$ (70:30) and submitted an acid-basic extraction procedure in agreement with described in Figure 3. The basic and acid chloroform extracts were weighted (30 mg), which were diluted in 9 mL of a solution of $CH_2Cl_2:MeOH$ (80:20, v/v). This mixture were homogenized and submitted to a filtration in a small column of silica-gel 60 Merck (1g) and were eluted with a solution of $CH_2Cl_2:MeOH$ (80:20, v/v). The filtrate was evaporated and the residue was dissolved in 2.5 mL of this solution and subsequently analyzed by GC or GC-MS. Standard compounds were obtained from *Aspidosperma* (Oliveira, 1999) using classic phytochemical extraction and separation methods and were identified by comparison of physical (mp, $[α]_D$) and spectroscopy data (UV, $^1H$ and $^{13}C$ NMR) with literature values. The purity of the compounds was determined by gas chromatography using FID and MS detector. The GC analyses were done using a Hewlett-Packard 3396A gas chromatograph equipped with a FID. A DB-5 capillary column (12 m, 0.25 mm i.d. x 0.2 mm film thickness, J & W Scientific, CA) was used for this analysis. Hydrogen was utilized as the carrier gas, and the optimized oven temperature program was 100-225 °C, linear increase 15 °C/min, after 225-260 °C in linear increase 2.5 °C/min, and 260 °C held for 10 min. Temperatures of the injector port and detectors were held at 270 °C. The injectors were operated in split mode (1:50). One microliter of all solutions was injected in all analysis. GC-MS analysis was carried out on HP 5970B gas chromatograph equipped with a MS detector. The temperature program was the same used for GC-FID analysis.

### 3.4 High Performance Liquid Chromatography

Despite the large number of *Aspidosperma* alkaloids isolated and their biological importance, there are very few reports on the HPLC analysis of extracts of species from this genus (Jacome et al., 2003; Jacome et. al., 2004). This can be explained due to HPLC that has been used in great extent to the analysis of different types of alkaloids, problems still exist, because analysis of this compound still occur with a low efficiency, large and asymmetric peaks, attributed to the dual retention mechanism (Philipson et al., 1982; Verpoorte & Baherheim, 1984; Stockigt et al., 2002). For *A. ramiflorum* analysis by HPLC, the three majors indole alkaloids were isolated and characterized as follows: 10-methoxy-geissoschizol (**3**), ramiflorine A(**1**) and ramiflorine B (**2**) from stem bark and β-yohimbine (**4**) from seeds, which were used as standards in HPLC analysis, as shown in Figures 9A and B. To minimize the negative effects attributed to the dual retention mechanism, a methodology described by Giroud el al. (1991) to *Chinchona legderiana* was adapted to *A. ramiflorum* alkaloidal extract analysis, with some minor modifications, which allowed a good separation of it majority compounds (Figure 9B). The main modification made was the addition of octane sulfonic acid to mobile phase. The presence of this component allowed that the separation to be mediated by more types of interaction, because it acts by an ion pair formation mechanism, increasing the resolution and efficiency of analysis.

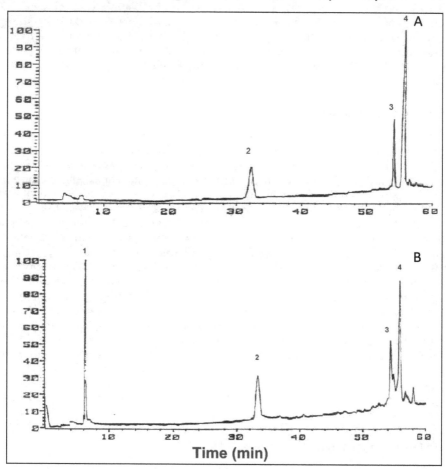

Fig. 9. A: high performance liquid chromatography (HPLC) chromatogram of standards mixture isolated from *Aspidosperma ramiflorum*; B: HPLC chromatogram of alkaloidal extract. Peaks – 1: internal standard (tryptophol); 2: 10-methoxy-geissoschizol; 3: ramiflorine A; 4: ramiflorine B.

### 3.4.1 High Performance Liquid Chromatography analysis

For HPLC analysis, the crude extract of *A. ramiflorum* (Ferreira et al., 2007) was dissolved in $CH_2Cl_2$:MeOH (80:20 v/v) and 10 µL were injected into a Waters µ-Bondapak RP-18 (reverse phase, 4.6 mm x 250 mm) column at 40°C. Solvent A was 100 mmol/L ammonium formate in 0.12% octane sulfonic acid (v/v), formic acid and acetonitrile (88:4:8, v/v), while solvent B

consisted of 100 mmol\L aqueous ammonium formate containing 0.12% octanesulfonic acid
(v/v)/formic acid/acetonitrile (64:4:32, v/v). The separation was carried out using a
mixture of solvent A and, a progressively increasing amount of B (0, 10, 40, 90, 100%) during
60 min. The flow rate was 1.3 ml min-1. The effluent was monitored with a photodiode-
array detector with windows at 222 nm and 254 nm and also by mass spectral analysis of
isolated eluates.

## 4. Conclusions

The development of the TLC solvent system for *Aspidosperma ramiflorum* alkaloids at the
beginning of this article clearly shows that TLC and PTLC as a routine and classical method
appreciated in alkaloid separation techniques, can still be optimized with introduction of
new methodologies as desorption electrospray ionization (DESI) which can permit coupled
TLC with mass spectrometry (Jackson et al., 2009). This is not necessarily true for alkaloid
determination by GC, because decomposition can occur due to the high temperatures
applied and however widespread application of this technique is limited by volatility of
more polar alkaloids. HPLC is a classical method for alkaloids separations and allowed a
good separation of majority compounds in *A. ramiflorum* alkaloids but can still be optimized.
However, the pre-purification of crude alkaloid extract will be a crucial step for all analytical
procedure described. The applicability of different chromatographic techniques for the
separation and identification of crude mixtures of bioactive *A. ramiflorum* indole alkaloids
has thus been demonstrated.

## 5. References

Barbosa, L. F.; Mathias, L.; Braz-Filho, R. & Vieira, I. J. C., (2010). Chemical constituents from
    *Aspidosperma illustre* (Apocynaceae). *J. Braz. Chem. Soc.*, 21, 1434-1438.
Biemann, K., (2002). Four Decades of Structure Determination by Mass Spectrometry: From
    Alkaloids to Heparin. *J Am Soc Mass Spectrom*, 13, 1254-1272.
Cardoso, C. A. L.; Vilegas, W. & Honda, N. H. (1998).Qualitative determination of indole
    alkaloids, triterpenoids and steroids of *Tabernaemontana hilariana J. Chromatogr., A.*
    *808*, 264.
Cardoso, C. A. L.; Vilegas, W. & Pozetti, G. L. (1997).Gas chromatographic analysis of indole
    alkaloids from *Tabernaemontana hilariana. J. Chromatogr.*,A, *788*, 204.
Dagnino, D., J. Schripsema, A. Peltenburg, R. Verpoorte & K. Teunis. (1991). Capillary gas
    chromatographic analysis of indole alkaloids: investigation of the indole alkaloids
    present in *Tabernaemontana divaricata* cell suspension culture. *J. Nat. Prod.* 54:1558–
    1563.
Ferreira, I., Leon, L., Gobbi Filho, L., Lonardoni, M., Silveira, T., Machado, G. & Oliveira, A.,
    (2004). Antileishmanial activity of alkaloidal extract from Aspidosperma
    ramiflorum. *Mem. Inst. Oswaldo Cruz.* Vol. 99, pp. 325–327.
Gallagher C. A.; Hough L. B.; Keefner S. M.,; Seyed-Mozaffari A.; Archer S.; & Glick S. D.,
    (1995). *Biochemical Pharmacology*, Vol. 49, No. 1, pp. 73-79.
Giroud, C.; Vanderleer, T.; Vanderheijden, R.; Verpoorte, R.; Heeremans, C. E. M.; Niessen,
    W. M. A. & vandergreef, J. (1991). Thermospray liquid-chromatography mass-
    spectrometry (TSP LC MS) analysis of the alkaloids from Cinchona in vitro
    cultures. *Planta Medica*, 57, 142-148.

Henrique, M. C.; Nunomura, S. M. & Pohlit, A. M. (2010). Alcaloides indólicos de cascas de *Aspidosperma vargasii* e *A. desmanthum. Química Nova*, 33, 284-287.

Jackson, A. U., Tata, A., Wu, C., Perry, R. H., Haas, G., West, L., & Cooks, R. G. (2009). Direct analysis of *Stevia* leaves for diterpene glycosides by desorption electrospray ionization mass spectrometry. *Analyst*, 134, 867-874.

Jácome, R. L. R. P.; Oliveira, A. B.; Raslan, D. S. & Wagner, H. (2004).Estudo químico e perfil cromatográfico das cascas de *Aspidosperma parvifolium* A. DC. ("Pereira"). *Quim. Nova*, Vol. 27, No. 6, 897-900.

Jácome, R. L. R. P.; Souza, R. & Oliveira, A.B. (2003).Comparação cromatográfica entre o extrato de *Aspidosperma parvifolium* e o fitoterápico "Pau-Pereira". *Rev. Bras. Farmacogn.*, v. 13, supl., p. 39-41, 2003. ISSN: 0102-695X

Kobayashi, J; Sekiguchi, M.; Shimamoto, S.; Shigemori, H.; Ishiyama, H. & Ohsaki, A., (2002). Subincanadines A-C, Novel Quaternary Indole Alkaloids from *Aspidosperma subincanum. J. Org. Chem.*, 67, 6449-6455.

Marques, M., Kato, L., Filho, H. & Reis, F., (1996). Indole alkaloid from Aspidosperma ramiflorum. *Phytochemistry*, v. 41, n. 3, pp.963-967.

Oliveira, A., (1999). Estudo de seis espécies do gênero Aspidosperma, utilizando CG, CG/MS e HPLC: Análise Qualitativa e Quantitativa. Teste Bioautográfico; Cultura de Tecidos e Células Vegetais e rota de Preparação dos Compostos Diméricos Ramiflorina A e Ramiflorina B. Universidade de Campinas, Campinas. *Tese (doutorado em química orgânica)*.

Oliveira, V. B.; Freitas, M. S. M.; Mathias, L.; Braz-Filho, R. &Vieira, I. J. C.; (2009).Atividade biológica e alcalóides indólicos do gênero Aspidosperma (Apocynaceae): uma revisão.*Rev. Bras. Pl. Med.*, *11*, 92.

Phillipson, J. D.; Supavita, N. & Anderson, L. A. (1982). Separation of heteroyohimbine and oxindole alkaloids by reversed-phase high-performance liquid chromatography. *Journal of Chromatography A*, 244, 91-98.

Stockigt, J.; Sheludko, Y.; Unger, M.; Gerasimenko, I.; Warzecha, H. & Stockigt, D., (2002). High-performance liquid chromatographic, capillary electrophoretic and capillary electrophoretic–electrospray ionisation mass spectrometric analysis of selected alkaloid groups. *Journal of Chromatography A*, 967, 85-113.

Tanaka J.C., Silva C.C., Ferreira I.C.P., Machado G.M.C., Leon L.L. & Oliveira A.J.B., (2007). Antileishmanial activity of indole alkaloids from *Aspidosperma ramiflorum. Phytomedicine* 14, 377-380.

Tanaka JCA, Silva C.C., Dias F., B.P., Nakamura C.V. & Oliveira A.J.B. (2006). Antibacterial activity of indole alkaloids from *Aspidosperma ramiflorum. Braz J MedBiol Res 39*: 387-391.

Verpoorte, R. & Baherheim, A. S. (1984).Chromatography of Alkaloids. Amsterdam, Netherlands, *Elselvier*, vol. 23B, p.31.

Zocoler, M.A., Oliveira, A.J.B., Sarragiotto, M.H., Grzesiuk, V.L. & Vidotti, G.J., (2005). Qualitative determination of indole alkaloids of *Tabernaemontana fuchsiaefolia* (Apocynaceae). *J. Braz. Chem. Soc.* 16, 1372–1377.

# 8

# Biomarkers

Yasser M. Moustafa and Rania E. Morsi
*Egyptian Petroleum Research Institute,*
*Egypt*

## 1. Introduction

Generally, biomarkers are naturally occurring, ubiquitous and stable complexes that are objectively measured and evaluated as an indicator of a certain state. It is used in many scientific fields; medicine, cell biology, exposure assessment, astrobiology, geology and petroleum.

Biomarkers "biological markers" in medicine are complex compounds that can be used as an indicator of a particular disease state or some other physiological state of an organism. Biomarkers have been defined also as cellular, biochemical or molecular alterations that are measurable in biological media such as human tissues, cells or fluids (Hulka et al., 1990). Broader definitions include biological characteristics that can be objectively measured and evaluated as an indicator of normal or abnormal biological processes, pathogenic processes or pharmacological responses to a therapeutic intervention (Naylor, 2003).

Biomarkers can also reflect the entire spectrum of disease from the earliest manifestations to the terminal stages. Characterization of the healthy and diseased cells, when identifying a specific biomarker as an indicator of cancer, is a needed research strategy for validating biomarkers. For the nervous system, as an example, there is a wide range of techniques used to gain information about the brain in both the healthy and diseased state. These may involve measurements directly on biological media (e.g. blood or cerebrospinal fluid) or measurements such as brain imaging which do not involve direct sampling of biological media but measure changes in the composition or function of the nervous system (Mayeux , 2004). Moreover, the pharmaceutical industry is beginning to rely on biomarkers information and their importance to the future of drug discovery and development.

In practice, biomarkers science includes tools and technologies that can aid in understanding the prediction, cause, diagnosis, progression, regression, and/or the outcome of disease treatment which provide a dynamic and powerful approach to understand the spectrum markers and offer the means for homogeneous classification of a disease and risk factors which can extend the base information about the underlying pathogenesis of disease.

## 2. Petroleum biomarkers

Due to the variety of geological conditions and ages under which oil was formed, every crude oil exhibits a unique biomarker fingerprint. Crude oils compositions vary widely depending on the oil sources, the thermal regime during oil generation, the geological migration and the reservoir conditions. Crude oils can have large differences in:

1- distribution patterns of the n- alkanes, iso-alkanes and cyclic- alkanes as well as the unresolved complex mixture (UCM) profiles 2- relative ratios of isoprenoid to normal alkanes 3- distribution patterns and concentrations of alkylated polynuclear aromatic hydrocarbons (PAHs) homologues. Most of these constituents undergo changes in their chemical structure by time as an effect of several factors among which are the biodegradation and weathering. Relative to other hydrocarbon groups in oil, there are some compounds that are more degradation-resistant in the environment as for example; *Pristane, phytane, steranes, triterpanes and porphyrins*. These undegradable compounds are known as *Biomarkers*.

Trebs (1934) was the first one to develop the biomarkers concept, with his pioneering work on the identification of porphyrins in crude oils suugesting that these porphyrins are generated from chlorophyll of plants. Blumer et al. (1963) and Blumer & Thomas (1965) isolated pristane from recent marine sediments and concluded that it was derived from the phytol side chain of chlorophyll. Later, other workers reported the present of various classes of degradation-resistant organic compounds and recognized their biomarker implementations.

Petroleum biomarkers can thus be defined as complex organic compounds derived from formerly living organisms found in oil (Mobarakabad et al., 2011). They show little or no changes in their structure from the parent organic molecules and this distinguishes biomarkers from other compounds (Maioli et al., 2011). Various biomarkers formed under different geological conditions and ages can occur in different carbon ranges exhibiting different biomarker fingerprints.

From the identification point of view, biomarkers are the most important hydrocarbon groups in petroleum because they can be used for chemical fingerprinting which provides unique clues to the identity of source rocks from which petroleum samples are derived, the biological source organisms which generated the organic matter, the environmental conditions that prevailed in the water column and sediment at the time, the degree of microbial biodegradation and the thermal history (maturity) of both the rock and the oil. The information from biomarker analysis can be used also to determine the migration pathways from a source rock to the reservoir for the correlation of oils in terms of oil-to-oil and oil-to-source rock and the source potential. Also chemical analysis of biomarkers generates information of great importance to environmental forensic investigations in terms of determining the source of spilled oil, differentiating and correlating oils and monitoring the degradation process and weathering state of oils under a wide variety of conditions.

Terpanes and steranes are highly resistant to biodegradation but few studies have shown that they can be degraded to certain degree under severe weathering conditions i.e, extensive microbial degradation (Chosson, 1991).

## 2.1 Biomarkers analysis

The commercial availability of Gas Chromatography- Mass Spectroscopy (GC-MS) and associated data systems in the mid-1970s led to use the biomarkers for a wide variety purposes. The complex structure of biomarkers and the possible presence in low concentrations make a pressing need for more sensitive and precise analysis. The development of analytical methodologies and the combination between these methods are of great importance to separate, monitor and detect the absolute concentrations and structure of petroleum biomarkers. The use of "hybrid" or "hyphenated" techniques, which are a combination of different separate techniques, increases the analytical power of the used methods. GC-MS can be considered as the most popular method used in the characterization of major biomarker groups. GC provides the significant advantage of the separation of different structures of biomarkers while MS can accurately detect and identify these structures. The concept and the development of these instrumentations will be briefly mentioned.

### 2.1.1 Separation by chromatographic techniques

Chromatography is the separation of a mixture of compounds into their individual components primarily according to their volatilities. There are numerous chromatographic techniques but gas chromatography (GC) is the most important one. It has a number of advantages over other separation techniques. It can identify (qualitate) and measure the amount (quantitate) various sample components.

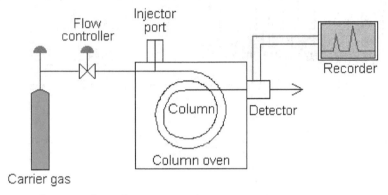

Fig. 1. Schematic diagram of gas chromatograph.

The GC column is the heart of the system; the structure of the stationary phase and the packed material greatly influence the separation of the compounds and affect the time of separation (retention time). Two types, packed and capillary columns, have been used. The advantages in capillary columns over packed columns are in obtaining practically improved resolution in order to give fine structured chromatographic fingerprints. The column is placed in an oven where the temperature can be controlled very accurately over a wide range of temperatures. As compounds come off the column, they enter a detector for identification. Figure 2 represent the carbon number range distribution of common hydrocarbons in crude oil and petroleum products.

### 2.1.2 Identification by spectroscopic techniques

There are different types of detectors that can be employed depending on the compounds to be analyzed. Mass spectrometry (MS) has very common use in analytical laboratories that study a great variety of compounds and provides a satisfactory tool for obtaining specific fingerprints for classes and homologous series of compounds resolved by gas chromatography. The technique has both qualitative and quantitative uses include identifying and determining the structure of a compound by observing its fragments.

The mass spectrometer has long been recognized as the most powerful detector for gas chromatography. Typical MS instruments consist of three modules; an ion source: which can

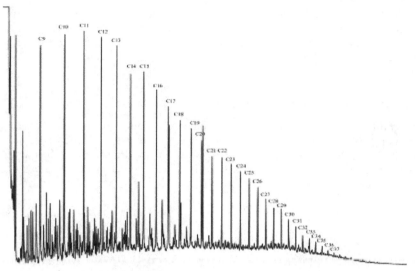

Fig. 2. Representative model of carbon number distribution in petroleum hydrocarbons.

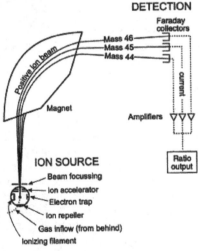

Fig. 3. Schematic diagram of mass spectrometer.

convert the separated constituent into ions, a mass analyzer: which sorts and separates ions by their masses by applying electromagnetic fields and a detector: which calculate the abundances of each ion present by a quantitative method to generate signals. The size of the signals corresponds to the amount the compound present in the sample.

Characterization of some major biomarker groups is largely achieved using the following MS fragment ions:

- alkyl-cyclohexanes: m/z 83
- methyl-alkyl-cyclohexanes: m/z 97

- isoalkanes and isoprenoids: m/z 113, 127, 183
- sesquiterpanes: m/z 123
- adamantanes: m/z 135, 136, 149, 163, 177, and 191
- diamantanes: m/z 187, 188, 201, 215 and229
- tri-, tetra-, penta-cyclic terpanes: m/z 191
- 25-norhopanes: m/z 177
- 28,30-bisnorhopanes: m/z 163, 191
- steranes: m/z 217, 218
- 5a(H)-steranes: m/z 149, 217, 218
- 5β(H)-steranes: m/z 151, 217, 218
- X diasteranes: m/z 217, 218, 259
- methyl-steranes: m/z 217, 218, 231, 232
- monoaromatic steranes: m/z 253
- triaromatic steranes: m/z 231

The m/z 191 fragment is often the base peak of mass spectra of biomarkers. In general, GC/MS chromatograms of terpanes (m/z 191) are characterized by the terpane distribution in a wide range from C19 to C35 with C29 αβ- and C30 αβ-pentacyclic hopanes and C23 and C24 tricyclic terpanes being often the most abundant. As for steranes (at m/z 217 and 218), the dominance of C27,C28 and C29 homologues. Figure 4, as a representative model, shows

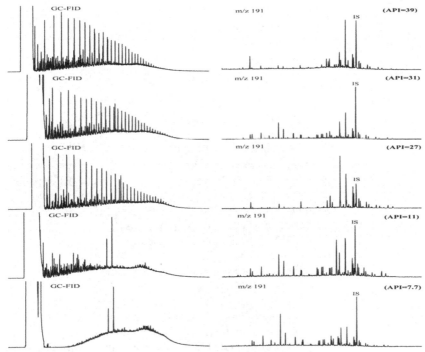

API: American Petroleum Institute. The larger the API, the greater the amount of the light components the oil contains and with decreasing the API, the amounts of medium and heavy weight components increase.

Fig. 4. GC-MS chromatograms at m/z 191 for light to heavy crude oils.

GC-MS chromatograms at m/z 191 for light (API > 35), medium (API: 25–35), and heavy (API < 25) crude oils (Wang et al., 2006).

## 3. Diagnostic ratios of biomarkers

Biomarker diagnostic parameters have been long established and are widely used by geochemists for oil correlation; determination of organic input and precursors, depositional environment, assessment of thermal maturity and evaluation of in-reservoir oil biodegradation (Peters et al., 2005). Diagnostic ratios (DRs) can either be calculated from quantitative (i.e., compound concentrations) or semi-quantitative data (i.e., peak areas or heights). Also, many diagnostic ratios currently used in oil spill and environmental studies. Oil–oil correlations are based on the concept that the composition of biomarkers in spill samples does not differ from those of the candidate source oils. Most biomarkers in spill samples and source oils, in particular those homologous series of biomarkers with similar structure, show little or no changes in their diagnostic ratios. An important benefit of comparing diagnostic ratios of spilled oil and suspected source oils is that concentration effects are minimized. In addition, the use of ratios (rather than absolute values) tends to induce a self-normalizing effect on the data because the variations due to the fluctuation of instrument operating conditions day-to-day, operator, and matrix effects are minimized. Therefore, comparison of diagnostic ratios reflects more directly differences of the target biomarker distribution between samples.

**Biomarker classes**          **Diagnostic ratios**
*Acyclic Isoprenoids*

                                 pristane/phytane
                                 pristane/n-C17
                                 phytane/n-C18

*Terpanes*

                                 C21/C23 tricyclic terpane
                                 C23/C24 tricyclic terpane
                                 C23 tricyclic terpane/C30αβ hopane
                                 C24 tricyclic terpane/C30αβ hopane
                                 C24 tertracyclic/C26 tricyclic (S)/C26 tricyclic (R) terpane
                                 C27 18α, 21β-trisnorhopane/C27 17α, 21β-trisnorhopane
                                 C28 bisnorhopane/C30αβ hopane
                                 C29αβ-25-norhopane/C30αβ hopane
                                 C29αβ-30-norhopane/C30αβ hopane
                                 oleanane/C30 αβ hopane
                                 moretane(C30 βα hopane)/C30αβ hopane
                                 gammacerane/C30αβ hopane
                                 tricyclic terpanes (C19-C26)/C30 αβ hopane
                                 C31 homohopane (22S)/C31 homohopane (22R)
                                 C32 bishomohopane (22S)/C32 bishomohopane (22R)
                                 C33 trishomohopane (22S)/C33 trishomohopane (22R)
                                 Relative homohopane distribution
                                 (C31-C35)/C30 αβ hopane
                                 homohopane index

| Biomarker classes | Diagnostic ratios |
|---|---|
| *Steranes* | |
| | Relative distribution of regular C27-C28-C29 steranes |
| | C27 αββ/C29 αββ steranes (at m/z 218) |
| | C27 αββ/(C27 αββ + C28 αββ + C29αββ) (at m/z 218) |
| | C28 αββ/C29 αββ steranes (at m/z 218) |
| | C28 αββ/(C27 αββ + C28 αββ + C29 αββ) (at m/z 218) |
| | C29 αββ/(C27 αββ + C28 αββ + C29 αββ) (at m/z 218) |
| | C30 sterane index: C30/(C27 to C30) steranes selected |
| | diasteranes/regular steranes |
| | Regular C27-C28-C29 steranes/C30αβ-hopanes |
| *Monoaromatic steranes* | |
| | C27-C28-C29 monoaromatic steranes (MA) distribution. |
| *Triaromatic steranes* | |
| | C20 TA/(C20 TA + C21 TA) |
| | C26 TA (20S)/sum of C26 TA (20S) through C28 TA (20R) |
| | C27 TA (20R)/C28 TA (20R) |
| | C28 TA (20R)/C28 TA (20S) |
| | C26 TA (20S)/[C26 TA (20S) +C28 TA (20S)] |
| | C28 TA (20S)/[C26 TA (20S) +C28 TA (20S)] |

Table 1. Examples of some diagnostic ratios of biomarkers frequently used for the environmental forensic studies.

# 4. Examples of parameters used in fingerprinting

## 4.1 Normal -alkanes characteristics

The distribution of n-alkanes in crude oils can be used to indicate the organic matter source (Duan and Ma, 2001). For example, the increase in the n-C15 to n-C20 suggests marine organic matters with contribution to the biomass from algae and plankton (Peters and Moldowan, 1993). Oil samples characterized by uniformity in n-alkanes distribution patterns suggest that they are related and have undergone similar histories with no signs of biodegradation (Ficken et al. 2000 and Duan and Ma, 2001).

## 4.2 Carbon preference index (CPI)

Carbon preference index, obtained from the distribution of n-alkanes, is the ratio obtained by dividing the sum of the odd carbon-numbered alkanes to the sum of the even carbon-numbered alkanes. CPI is affected by both source and maturity of crude oils (Tissot and Welte, 1984). CPI of petroleum oils ranging about 1.00 generally shows no even or odd carbon preference indicates mature samples. Also, it can be used in source identification; petroleum origin contaminants characteristically have CPI values close to one (Maioli et al., 2011).

## 4.3 Degree of waxiness

The degree of waxiness can be expressed by the $\Sigma C21$-$C31/\Sigma C15$-$C20$ ratios. The oils characterized by high abundance of n-C15to n-C20 n-alkanes in the saturate fractions reflecting low waxy (Moldowan et al., 1994). Generally, the degree of waxness < 1 reveals low waxy nature and suggests marine organic sources (Peters and Moldowan, 1993) mainly of higher plants deposited under reducing condition.

## 5. Examples of parameters used in biomarker fingerprinting

### 5.1 Pristane/phytane ratio

Both pristine (2,6,10,14- tetramethyl pentadecane) and phytane (2,6,10,14- tetramethyl hexadecane) are derived from the phytol side chain of chlorophyll, either under reducing conditions (phytane) or oxidizing conditions (pristane). Also both pristine and phytane became dominant saturated hydrocarbon components of highly weathered crude oils until they are degraded (Moustafa et al., 2004).

The pristane/phytane (Pr/Ph) ratio is one of the most commonly used correlation parameters which have been used as an indicator of depositional environment (Peters et al., 2005). It is believed to be sensitive to diagenetic conditions; Pr/Ph ratios substantially below unity could be taken as an indicator of petroleum origin and/or highly reducing depositional environments. Very high Pr/Ph ratios (more than 3) are associated with terrestrial sediments. Pr/Ph ratios ranging between 1 and 3 reflect oxidizing depositional environments (Hunt, 1996).

According to Lijmbach (1975) low Pr/Ph values (<2) indicate aquatic depositional environments including marine, fresh and brackish water (reducing conditions), intermediate values (2–4) indicate fluviomarine and coastal swamp environments, whereas high values (up to 10) are related to peat swamp depositional environments (oxidizing conditions).

### 5.2 Isopreniods/n-alkanes

Waples (1985) stated that by increasing maturity, n-alkanes are generated faster than iosprenoids in contrast to biodegradation. Accordingly, isopreniods/n-alkanes (Pr/$n$-C17 and Ph/$n$-C18) ratios provide valuable information on biodegradation, maturation and diagenetic conditions. The early effect of microbial degradation can be monitored by the ratios of biodegradable to the less degradable compounds. Isoprenoid hydrocarbons are generally more resistant to biodegradation than normal alkanes. Thus, the ratio of the pristane to its neighboring n-alkane C17 is provided as a rough indication to the relative state of biodegradation. This ratio decreases as weathering proceeds.

### 5.3 Steranes (m/z 217) distribution

The distribution of steranes is best studied on GC/MS by monitoring the ion m/z=217 which is a characteristic fragment in the sterane series. It is agreed that the relative amounts of C27-C29 steranes can be used to give indication of source differences (Lijmbach, 1975). For example, predominance of C28, C29 and C30 steranes indicate an origin of the oils derived mainly from mixed terrestrial and marine organic sources, while oils show slightly low abundance of C28 and C29 and relatively higher concentrations of C27 steranes indicate more input of marine organic source.

## 5.4 Triterpanes (m/z 191) distribution

Together with steranes, triterpanes belong to the most important petroleum hydrocarbons that retain the characteristic structure of the original biological compounds. Tricyclic, tetracyclics hopanes and other compounds contribute to the terpane fingerprint mass chromatogram (m/z=191) are commonly used to relate oils and source rocks (Hunt, 1996). Mass fragmentogram at m/z=191 can be used to detect triterpanes in the saturate hydrocarbon fraction.

### 5.4.1 Tricyclic terpanes

Aquino et al. (1983) indicated that tricyclic terpanes are normally associated with marine source. In addition it has been used as a qualitative indicator of maturity (Van Grass, 1990). In high mature oils, the tricyclic terpanes is dominated more than in low mature oils (Hunt, 1996).

### 5.4.2 Homohopanes

The homohopanes (C31 to C34) are believed to be derived from bacteriopolyhopanol of prokaryotic cell membrane. C35 homohopane may be related to extensive bacterial activity in the depositional environment (Ourisson et al., 1984). Homohopane index can be used as an indicator of the associated organic matter type, as it can also be used to evaluate the oxic/anoxic conditions of source during and immediately after deposition of the source sediments (Peters and Moldowan, 1991). Low C35 homohopanes is an indicator of highly reducing marine conditions during deposition whereas high C35 homohopane concentrations are generally observed in oxidizing water conditions during deposition, consistent with the oxic conditions (Peters and Moldowan, 1991).

### 5.4.3 Gammacerane

Gammacerane, originally thought to be as hypersalinity indicator (Sinninghe-Damste et al., 1995), is associated with both marine and lacustrine environments of increasing salinity (Waples and Machihara, 1991; and Peters and Moldowan, 1993).

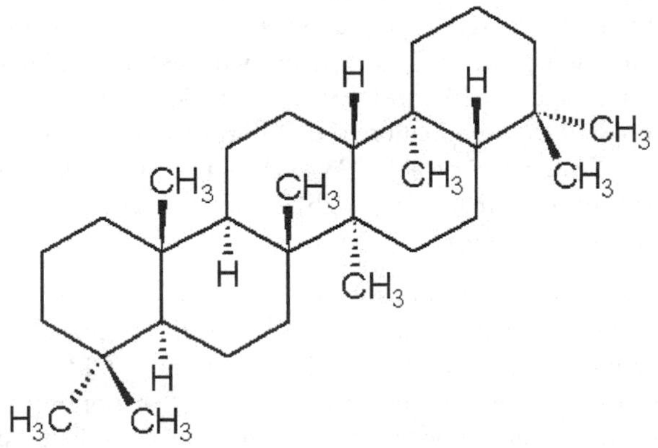

Fig. 5. Gammacerane chemical structure.

### 5.4.4 Ts/Tm

The ratio of Ts (trisnorneohopane) to Tm (trisnorhopane) more than (0.5) was found to increase as the portion of shale in calcareous facies increases (Hunt, 1996). Van Grass (1990) stated that Ts/Tm ratios begin to decrease quite late during maturation but Waples and Machihara (1991) reported that Ts/Tm ratio does not appear to be appropriate for quantitative estimation of maturity.

### 5.4.5 C29/C30 hopanes ratios

C29/C30 hopanes ratios are generally high (>1) in oils generated from organic rich carbonates and evaporates (Connan et al., 1986).

### 5.4.6 Steranes/17α (H)-hopanes ratio

The regular steranes /17α(H)-hopanes ratio reflects input of eukaryotic (mainly algae and higher plants) versus prokaryotic (bacteria) organisms to the source rock. The sterane/hopane ratio is relatively high in marine organic matter with values generally approaching unity or even higher. In contrast, low steranes and sterane/hopane ratios are more indicative of terrigenous and/or microbially reworked organic matter (Suzuki et al.,1996).

### 5.4.7 Bisnorhopanes

It is believed that sediments containing large amounts of bisnorhopane were deposited under anoxic conditions (Mello et al., 1988). Bisnorhopanes are types of pentacyclic triterpanes present in significant concentrations in oil. Bisnorhopanes are observed in Guatemalan evaporites (Connan et al., 1986) and frequency reported in other biogenic siliceous rocks of the circum-Pacific region (Katz and Elrood, 1983).

### 5.5 Metalloporphyrins

Porphyrins are the tetrapyrole compounds; the porphyrin nucleus consists of four pyrrole rings joined by four methine bridges giving a cyclic tetrapyrrole structure. The majority of these compounds are thought to originate from various chloropigments produced by phototrophic organisms of the geological past (Yui et al., 2007). Metalloporphyrins has become a valuable tool in the determination of the origin and maturity of the organic matter (Doukkali et al., 2002; Chikaraishi et al., 2005 and Ohkouchi et al., 2006). The porphyrin structure consists of a porphyrin nucleus with various groups of side chains occupying some or all of its peripheral positions.

Metalloporphyrins were extracted from asphaltene and maltene fractions using adsorption column chromatography (Faramawy et al., 2010). Porphyrins occur as etioporphyrin (Etio), Benzo-etio, deoxophylloerythroetioporphyrin (DPEP), Benzo-DPEP and tetrahydrobenzo-DPEP (THBD). The distribution of different types of metalloporphyrins is useful for interpreting transformation of kerogen into bitumen, depositional environments and maturation levels of deposited organic matters.

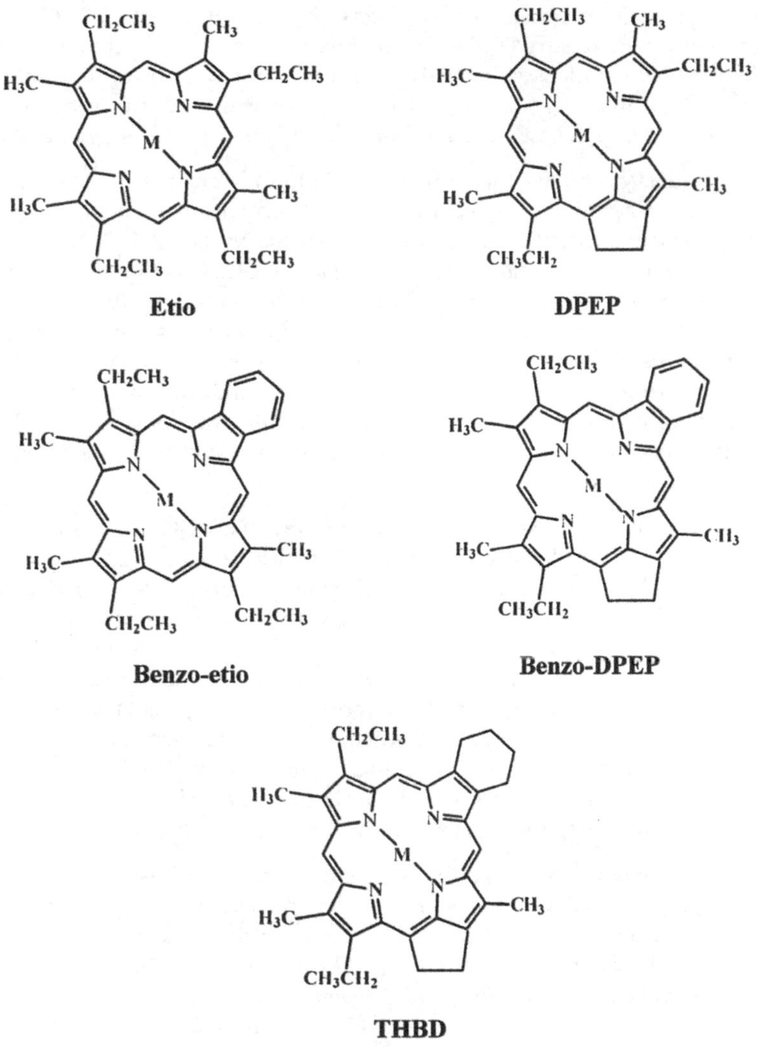

Fig. 6. Structures of different types of metalloporphyrins.

## 6. Developments in GC-MS instrumentation

The low biomarker concentrations in oils (often in the range of several parts per million) in the presence of a highly complex petroleum hydrocarbon matrix especially weathered oils, the variety of chemical classes present in oils and the possible co-elutions in conventional chromatographic separations make the identification of biomarkers a more difficult task.

The development of more reliable, highly selective, fast and sensitive separation and identification tools for biomarker analysis purposes can be considered as one of the most important research points in this field for a meaningful biomarker analysis.

The use of comprehensive two-dimensional gas chromatography (GC × GC) coupled to time-of-flight mass spectrometry (TOFMS) was found to be a powerful tool for overcoming some problems and limitations since it (i) separates substances using two interconnected capillary columns containing different stationary phases and (ii) uses the fast data acquisition of time-of-flight analyzer as a robust registry for GC × GC (Aguiar et al., 2001).

In their work, Aguiar et al. (2001) used this technique to overcome the co-elution between tri- and pentacyclic terpanes separated by extracted ion chromatograms (EIC) for ions of mass-to-charge ratio (m/z) 191. The biomarker analysis by GC × GC–TOFMS was much better than in previous works using one-dimensional GC. Co-elutions between tri- and pentacyclic terpanes were clearly resolved in the second column. Noteworthy separation between the C30 hopane and C30 dimethylated homohopane was achieved and overlap of hopanes with steranes in the m/z 217 was eliminated. Besides hopanes, dimethylated tri- and tetracyclic terpanes were identified. These findings indicate the superiority of GC × GC–TOFMS as a technique for separation and identification of biomarkers in oils due to its high sensitivity, specificity and capability to elucidate compounds structure with high spectral resolution.

Comprehensive two-dimensional gas chromatography (GC×GC) has also been used to separate and identify alkylated aromatics (naphthalenes, biphenyls, fluorenes, phenanthrenes and chrysenes), sulfur-containing aromatics (dibenzothiophenes, benzonaphthothiophenes), steranes, triterpanes, and triaromatic steranes. These biomarkers were separated into easily recognizable bands in the GC×GC chromatogram. Methods used to identify the bands included peak matching with chemical standards and comparison with GC/MS extracted ion chromatograms (Frysinger and Gaines, 2001). By designing mass spectrometers that can determine m/z values accurately to four decimal places, it is possible to distinguish different formulas having the same nominal mass. Since a given nominal mass may correspond to several molecular formulas, lists of such possibilities are especially useful when evaluating the spectrum of an unknown compound.

GC/MS/MS is an operation based on the covariant scan of electrostatic magnetic fields on the trisector double focusing mass spectrometer providing more accurate data. The quadraupole is a common mass separator gives a sufficient sensitivity and selectivity however, high resolution mass spectrometry (HRMS) is also used due to its ability to provide quantitative data for compounds present in complex mixtures for biomarkers analysis. Triple quadrupole GC/MS offers a viable alternation for the rapid, routine analysis providing excellent precision, sensitivity, selectivity, and dynamic range (Thermo-application note 10261).

Fourier transform ion cyclotron resonance mass spectrometry (FT-ICR MS) benefits from ultra-high mass resolving power (greater than one million), high mass accuracy (less than 1 ppm) and rapid analysis which make it an attractive alternative for the analysis of different and wide range of petroleum products (Klein et al., 2003).

It should be noted, however, that there is no single fingerprinting technique that can fully and readily meet the objectives of biomarkers investigation and quantitatively allocate hydrocarbons to their respective sources, particularly for complex hydrocarbon mixtures or extensively weathered and degraded oil residues. Combined and integrated multiple tools are often necessary under such situations.

## 7. Data analysis by computerized techniques

Data analysis is an important part of chemical fingerprinting and a broad collection of statistical techniques has been used for evaluation of data.

After separation and identification of biomarkers, principal component analysis PCA, a mathematical procedure, can be used for analyses of chromatograms using a fast and objective procedure with more comprehensive data usage compared to other fingerprinting methods. The discriminative power of PCA can be enhanced by deselecting the most uncertain variables or scaling them according to their uncertainty.

For example, preprocessing of GC-MS chromatograms followed by principal component analysis (PCA) of oil spill samples collected from the coastal environment in the weeks after the Baltic Carrier oil spill and from the tank of the Baltic Carrier (source oil) was carried out (Christensen et al., 2005). The preprocessing consists of baseline removal by derivatization, normalization, and alignment using correlation optimized warping. The method was applied to chromatograms of $m/z$ 217 (tricyclic and tetracyclic steranes) of oil spill samples and source oils. The four principal components were interpreted as follows: boiling point range (PC1), clay content (PC2), carbon number distribution of sterols in the source rock (PC3), and thermal maturity of the oil (PC4). The method allows for analyses of chromatograms using a fast and objective procedure and with more comprehensive data usage compared to other fingerprinting methods.

## 8. Previous studies in Egypt

### 8.1 Egyptian Western Desert

The Western Desert covers about 700,000 square kilometers (equivalent in size to Texas) and accounts for about two-thirds of Egypt's land area. This immense desert to the west of the Nile spans the area from the Mediterranean Sea south to the Sudanese border. The chemical fingerprinting of oils in this area is a great interesting research area for many proposes as for example identifying the sources of petroleum oil or complex environmental pollutants. The original sources of complex mixtures can often be identified by the relative abundance of some major individual compounds (e.g n-alkanes) forming a chemical pattern by ratios of specific constituents or by identifying source-specific compounds or markers (e.g triterpanes) in the environmental sample being investigated (Peters et al., 2005). These parameters depend mostly on the preburial environments of the living organisms, the depositional environments of the organic matter and the diagenetic processes in the source rocks.

In their work, Roushdy et al. (2011) utilize biomarkers characteristics together with bulk geochemical parameters to identify and characterize the crude oils and to assess the respective depositional environments and maturation. Variation of crude oil-gravities in the Western Desert reflects different stages of oil migration and accumulation as well as different oil source rocks in the same and different ages (Zein El Din et al., 1990). The authors attempt to assess the correlation between the crude oil samples and the potential source rocks to confirm the indigenous sources for the petroleum generation of some oilfields of the North Western Desert. This target was made throughout the study in detail of the analytical results for three crude oil samples collected from three oilfields in the North Western Desert oilfields (Meleiha, Misaada and Qarun) as well as three extract samples (Baharia, Kharita and Khtataba) from formations ranging in age from Upper Cretaceous to Middle Jurassic.

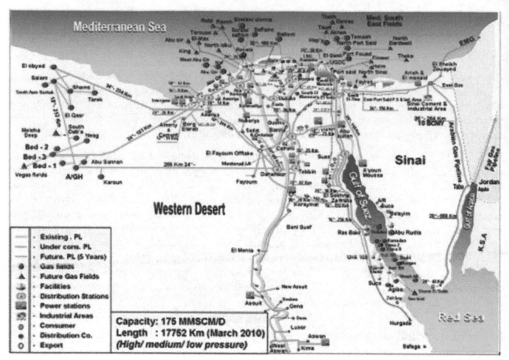

Fig. 7. Map of Egypt showing the main oil and gas fields.

The specific geochemical parameters have been assessed by the aid of gas chromatography and gas chromatographic-mass spectrometric analyses of the saturated fractions. The degree of the correlation between crude oils and the extracted samples was determined by studying the correlation scores for both oils and extracts. Eight correlation parameters have been studied for this purpose includes: saturates%, saturates/aromatics ratio, $C_{max}$, $C21+C22/C28+C29$, CPI, pristane/phytane, pristane/$n$-C17 and pristane+$n$-C17/phytane+$n$-C18. An overall correlation score was obtained for each oil and extract by summing up the contribution from each parameter. The GC/FID chromatogram of the Meleiha crude oil sample is characterized by a monotonically decreasing homologous series of heavy normal alkanes ($n$-C25 to $n$-C30) and display odd carbon preference at $n$-C15 which reflects mature oils originated mainly from non-marine origin mainly terrestrial organic matters deposited under slightly oxidizing environment and slightly mixed with inputs from marine source (Hunt, 1996). The mode of distribution of $n$-paraffins in the crude oils of Misaada and Qarun oilfields show that the maximum abundance is at $n$- C15 to $n$-C25 reflecting marine origin.

The steranes distribution of crude oils was studied. Meleiha crude oil was found to be characterized by low predominance of C27 steranes and slightly high abundance of C28 and C29 indicating that the Meleiha oil is believed to be generated from both marine shales and carbonates enriched in marine algae with more contribution from terrestrial organic sources deposited under saline conditions. It reveals that the Meleiha oil is derived mainly from terrestrial organic sources.

The high concentrations of C27 - C29 diasteranes in case of Meleiha oil indicate input of marine organic source with more contribution from terrestrial organics (Waples and Machihara, 1992). The high diasteranes concentration compared to regular steranes suggest a clay rich source rock because the clay is required to catalyze the steroids transformation to diasteranes (Peters and Moldowan, 1991). Misaada and Qarun oils are characterized by slightly lower predominance of C27 steranes and higher abundance of C28 and C29 steranes indicating inputs from marine organic sources (Waples and Machihara, 1992). Moreover, the distributions of regular steranes C29, C27 and C28 on the ternary diagram reveal also more contribution from marine organic sources. The diasteranes concentrations compared to regular steranes is low, suggesting a clay rich source rock.

The crude oil samples of Qarun and Misaada oilfields have Pr/$n$-C17 and ph/$n$-C18 ratios 0.28, 0.47 and 0.1, 0.1, respectively reflecting mostly mature and originated mainly from marine organic sources deposited under reducing environment. The crude oil of Meleiha oilfield has Pr/$n$-C17 and ph/$n$-C18 of 0.40 and 0.28 indicating mixed organic sources.

Terpanes biomarkers distributions derived from the m/z 191 mass chromatograms show that the C21-C25 tricyclic terpanes of Meliha oil appear to be the largest components which may support that the oil of Meleiha oilfield is more mature and sourced mainly from marine carbonate source rocks. At the same time, the C23, C24 and C25 tricyclic terpanes are generally of lower values compared with C22 indicating that the oil has some inputs from terrestrial organic materials (Hunt, 1996). The unusual low amounts of C30 extended hopanes seem to be associated with mixed organic sources (Moldowan et al., 1985). This phenomenon can be displayed by the low ratio of C29/C30 extended hopanes.

On the other hand, The C30 hopanes are the largest components in the series C27-C34 in oil samples from Misaada and Qarun oilfields. This indicates that the organic materials in these oils were originated mainly from saline and hypersaline environments (Peters and Moldowan, 1993). The extended hopanes are available as paleo environmental indicator (Waples and Machihara, 1992). The unusual large amounts of C30 hopanes seem to be associated with marine sources (Moldowan et al., 1985).

Bisnorhopanes are types of pentacyclic triterpanes present in significant concentrations in oil. Bisnorhopanes are observed in Guatemalan evaporites and frequency reported in other biogenic siliceous rocks of the circum-Pacific region (Connan et al., 1986). It is believed that sediments containing large amounts of bisnorhopane were deposited under anoxic conditions (Mello et al., 1988). The crude oils of Misaada oil field have relatively higher amounts of C28 bisnorhopane indicating more anoxic environment than Qarun and Meleiha oils.

Carbon preference index (CPI) values of the studied crude oils are close to unity, ranging from (0.94 to 1.04) indicating mature crude oils.

Pr/Ph ratio of the oil sample from Meleiha oilfield is 3.0 indicating oxidizing depositional environment of the crude oil while the crude oils from Qarun and Misaada oil fields have Pr/Ph ratios of 0.63 and 2.00 respectively reflecting that these crude oils were deposited under transitional (reducing- oxidizing) environments. These results indicate good correlation between crude oils from Qarun and Misaada oilfields with slight correlation to crude oil from Meleiha oilfield.

Oil: source correlation reflect a good correlation between the extract samples of Kharita and Khatatba source rocks and crude oils from Meleiha and Qarun oilfields. The extract of

Bahariya source rock shows slight correlation with Meleiha oil and differ from the other oil samples. These evidences indicate that Kharita and Khtataba source rocks seem to act as sources and reservoirs for oil generation in the Qarun and Misaada oilfields while the oil generation of Meleiha oilfield seems to be migrated from Bahariya source rocks.

## 8.2 Suez Gulf

The Gulf of Suez occupies the northwestern arm of the Red Sea between Africa proper (west) and the Sinai Peninsula (east) of Egypt. The length of the gulf, from its mouth at the Strait of Jubal to its head at the city of Suez, is 195 miles (314 km) and it varies in width from 12 to 20 miles (19 to 32 km). Because the importance occurrence of crude oil in the Gulf of Suez, the biological markers was analyzed to evaluate the geochemical relationships between the oils recovered from some oil fields within the Gulf of Suez to assess and investigate oil characterization, maturation, source depositional environments and oil families.

Roushdy et al. (2010) evaluate the geochemical relationships between the oils recovered from some oil fields within the Gulf of Suez. This target was achieved through analytical results of GC and GC-MS analysis for seven crude oil samples collected from seven oilfields namely: Ras Badran, Belayim marine, Belayim Land, Rahmi, West Bakr, Esh El Mellaha and Geisum distributed within the Gulf of Suez. These samples are representative for the producing horizon zones (Belayim, Rudies and Nuhkul formations.) of Upper- Lower Miocene age characterized by limestone facies with depths ranging from 2250 to 8286 ft. Geochemical parameters based upon acyclic isoprenoids, steranes and terpanes coupled with bulk geochemical parameters indicated whether the crude oils are of marine, terrestrial or mixed marine-terrestrial origin.

Biomarkers analyses of crude oils from the Gulf of Suez suggest that oils are more mature and derived mainly from mixed organic sources from terrestrial and marine inputs contribution to the biomass from algae and plankton in different saline environments.

A few discrepancies that appear between the results obtained by using the different parameters can be related to the alteration caused by the number of processes (physical, chemical and/or biological) affecting part of the source related biomarkers pattern of the oil after generation and/or primary migration from the source rock.

In another study, two genetic families based on biomarker analyses of oils were isolated from the Gulf of Suez, Egypt. Oils from Ras Fanar and East-Zeit wells have high gammacerance, low diasterances and high C33/C34 hopanes, consistent with an origin from the Brown Limestone. Oils from the Gama and Amal-9 wells have low gammacerance, high diasterances and oleanane indices > 20 %, indicating an angiosperm-rich Tertiary siliciclastic source rock, probably the Lower Miocene Rudeis Formation (Peters et al., 2005).

(Younes et al., 2004) evaluated the depositional environments and maturation assessments of source rocks from the central Gulf of Suez, Egypt utilizing the biomarker distributions in nine crude oils derived from a synrift tectonic sequence of the central Gulf of Suez province. No obvious variations were observed amongst the studied crude oils, suggesting that these oils are all of the same genetic type. These oils features, a predominance of oleanane, reaching 24%, and a relatively low gammacerane concentration of 10%, suggested that these oils were derived from a terrigenous organofacies source rock with a significant angiosperm higher land plants input deposited within the marginally mature syn-rift shale of Lower Miocene Nukhul, Rudeis and Kareem formations of mixed kerogen types II-III. Maturity parameters based on various sterane isomerisation distributions and polycyclic aromatic compounds indicate a low thermal maturation level for the generated hydrocarbons within the syn-rift lithostratigraphic succession. These similarities in geologic occurrences and biomarker characteristics suggest the possibility that the hydrocarbon expulsion could have been initiated from deeply buried Miocene source rocks and trapped within the syn-rift structures throughout the extensional faults of the central Gulf of Suez province.

(Barakat et al., 2000) studied the aliphatic and aromatic fractions of a beach tar sample from the Mediterranean coast of Sidi Kreir, 37 Km west of the city of Alexandria by GC and GC/MS techniques. A complete analysis was carried out to investigate chemical composition changes, fate of weathered oil residue and possible source identification. The distribution of sterane, hopane, mono-and triaromatic steroids, $C_2$ and $C_3$ phenanthrenes and dibenzothiophenes and chrysenes, however, had remained unaltered by weathering. The beach tar possessed geochemical features consistent with a marine carbonate or evaporite source depositional environment under normal saline reducing conditions.

## 9. Current work

### Applications of biomarkers in oil spill source identification

Although oil is the dominant energy source, oil spill occurs worldwide causing a severe global environmental problems (Abostate et al., 2011). Egypt is suffering from oil pollution owing to the increasing petroleum activities in the last decades. Environmental protection is currently an important subject of increasing public and research concern and as a result, special efforts have already been done so as to develop oil spill detection and fingerprinting. Therefore, to unambiguously characterize, identify, categorize, and quantify all sources of hydrocarbons entering the environment is very important for environmental damage assessment, evaluation of the relative risks to the ecosystem posed by each spill and selecting appropriate spill response and taking effective cleanup measures. Biomarkers are the most important hydrocarbon groups for chemical fingerprinting which play a very important role in source identification in environmental forensic investigations of oil spills. It was a useful analogy to explain this type of forensic analyses for spilled oil. However, it was recognized then, and remains true today, that the analyses of spilled oils do not have the statistical discriminating power of the human fingerprint in the sense that each human has an individual fingerprint. Analyses of spilled oils and potential sources are usually undertaken by increasingly sophisticated chemical analyses until either all but one potential source oil remains that cannot be distinguished from the spilled oil, or all potential sources have been eliminated and the spill is then a "mystery". The presumption for success using fingerprinting is that a complete collection of possible sources has been secured for the matching analyses. The term "passive tagging" has been used in place of fingerprinting in the past to describe the chemical analyses of oils. The term derives from the process of using the chemicals naturally present in the oil as "tags". The "passive" part of the term was used because there were proposals and some experiments conducted in the late 1960s and early 1970s to introduce "active tags" into various oil cargos to allow for identifying the oils if they were spilled (Adlard, 1972). Various chemicals were proposed as active tags, but the obvious international administrative and logistical effort needed to keep track of such "active tags" prevented operational use of active tagging systems.

Nothing sparks concerns about contaminates in the environment quite like a petroleum release. Unfortunately, the events of 2010 served to heighten the awareness and need to have the capability to monitor and characterize the extent and breadth of the impact of these events. Using petroleum biomarker analysis make it possible to accurately identify the source of contaminates back to the specific origin as well as determining the absolute concentrations of priority pollutant PAHs.

Generally, gas chromatograms of two oil samples are compared by comparing the envelop shapes of the n-alkanes, the unresolved backgrounds and individual peak intensities. By means of GC/MS, a big number of compound classes of oils may be separately detected and compared. Computerized oil spill identification (COSI) may highly support analysts in GC and GC/MS results evaluation and adds a new dimension to forensic oil spill identification. It is greatly increases the possibilities for finding the sources of oil pollution. The patterns of the biomarkers and a set of parameters based on the literature findings was chosen to be investigated for most Egyptian crude oil and stored in the database in order to construct an Egyptian computerized oil spill identification database of local crude oils. Gas-chromatograms and mass-fragmentograms are rapidly produced from raw GC- and GC/MS-data for comparing an unknown pollutant sample with any oil sample stored in the database then simultaneously a much stronger connection between a distinct oil spill and its actual source accurately established than before, as shown clearly in Figure 8. These parameters allow a more objective, provable and defensible result

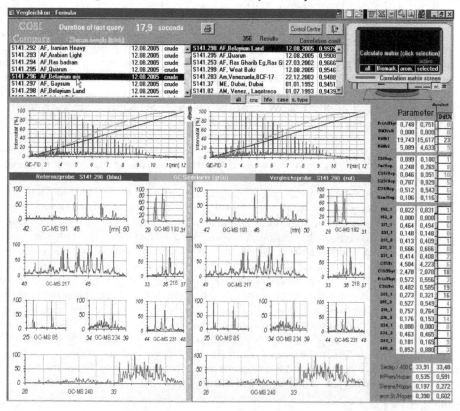

Fig. 8. Representative model of computerized oil spill identification matching.

evaluation than the mere visual comparison of the chromatograms. In addition, these parameters may also be used for finding oils in the database, which are similar to the spill sample. The system is fast and greatly saves laboratory resources and reliable and comfortable.

## 10. Conclusion

Biomarkers are naturally occurring, ubiquitous and stable complexes that are objectively measured and evaluated as an indicator of a certain state. It is used in many scientific fields; medicine, cell biology, exposure assessment, geology and astrobiology.

Due to the variety of geological conditions and ages under which oil was formed, every crude oil exhibits a unique biomarker fingerprint. From the identification point of view, biomarkers are the most important hydrocarbon groups in petroleum because they can be used for chemical fingerprinting which provides unique clues to the identity of source rocks from which petroleum samples are derived and the biological source organisms which generated the organic matter, the environmental conditions that prevailed in the water column and sediment at the time, the thermal history (maturity) of both the rock and the oil, and the degree of microbial biodegradation.

GC-MS is considered the most widely used method for biomarkers detection and identification which is a true combination of its separate parts (gas chromatography, GC and mass spectrometry, MS). The mass spectrometer has long been recognized as the most powerful detector for gas chromatography due to its high sensitivity, specificity and capability to elucidate compound structure. Mass fragmentography provides a satisfactory tool for obtaining specific fingerprints for classes and homologous series of compounds resolved by gas chromatography. The development of more sensitive and selective identification tool for biomarker analysis purpose especially for crude oils containing low concentration biomarkers as weathered and light oils can be considered as one of the most important research points in this field. After separation and identification of biomarkers, principal component analysis PCA, a mathematical procedure, can be used for analyses of chromatograms using a fast and objective procedure with more comprehensive data usage compared to other fingerprinting methods. The discriminative power of PCA was enhanced by deselecting the most uncertain variables or scaling them according to their uncertainty. Chemical analysis of biomarkers generates information of great importance to environmental forensic investigations in terms of determining the source of spilled oil. The patterns of the biomarkers and a set of parameters were used to construct an Egyptian computerized oil spill identification database. This can greatly increase the possibilities for finding the sources of oil pollution by comparing an unknown pollutant sample with any similar oil sample stored in the database. A much stronger connection between a distinct oil spill and its actual source may be established than before.

## 11. References

Abostate, M.A.; Moustafa, Y.M. & Mohamed, N. H. (2011). Biodegradation of slop wax byBacillusspecies isolated from chronic crude oil contaminated soils, *Fuel Processing Technology,* Vol. 92, 2348-2352.

Adlard, E. (1972). Review of Methods for Identification of Persistent Hydrocarbon Pollutants on Seas and Beache, *Journal of Institute petroleum,* Vol. 58 (560) 63-74.

Aguiar, A.; Ademário, I.; Azevedo, A. & Aquino-Neto, R. (2010). Application of comprehensive two-dimensional gas chromatography coupled to time-of-flight mass spectrometry to biomarker characterization in Brazilian oils, *Fuel*, 2760-2768.

Aquino, F.; Trendel, J.; Restle, A.; Connan, J. & Albrecht, P. (1983). Occurrence and formation of tricyclic and tetracyclic terpanes in sediments and petroleums, *Advances in organic geochemistry*, Wiley, chichester, 659-667.

Barakat, O.; El-Gayar, M. & Mostafa, A. (2000). Geochemical significance of fatty acids in crude oils and related source rocks from Egypt, *Petroleum Science and Technology*, Vol. 18, 635-655.

Blumer, M. & Thomas, D. (1965). Phytadienes in zooplankton, *Science*, Vol. 149, 1148-1149.

Blumer, M.; Mullin, M. & Thomas, D. (1963). Pristan in zooplankton, *Science* , Vol. 140, 974.

Chikaraishi, Y.; Matsumoto, K.; Ogawa, N.; Suga, H.; Kitazato, H. & Ohkouchi, N. (2005). Hydrogen, carbon and nitrogen isotopic fractionations during chlorophyll biosynthesis in higher plants, *Phytochemistry*, Vol. 66, 911.

Chosson, P.; Lanau, C.; Connan, J. & Dessort, D. (1991). Biodegradation of refractory hydrocarbon biomarkers from petroleum under laboratory conditions, *Nature*, Vol. 351, 640–642.

Christensen, H.; Giorgio, T. & Hansen, B. (2005). Chemical Fingerprinting of Petroleum Biomarkers Using Time Warping and PCA, *Environ. Sci. Technol.*, Vol. 39, 255–260.

Connan, J.; Bouroullec, J.; Dessort, D. & Albrecht, P. (1986). The microbial input in carbonate-anhydrite facies of sabkha palaeoenvironment from Guatemala, A molecular approach. In: *advances in organic geochemistry* 1985. org. geochem., Vol. 10, 29-50.

Doukhali, A.; Saoiabi, A.; Zrineh, A.; Hamad, M.; Ferhat, M.; Barbe, J. & Guilard, R. (2002). Sparation and identification of petroporphyrins extracted from the oil shales of tarfaya, *Fuel*, Vol. 81, 467-472.

Duan,Y.; Ma, L. (2001). Lipid geochemistry in a sediment core from Ruoergai Marsh deposit (Eastern Qinghai-Tibet Plateau, China), *Organic Geochemistry*, Vol. 32, 1429-1442.

Faramawy, S.; El-Sabagh, S.; Moustafa, Y. & El-Naggar, A. (2010). Mass Spectrometry of Metalloporphyrins in Egyptian Oil Shales from Red Sea Area, *Petroleum Science and Technology*, Vol. 28, 603 — 617.

Ficken, K.; Li, B.; Swain, D. & Eglinton, G. (2000). N-alkane proxy for the sedimentary input of submerged/floating freshwater aquatic macrophytes, *Organic geochemistry*, Vol. 31, 745-749.

Frysinger, S. & Gaines, B. (2001). Separation and identification of petroleum biomarkers by comprehensive two-dimensional gas chromatography, *Journal of Separation Science*, Vol. 24, 87–96.

Huebschmann, H. Analysis of Molecular Fossils: Crude Oil Steroid Biomarker Characterization Using Triple Quadrupole GC-MS/MS Frank Theobald, Environmental Consulting, Cologne, *Thermo Fisher Scientific*, Austin, TX, USA, application note 10261.

Hulka, B.; Griffith, J. & Wilcosky, T. (1990). Overview of biological markers, In: *Biological markers in epidemiology*, New York: Oxford University Press 3–15.

Hunt, J. (1996). Petroleum geochemistry and geology, 2nd ed., Freeman and Company, New York, 743.

Lijmbach, G. (1975). On the origin of petroleum: proceedings of the 9th world petroleum congress, *Applied science publishers*, London, Vol. 2, 357-369.

Maioli, O. L.; Rodrigues, K. C.; Knoppers, B. A. & Azevedo, D. A. (2011). Distribution and sources of aliphatic and polycyclic aromatic hydrocarbons in suspended particulate matter in water from two Brazilian estuarine systems, Continental Shelf Research, Vol. 31, 1116-1127.

Mayeux, R. (2004). Biomarkers: Potential Uses and Limitations, *The American Society for Experimental NeuroTherapeutics* , Vol. 1, 182–188.

Mello, M.; Gaglianone, P.; Brassell, S. & Maxwell, J. (1988). On geochemical and biological marker assessment of depositional environments using Brazilian offshore oils, *Marine and petroleum Geology*, Vol. 5, 205-223.

Mobarakabad, A.; Bechtel, A.; Gratzer, R.; Mohsenian, E. & Sachsenhofer, R. F. (2011). Geochemistry and origin of crude oils and condensates from the central Persian Gulf, offshore Iran, Journal of Petroleum Geology, Vol. 34, 261–275.

Moldowan, J.; Dahl, J.; Huizinga, B.; Fago, F.; Hickey, L.; Peakman, T. & Taylor, D. W. (1994). The molecular fossil record of oleanane and its relation to angiosperms, *Science*, Vol. 265, 768-771.

Moustafa Y. (2004). Environmental assessment of  petroleum contamination of Gamasa-Damiette Beaches, *Oriental journal of chemistry*, Vol. 20, No. 2, 219-226.

Naylor, S. (2003) Biomarkers: current perspectives and future prospects, *Expert Rev Mol Diagn*, Vol. 3, 525–529.

Ohkouchi, N.; Kashiyama,Y.; Kuroda, J.; Ogawa, N. & Kitazato, H. (2006). The importance of diazotrophic cyanobacteria as primary producers during Cretaceous Oceanic Anoxic Event, *Biogeosci. Discuss.*, Vol. 3, 575.

Ourisson, G., Albrecht, P. & Rohmer, M. (1984). The microbial origin of fossil fuels, *Scientific American*, Vol. 251, 44-51.

Peters, K. & Moldowan, J. (1991). Effects of source, thermal maturity and biodegradation on the distribution and isomerization of homohopanes in petroleum, *Organic geochemistry*. Vol. 17, 47-61.

Peters, K. & Moldowan, J. (1993). The biomarker Guide: Interpreting molecular fossils in petroleum and ancient sediments, Prentice hall, Englewood cliffs, NJ, 363.

Peters, K.; Walters, C. & Moldowan, J. (2005). *The Biomarker Guide*:  Biomarkers and isotopes in petroleum systems and Earth history Vol. 2, 2nd edition, Cambridge university press, New York.

Rodgers, P.; Klein, C.; Wu, Z. & Marshall, G. (2003). Environmental applications of ESI ft-ICR-mass spectrometry: the identification of polar n, s and o containing PAH species in crude oil and coal extracts, *Am. Chem. Soc.*, Div. Fuel Chem., Vol. 48,  758.

Roushdy, M.; El-Nady, M.; Moustafa, Y.; El-Gendy, N. & Ali, H. (2010). Biomarkers Characteristics of Crude Oils from some Oilfields in the Gulf of Suez, Egypt, *Journal of American Science*, Vol. 6, 911-925.

Roushdy, M.; Hashem, A.; El Nady, M.; Mostafa, Y.; El Gendy, N. & Ali, H. (2011). Specific geochemical parameters and oil: source rock correlations of some oilfields in the North Western Desert, Egypt, *Journal of American Science* Vol. 7, 715-729.

Sinninghe-Damste, J.; Kenig, F.; Koopmans, M.; Koster, J.; Schouten, S.; Hayes, J. & De-Leeuw, J. (1995). Evidence for gammacerane as an indicator of water column stratification, In: *Geochimica et Cosmochimica Acta* , Vol. 59, 1895-1900.

Tissot, P. & Welte, D. (1984). Petroleum formation and occurrence, 2nd ed., Springer Verlag, Berlin, 699.

Treibs, A. (1934). The occurrence of chlorophyll derivatives in an oil shale of the upper Triassic, *Annalen*, Vol. 517, 103-114.

Van Graas, G. (1990). Biomarker maturity parameters for high maturities: calibration of the working range up to the oil/condensate threshold, *Organic Geochemistry*, Vol. 16, 1025-1032.

Van Graas, G. (1990). Biomarker maturity parameters for high maturities: calibration of the working range up to the oil/condensate threshold, *Organic Geochemistr*, Vol. 16, 1025-1032.

Wang, Z.; Stout, S. & Fingas, M. (2006) Forensic Fingerprinting of Biomarkers for Oil Spill Characterization and Source Identification, *Environmental Forensics*, Vol. 7, 105-146.

Waples, D. (1985). Geochemistry in petroleum exploration, International Human Resources development corporation, Boston. 232.

Younes, M.; Hegazi, A.; El-Gayar, M. & Andersson, J. (2004). Petroleum biomarkers as environment and maturity indicators for crude oils from the central Gulf of Suez, Egypt, *Oil Gas European Magazine. Chemosphere*, Vol. 55, 1053-1065.

Yui, C.; Hiroshi, K. & Naohiko, O. (2007). An improved method for isolation and purification of sedimentary porphyrins by high-performance liquid chromatography for compound-specific isotopic analysis, *J. Chromatography. A* , 1138:73. Vol. 1138, 73-83.

Zein El Din, M.; Abd El Khalik, M.; Matbouly, S. & Moussa, S. (1990). Geochemistry and oil-oil correlation in Western Desert, Egypt, EGPC 10th Petrol. Explor. and Prod. Confer, Cairo, Vol. 2, 107-137.

# 9

# Secondary Metabolites

Tânia da S. Agostini-Costa[1], Roberto F. Vieira[1],
Humberto R. Bizzo[2], Dâmaris Silveira[3] and Marcos A. Gimenes[1]
[1]*Embrapa Genetic Resources and Biotechnology, Brasília*
[2]*Embrapa Food Technology, Rio de Janeiro,*
[3]*Health Sciences Quality, University of Brasília, Brasília,*
*Brazil*

## 1. Introduction

Secondary metabolites are organic molecules that are not involved in the normal growth and development of an organism. While primary metabolites have a key role in survive of the species, playing an active function in the photosynthesis and respiration, absence of secondary metabolites does not result in immediate death, but rather in long-term impairment of the organism's survivability, often playing an important role in plant defense. These compounds are an extremely diverse group of natural products synthesized by plants, fungi, bacteria, algae, and animals. Most of secondary metabolites, such as terpenes, phenolic compounds and alkaloids are classified based on their biosynthetic origin. Different classes of these compounds are often associated to a narrow set of species within a phylogenetic group and constitute the bioactive compound in several medicinal, aromatic, colorant, and spice plants and/or functional foods.

Secondary metabolites are frequently produced at highest levels during a transition from active growth to stationary phase. The producer organism can grow in the absence of their synthesis, suggesting that secondary metabolism is not essential, at least for short term survival. A second view proposes that the genes involved in secondary metabolism provide a "genetic playing field" that allows mutation and natural selection to fix new beneficial traits via evolution. A third view characterizes secondary metabolism as an integral part of cellular metabolism and biology; it relies on primary metabolism to supply the required enzymes, energy, substrates and cellular machinery and contributes to the long term survival of the producer (Roze et al, 2011).

A simple classification of secondary metabolites includes tree main groups: terpenes (such as plant volatiles, cardiac glycosides, carotenoids and sterols), phenolics (such as phenolic acids, coumarins, lignans, stilbenes, flavonoids, tannins and lignin) and nitrogen containing compounds (such as alkaloids and glucosinolates). A number of traditional separation techniques with various solvent systems and spray reagents, have been described as having the ability to separate and identify secondary metabolites. This chapter proposes to discuss major secondary metabolites classes (terpenoids, phenolic compounds and alkaloids) with different chemical structures and functions being screened, separated, fractionated, purified or analyzed using various adsorbents and eluents through column chromatography (CC) and thin layer chromatography (TLC).

## 2. Terpenoids

Terpenoids are the largest and most diverse family of natural products, ranging in structure from linear to polycyclic molecules and in size from the five-carbon hemiterpenes to natural rubber, comprising thousands of isoprene units. All terpenoids are synthesized through the condensation of isoprene units ($C_5$) and are classified by the number of five-carbon units present in the core structure (Mahmoud et al. 2002). Many flavor and aromatic molecules, such as menthol, linalool, geraniol and caryophyllene are formed by monoterpenes ($C_{10}$), with two isoprene units, and sesquiterpenes ($C_{15}$), with three isoprene units. Other bioactive compounds, such as diterpenes ($C_{20}$), triterpenes ($C_{30}$) and tetraterpenes ($C_{40}$) show very special properties and will be also discussed in this chapter.

### 2.1 Monoterpenes and sesquiterpenes (Plant volatiles)

Plant volatiles are typically lipophilic liquids with high vapor pressures. Non-conjugated plant volatiles can cross membranes freely and evaporate into the atmosphere when there are no barriers to diffusion. The number of identified volatile chemicals synthesized by various plants exceeds 1000 and is likely to grow as more plants are examined with new methods for detecting and analyzing quantities of volatiles that are often minute (Pichersky et al. 2006; Dudareva et al., 2004). Studying the volatile fraction requires analytical methods and technologies that not only evaluate its composition exhaustively but also monitor variations in its profile and detect trace components characterizing the plant being investigated (Bicchi et al., 2011).

The gas chromatography (GC and GC-MS) is a very powerful analytical tool for the identification of essential oil components. However, GC-MS has its limitations. Isomers usually give very similar mass spectra. This is particularly true for terpenes and even more for sesquiterpenes. Therefore, a favorable match factor between mass spectra is not sufficient for identification (Zellner et al., 2010). Retention indices have been used, together with mass spectrometry, for the proper identification of essential oils composition. Misidentification is not rare, however, either if a non-authentic mass spectra library is used, which means a database built with data from the literature, not from the analysis of real standards, or by misuse of retention indices (Joulain & König, 1998). CC has been applied to solve this problem. After isolation, a NMR analysis can be performed for the unknown or suspect compound, so that the correct mass spectrum and retention index can be recorded.

Despite the advances in analytical methods to evaluate the composition of essential oil, CC remains a powerful technology for separation and characterization of specific compounds of interest. Column chromatography has been largely used and reported for searching and identification of new molecules, sometimes associated with their antimicrobial, antibacterial and antifungal activities. Also, this technique has been successfully used to obtain sufficient amounts of a substance for the investigation of its biological properties and allowing the detection of its olfactory properties. Isolation is also applied to and, very important for volatiles, to evaluate its odor.

## 2.1.1 Compound identification

The major constituents present in the *volatile oils* of different aromatic species may also be isolated or fractionated by silica gel on preparative TLC or CC. For example, the essential oil of *Tanacetum chiliophyllum* was subjected to silica gel CC, using a solvent mixture of n-hexane and ethyl acetate, to isolate one of its major components, dihydro-α-cyclogeranyl hexanoate (Salamci et al, 2007).

Djabou et al. (2010) have characterized the essential oils of *Teucrium massiliense* L. from Corsican and Sardinian islands, using a combination of capillary GC/retention indices, GC/MS and [13]CNMR spectroscopy after fractionation on CC. A mixture of all Corsican oil samples was submitted to flash chromatography [FC; silica gel, elution with n-pentane, then with diethyl ether]. The polar fraction was separated on silica gel and 14 fractions were eluted with a mixture of n-pentane and diethyl ether of increasing polarity to give the major components: 6-methyl-3-heptyl acetate, 3-octyl acetate, isobutyl isovalerate, germacrene D and linalool. Successive CC revealed the unknown compound 6-methyl-3-heptyl acetate.

Chemical investigations on the essential oil of *Lippia integrifolia* performed by CC (silica gel using n-hexane with increasing amounts of diethyl ether), followed by HPLC, GC-MS, [1]H and [13]C-NMR spectroscopy led to the identification of 78 components. A new sesquiterpenic alcohol, *trans*-africanan-1α-ol, was identified (Coronel et al. 2006).

Sutour et al. (2008) studied the essential oil of *Mentha suaveolens* ssp. *insularis* (Req.) Greuter, finding pulegone to be the major constituent. However, the second most abundant component of the essential oil remained unknown, being isolated by repeated chromatography on silica gel and submitted to a full set of NMR experiments, and identified as *cis-cis-p*-menthenolide. Fractions were eluted with a gradient of solvents of increasing polarity (pentane:diethyl ether 100:0–0:100).

The essential oil from the leaves of *Piper hispidinervum*, a native species from the Amazon, is very rich in safrole (Maia et al., 1987). During an extensive agronomical study with this species, a population produced an essential oil rich in a different phenylpropanoid (up to 78%). The mass spectrum was identical to that of myristicin (Figure 1). However, a difference of near 30 units in retention index between myristicin and the phenylpropanoid was strong evidence that it was another compound, probably a myristicin isomer. No retention indices were available for the two possible isomers, croweacin and sarisan. For identification, the unknown compound was isolated by CC on silica gel and eluted with hexane-ethyl acetate mixtures. After [1]H and [13]CNMR studies, it was identified as sarisan (Bizzo et al., 2001).

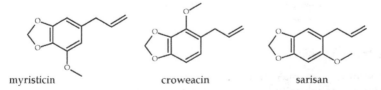

|  myristicin | croweacin | sarisan |

Fig. 1. Possible isomers of myristicin from the oil of a population of *Piper hispidinervum*; sarisan was the actual compound present.

Evaluating the essential oils from the leaves of 40 individuals of *Croton cajucara* Benth. from a germplasm bank in Manaus (State of Amazonas, Brazil), Quadros et al. (2011) observed some plants, instead of linalool, produced an oil rich in a hydroxylated sesquiterpene. The mass spectrum was very similar to that of 5-hydroxycalamenene. After the last edition of Adams' reference book (Adams, 2007), it was verified that the retention index of the sesquiterpenic alcohol did not correspond to that of 5-hydroxycalamenene. The oil was chromatographed over silica gel with hexane and hexane-ethyl acetate mixtures. The fraction containing the compound of interest was re-chromatographed on a silica gel preparative TLC plate with hexane-ethyl acetate (90:10). The purified compound was extracted from the silica with chloroform. Two clear singlets in the aromatic hydrogens at 6.56 and 6.94 ppm instead of a doublet pointed to a 7-hydroxy-substituted structure (Figure 2). Chemical shifts were in good agreement with published data for *cis*-7-hydroxycalamenene.

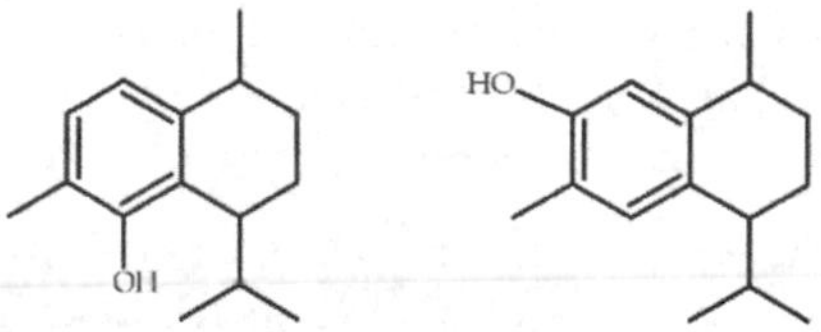

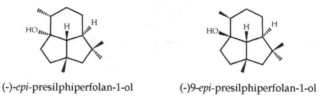

5-hydroxycalamenene          7-hydroxycalamenene

Fig. 2. Hydroxylated sesquiterpenes from *Croton cajucara*.

Sometimes, despite all careful efforts and extensive study, wrong structure assignment is published. For example, a new triquinane sesquiterpenic alcohol was isolated from the essential oil of *Anemia tomentosa* var. *anthriscifolia* by CC on silicagel with a hexane-ethyl acetate gradient. A new structural formula, namely (−)-*epi*-presilphiperfolan-1-ol, was initially proposed, after extensive 1D- and 2D-NMR analyses, as well as by GC–MS, chiral bidimensional GC, dehydration reactions, and a comparative (GIAO/DFT) theoretical study of the $^{13}C$ NMR chemical shifts (Pinto et al., 2009a). Further examination by X-ray diffraction and vibrational circular dichroism studies, however, led to its reassignment to (−)-9-*epi*-presilphiperfolan-1-ol (Figure 3), also a new compound. Its absolute configuration was established as 1S,4S,7R,8R,9S (Joseph-Nathan et al., 2010).

(-)-*epi*-presilphiperfolan-1-ol          (-)9-*epi*-presilphiperfolan-1-ol

Fig. 3. Isomers of (−)-presilphiperfolan-1-ol: the 9-*epi*-isomer was isolated from *Anemia tomentosa* var. *anthriscifolia* essential oil.

## 2.1.2 Olfactory analysis

Aroma concentrate separated from an aqueous solution of Haze honey by adsorptive CC had stronger aroma intensity than that separated by other methods. Fractions separated by preparative GC were sniffed to evaluate the sensory importance of the volatile compounds of Haze honey. Benzeneacetaldehyde, linalool, phenethyl alcohol, *p*-cresol, *p*-anisaldehyde, methyl-*p*-anisaldehydes, trimethoxybenzene, 5-hydroxy-2-methyl-4*H*-pyran-4-one, and lilac aldehydes seemed to contribute to Haze honey's aroma (Shimoda et al., 1996).

Shimoda et al. (1995) reported that a column extraction method was applied to the separation of volatile compounds from infusions of green tea and black tea, wherein their aromas degrade during distillation even under reduced pressure, and from sake, wherein ethyl alcohol disturbs a quantitative separation of volatile compounds.

Linalool is a very frequent terpenoid in essential oils, usually responsible for the scent of a specific plant, such as mint, basil, and other plant species. In order to evaluate the effect on humans of inhalation of linalool enantiomers (Figure 4), repeated flash chromatography was used to isolate R-(-)-linalool from lavender essential oil, while (S)-(+)-linalool was obtained from coriander oil and commercial linalool (Sugawara et al. 2000).

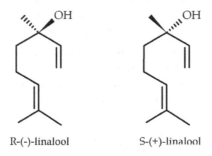

R-(-)-linalool          S-(+)-linalool

Fig. 4. Linalool enantiomers.

The Australian finger lime (*Citrus australasica*), an endemic Australian species, is an unique citrus fruits in terms of their size, shape and aroma. The fractionation of finger lime peel oil by chromatography on silica gel (with a gradient of pentane:ether) was able to identify a high number of terpenyl esters enriched in a medium polar fraction. This included citronellyl and geranyl esters, commonly found in lime peel oils, but also several less common (Z)-3-hexenyl esters and bornyl esters. In finger limes these compounds, together with isomenthone, menthol, carvone, *cis*-3-hexenol and methyl salicylate, may also contribute to the characteristic fresh green aroma of the fruits (Delort et al., 2009).

L-menthyl-L-lactate [(2*S*, 1'*R*, 2'*S*, 5'*R*)-2-isopropoyl-5-methylcyclohexyl 2-hydroxy-propanoate, LM-LL] is widely used as a cooling compound in mint flavors, fruit flavors, oral care products, confections and beverages. L-menthyl-L-lactate was identified in dementholized cornmint oil from India by comparing mass spectrum and retention time with a reference sample on two columns. Separation was achieved by increasing polarity of the hexane–methyl-tert-butyl-ether (MTBE) gradient on a flash chromatography column packed with silica gel. It is worth noting that L-menthyl-L-lactate was identified in *M. arvensis* oils from India but not in *M. piperita* oils from the USA. Model experiments proved that LM-LL is formed during water vapor distillation of mint leaves in the presence of lactic acid. Lactic acid is spontaneously formed when mint herb is stored for some days in a humid environment. A prolonged storage of damp mint herb may increase formation of lactic acid and thus give rise to the formation of LM-LL during distillation. The presence of lactic acid will be influenced by the agricultural practice employed (Gassenmeier, 2006).

Gas chromatography-olfactometry (GC-O), gas chromatography-mass spectrometry (GC-MS) and preparative CC were used to identify the key odorants present in laboratory-extracted clementine oil from Spain. Almost 50 odorants were identified using GC-O, many of which were unsaturated aldehydes with high odor spectrum values. α- and β-sinensal, *trans*-4,5-epoxy-(*E*)-2-decanal, (*E,Z*)-2,6-dodecadienal and linalool were found to dominate clementine oil aroma. Enrichment of the oxygenates using preparative CC provided further identification of a total of 50 aldehydes, which were originally present in the oil at concentrations not high enough to produce a response using GC-O (Chisholm et al., 2003).

### 2.1.3 Biological activities

Human enteropathogens (*Campylobacter jejuni, Campylobacter coli,* and *Campylobacter lari* strains) were inhibited by the essential oil of wild *Daucus carota* L. DCEO. Major components of the essential oil responsible for the antibacterial activity were isolated on CC and identified as (*E*)-methyl-isoeugenol and elemicin (Rossi et al., 2007).

Confertifolin (6,6,9a-trimethyl-4,5,5a,6,7,8,9,9a-octahydronaphtho[1,2-c] furan-3 (1*H*)-one), isolated from the essential oil of *Polygonum hydropiper* L. (Polygonaceae) leaves using CC, showed activity both against bacteria and fungi. The oil was chromatographed on a column packed with silica gel and eluted with hexane-ethyl acetate gradient solvent system with increasing polarity, yielding 117 fractions (Duraipandiyan et al., 2010).

The essential oil of *Daucus crinitus* Desf. was submitted to flash chromatography (silica gel, using petroleum ether with increasing amounts of diethyl ether as eluent), and fractions were analyzed by GC-MS. Those containing the isochavicol esters were gathered for the preparation of isochavicol. The antibacterial and antifungal activities of the whole essential oil, of these two esters, and of isochavicol itself, were investigated against a wide range of bacteria and fungi. Additionally, their antimalarial and antiradical properties were also evaluated, showing an interesting antiplasmodial activity of isochavicol (Lanfranchi et al., 2010).

Patchouli oil obtained from *Pogostemon cablin* (Blanco) Benth and its main constituent, patchouli alcohol, were tested for their repellency and toxicity against Formosan subterranean termites (*Coptotermes formosanus* Shiraki). Both were found to be toxic and repellent. The fraction containing patchouli alcohol was purified by silica CC, and its purity was determined by GC-MS. Unusual tissue destruction was noted inside the exoskeleton of the termite after patchouli alcohol was topically applied to the dorsum (Zhu et al., 2003).

A successful breeding program of *Cannabis sativa* cultivars with a low content of the hallucinogenous D-9-tetrahydrocannabinol (THC) is a prerequisite to avoid drug abuse of hemp. Four TLC systems for analyzing cannabinoids were compared, being able to identify the absence of THC in the hemp essential oils from five commercial fiber varieties. The system using HPTLC, RP-18 and acetonitrile-water (9:1) achieved the best separation of hallucinogenous and non-hallucinogenous cannabinoids (Novak et al., 2001).

Nine fractions of essential oil extracted from *Thymus praecox* were separated by silica gel CC and the fractions were demonstrated to have antioxidant activity (Ozen et al, 2011). The essential oil of *Coriandrum sativum* was submitted to silica gel on dry-CC, resulting in six fractions. Essential oil and its fractions presented antimicrobial potential against Candida yeast infections (Begnami et al, 2010). Essential oils, distilled from seeds of *Coriander sativum* and *Carum carvii* and from leaves of *Ocimum basilicum*, were fractionated by silica gel over CC and tested in the laboratory for volatile toxicity against three stored rice pests. Fractions, where combinations of products occurred with or without other minor compounds, were often more toxic than any one compound alone (Lopez et al, 2008).

The main component of the essential oil of *Anemia tomentosa* var. *anthriscifolia*, (−)-9-*epi-presilphiperfolan*-1-ol (Figure 3) was isolated by CC over silica gel, eluted with hexane and hexane-ethyl acetate (95:5), then tested against *Mycobacterium tuberculosis* (H37Rv) and *M. smegmatis*. A minimum inhibitory concentration of 120 μg/ml was recorded for *M. tuberculosis* (Pinto et al., 2009b).

## 2.2 Diterpenes and sesterterpenes

The diterpene compounds arise from geranylgeranyl diphosphate, and present 20 carbon units in their basic skeletal type. One of the simplest and most important of the diterpenes is phytol (Figure 5), a reduced form of geranylgeraniol, which forms the lipophilic side-chain of the chlorophylls (Vetter and Schröder, 2011).

Cyclization reactions of geranylgeranyl diphosphate led to many structural types of diterpenoids, presenting a large range of polarity nature, from apolar hydrocarbons such as cembrene (Villanueva and Setzer, 2010), a 14-membered ring, to fully oxidized skeleton of virescenoside (Figure 5), isolated from marine fungus *Acremonium striatisporum* (Ebel, 2010).

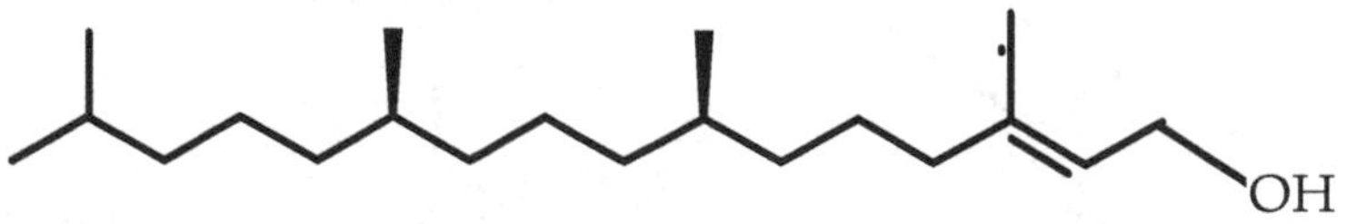

phytol

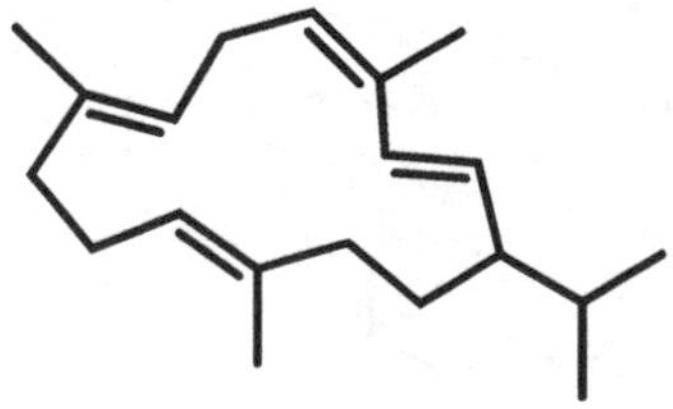

cembrene

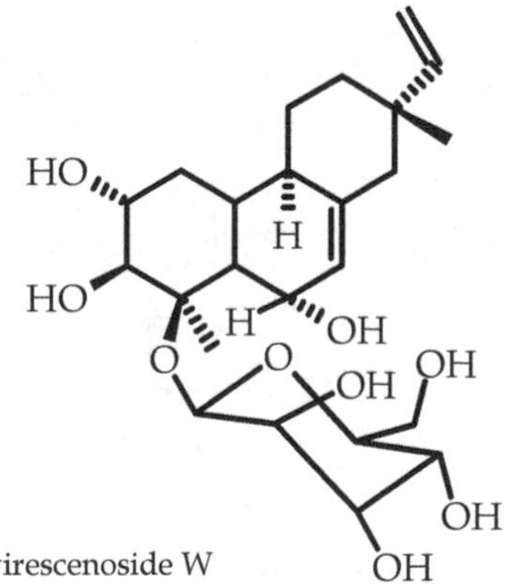

virescenoside W

Fig. 5. Diterpenes phytol, cembrene and virescenoside W.

Sesterterpenes ($C_{25}$) may be the least common group of terpenoids. This class of compounds arises from geranylfarnesyl diphosphate (Figure 6), which by cyclization can give rise to various skeletal types, presenting different oxidation levels and several biological activities.

Although many examples of these natural terpenoids are known, they are primarily isolated from fungi and marine organisms.

Considering the large range of polarity nature presented by both diterpene and sesterterpene, the isolation and purification techniques vary and can be classic TLC, preparative thin-layer chromatography (PTLC), CC, flash chromatography (FC), or modern high performance liquid chromatography (HPLC), multiflash chromatography, vacuum liquid chromatography (VLC), solid-phase extraction and others (Lanças, 2008). However, the most important factor that has to be considered before designing an isolation protocol is the nature of the target compound, such as solubility (hydrophobicity or hydrophilicity), acid–base properties, charge, stability, molecular size. Taking these factors into account, the choice of chromatographic methods and the stationary phases to be used are important for the design of a purification system.

Silica gel-based material is the most usual stationary phase. However, polyamide, Sephadex® LH-20, alumina (Al$_2$O$_3$), florisil and others are also used, mainly for isolation of more polar compounds (Ajit-Simh et al., 2003). The air-dried and ground roots of *Peltodon longipes* were extracted with hexane. The methanol soluble fraction of the hexane crude extract was subjected to open CC on silica gel 60 eluted with hexane/ethyl acetate step gradient. After subsequent purification, several abietane diterpenes were isolated, including royleanone, inuroylleanol, deoxyneocryptotanshinone and 7α-ethoxyroyleanone (Figure 6) (Fronza et al., 2011).

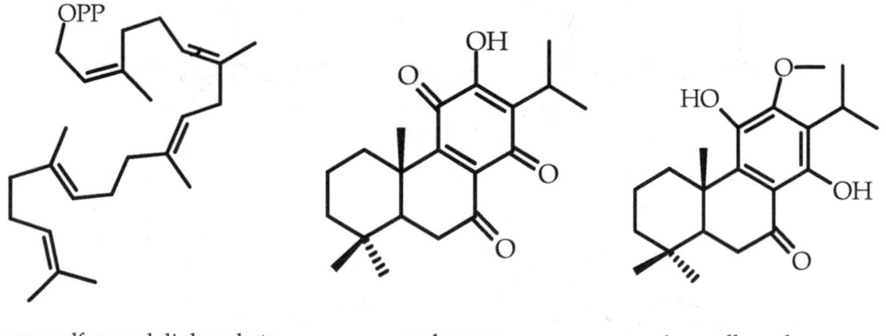

geranylfarnesyl diphosphate           royleanone              inuroylleanol

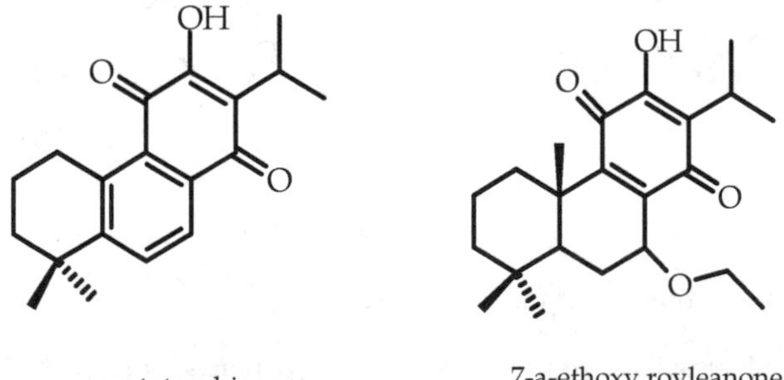

deoxyneocryptotanshinone              7-a-ethoxy royleanone

Fig. 6. Geranylfarnesyl diphosphate and abietane diterpenes from *Peltodon longipes*.

The sesterterpenes 24-O-methylmanaolide and 24-O-ethylmanaolide (Figure 7), besides others, were isolated from the sponge *Luffariella* cf. *variabilis*. Briefly, the frozen sponge tissue was extracted by methanol:chloroform (1:2). The crude extract was successively partitioned between equal volumes of aqueous methanol and a solvent series of hexane, carbon tetrachloride and chloroform. The hexane fraction was subjected to silica gel columns, using eluents of increasing polarity from hexane: ethyl acetate (Gauvin-Bialecki et al., 2008).

Olefins can form co-ordination products with silver ions, and this ability has been used for chromatographic separation of several terpenes, mainly those with a carbon skeleton higher than $C_{20}$. The procedure to prepare the modified silica was described as follows: Silver nitrate ($AgNO_3$) in water (1:2 w:v) is added to silica. After water evaporation, the residue is dried in oven at 130 °C for 15 h, resulting in an almost white powder that should be stored in a dark bottle. Column packing is carried out in the same way as ordinary silica gel column; however, should be wrapped in dark paper or built with amber glass (Norin and Westfelt, 1963).

Methyl ester from kaurenoic acid, iso-kaurenoic acid and grandiflorenic acid (Figure 7) were purified by using argentum silica (Batista et al., 2010). The diterpene acids were isolated from aerial parts of *Wedelia paludosa*. The plant material was extracted by percolation first using hexane, followed by ethanol. The crude ethanol extract was chromatographed over silica gel column. A mixture of these three compounds was obtained and further submitted to isocratic CC on silica gel impregnated with 20% of $AgNO_3$, eluting with hexane:diethyl ether 97:3 (Batista et al., 2010).

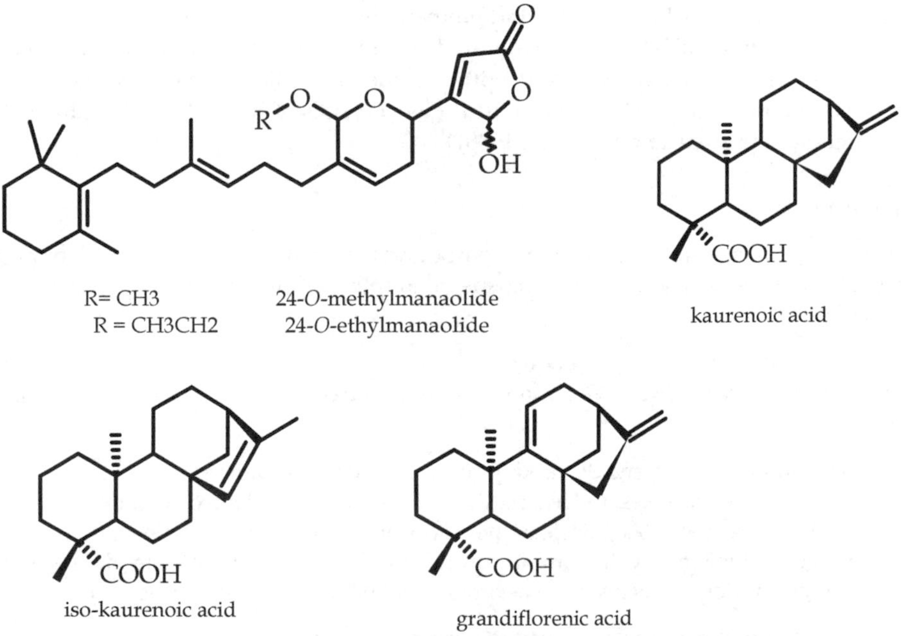

R= CH3          24-O-methylmanaolide
R = CH3CH2      24-O-ethylmanaolide

kaurenoic acid

iso-kaurenoic acid

grandiflorenic acid

Fig. 7. Sesterterpenes 24-O-methylmanaolide and 24-O-ethylmanaolide from *Luffariella* cf. *variabilis* and diterpene acids isolated from *Wedelia paludosa*.

Polyamide 6 (ε-aminopolycaprolactam) has been used in our laboratory to isolate polar natural products due to its low cost and re-usability. The pre-treatment of polyamide is carried out as follows: before column package, polyamide should be washed with methanol, rinsed with distilled water, and poured into the glass column, using an approximate proportion of 1 kg of polyamide to 10 L of each solvent, in order to eliminate oligomers and other impurities (Ye et al., 2011). After elution with $H_2O$ until column stabilization, elution is performed with $H_2O$ and a miscible polar solvent (usually methanol) gradient. In our experience, the same polyamide column can be re-used to obtain an additional amount of the obtained substance after the followed procedure: the column should be washed starting with 100% methanol to 100% $H_2O$. We did not observe any substantial reduction in quality of column performance for five consecutive recovery procedures, using the same sample.

A polyamide column was used in the first step to obtain diterpenes from aerial parts of *Euphorbia pannonica*. The plant material was extracted by methanol at room temperature. The obtained crude extract was partitioned by $H_2O$ and dichloromethane and the organic fraction was chromatographed on a polyamide column with mixtures of methanol and $H_2O$ (2:3, 3:2 and 4:1) as eluent. The fraction obtained with methanol–$H_2O$ (3:2) was subjected to silica gel VLC using a gradient system of cyclohexane, chloroform and acetone (3:2:0, 1:1:0, 2:3:0, 3:7:0, 1:4:0, 10:40:1, 4:16:1 and 1:4:1). After purification, two tigliane-type diterpenes, 4,12-dideoxy(4β)phorbol-20-benzoate-13-isovalerianate   and   4,12-dideoxy(4β)phorbol-20-benzoate-13-isobutyrate (Figure 8), were isolated (Sulyok et al., 2009). Sephadex LH-20 column was used to isolate halisulfate 1 (16) from dark brown sponge. The freeze-dried sponge was extracted by methanol. After the solvent had been filtered and removed, the residue was partitioned between dichloromethane and water. The organic extracts were dried over sodium sulfate and evaporated to obtain a dark brown oil that was chromatographed on Sephadex LH-20 with 1:1 methanol-dichloromethane as eluant to obtain two antimicrobial fractions. Fraction B gave crystals of halisulfate 1 (Figure 8) from dichloromethane (Kernan and Faulkner, 1988).

## 2.3 Triterpenes

Triterpenes ($C_{30}$) are a large class of compounds presenting a number of important biological activities; they arise from squalene, a coupling of two farnesil diphosphate units (Abe, 2007).

Cyclization of squalene, or squalene oxide leads to a large number of diverse structural triterpene skeletal types (with 30 carbons), such as lupane, oleane, ursane types (Tantillo, 2011).

Steroids are modified triterpenoids, lacking the three methyl groups at C-4 and C-14. Sterols are characteristic of eukaryotes. In bacteria they are sparsely distributed with limited array of products (Summons et al., 2006). Skeletal modifications, especially to the side-chain, originate a wide range of biologically important natural products, e.g. sterols, steroidal saponins, cardioactive glycosides, bile acids, corticosteroids, and mammalian sex hormones (Abe, 2007).

Saponins are a group of natural compounds presenting triterpenoidal or steroidal aglycone, designated genin or sapogenin, covalently linked to one or more sugar moieties (Augustin et al., 2011).

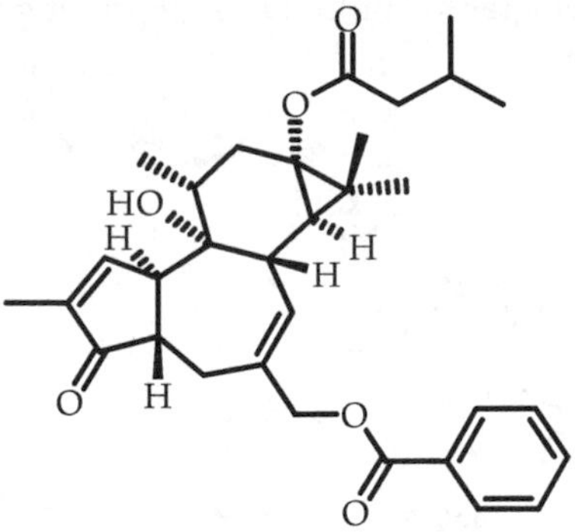

4,12-dideoxy(4β)phorbol-20-benzoate-13-
isovalerianate

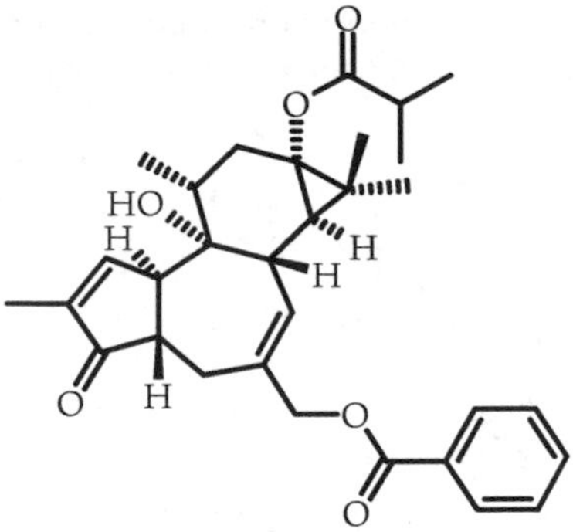

4,12-dideoxy(4β)phorbol-20-benzoate-13-
isobutyrate

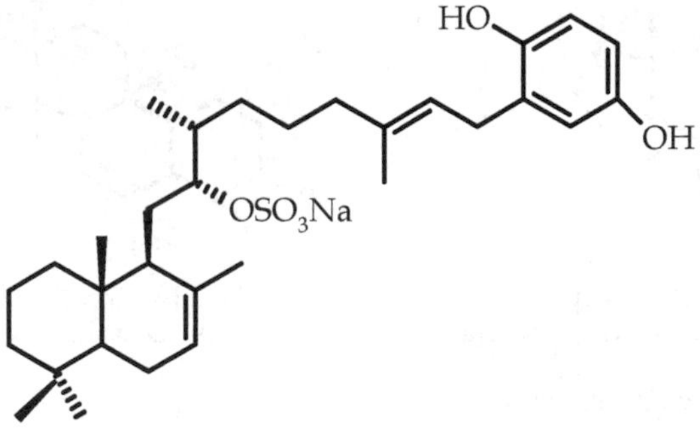

halisulphate 1

Fig. 8. Tigliane diterpenes from *Euphorbia pannonica* and halisulphate isolated from dark brown sponge.

The most usual technique to isolate triterpenes and steroids is by silica open column. As an example, six triterpenes were isolated from petroleum ether extract of *Azorella trifurcata* whole plant (Areche et al., 2009). The crude extract was chromatographed over silica gel using a petroleum ether–ethyl acetate gradient. The fraction eluted with 20% ethyl acetate was purified by CC on silica gel impregnated with 10% AgNO$_3$, furnishing four triterpenes: lanost-7-en-3β-ol, lanost-9(11)-en-3β-ol , lanosta-7,24-dien-3β-ol and cycloartenol (Figure 9). The acetylation product from the fraction eluted with 40% ethyl acetate, after purification by argentum silica gel CC, yielded the triterpenes lanosta-7,24-dien-3β-yl acetate and 28-acetoxycycloartenyl acetate (Areche et al., 2009).

Several cycloartane-type saponins were isolated from *Astragalus wiedemannianus* methanol crude extract (Polat et al., 2010). Plant material was extracted under reflux. After filtration and solvent removal, the methanol crude extract was solved in water and successively partitioned with hexane, dichloromethane and butanol saturated with water. The butanol fraction was submitted to VLC on reversed-phase material (Lichropep RP-18), employing a H$_2$O:methanol gradient. Fractions developed with H$_2$O:methanol 2:8 were rich in saponins. The first fraction was chromatographed over Lichropep RP-18 (H$_2$O:methanol gradient). The obtained sub-fractions were submitted to new chromatographic procedures, depending on the nature of target compounds furnishing the saponins.

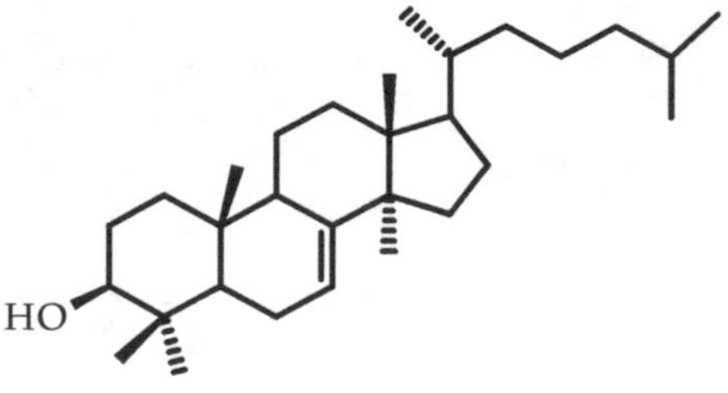

lanost-7-en-3β-ol

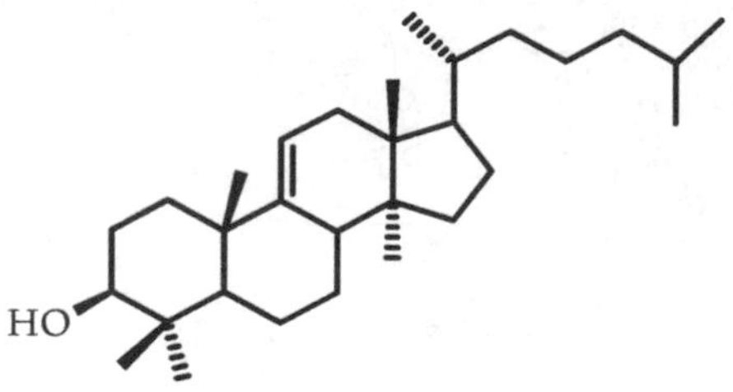

lanost-9(11)-en-3β-ol

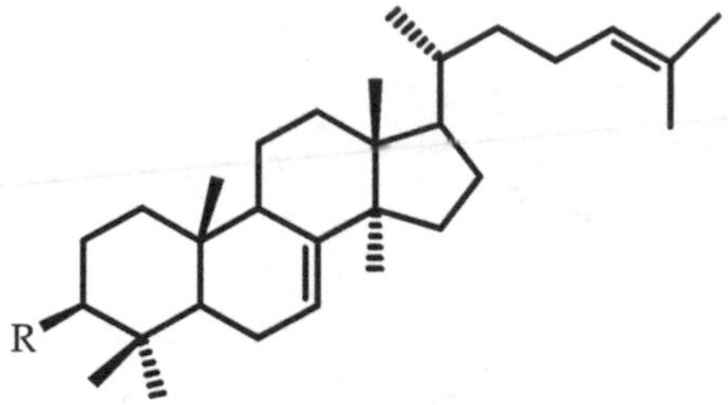

R= OH     lanosta-7,24-dien-3β-ol
R= OAc   lanosta-7,24-dien-3β-yl acetate

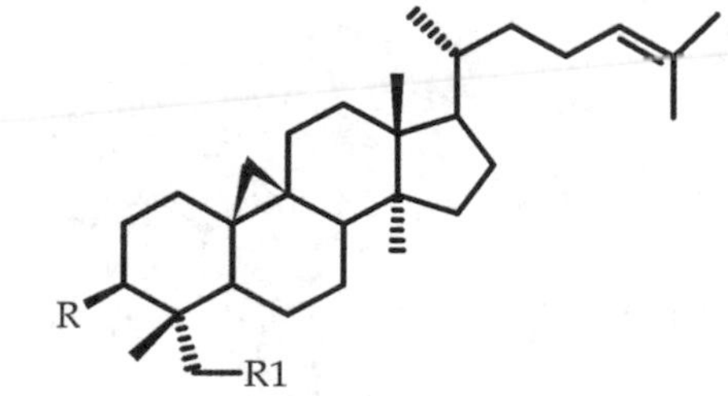

R= OH , R1= H cycloartenol
R=R1=OAc  28-acetoxycycloartenyl acetate

Fig. 9. Triterpenes from *Azorella trifurcata*.

## 2.4 Tetraterpenes (Carotenoids)

More than 650 carotenoids ($C_{40}$) are found in nature, constituting the largest group of natural dyes. The carotenoids are substances with very special properties possessed by no other group of substances; these form the basis of their many varied functions and actions in all kinds of living organisms (Britton, 1995). Carotenoids are biosynthesized by plants, algae, fungi, yeasts and bacteria.

The carotenoids are isoprenoid compounds, biosynthesized by tail-to-tail linkage of two geranylgeranyl diphosphate molecules. This produces the parent $C_{40}$ carbon skeleton from which all the individual variations are derived. This skeleton can be modified: a) by cyclization at one end or both ends of the molecule to give the seven different end groups; b) by changes in hydrogenation level, and c) by addition of oxygen-containing functional groups. Carotenoids that contain one or more oxygen functions are known as xanthophylls, the parent hydrocarbons as carotenes (Britton, 1995). After being absorbed through human diet, some carotenes, among them β-carotene (Figure 10), are pro-vitamin A; other, such as lycopene (Figure 10) , are important due to their antioxidant properties.

Carotenoid extracts have been screened by TLC and separated by CC, involving liquid-solid chromatography (adsorption). Various adsorbents have been applied in carotenoid analysis, including $Al_2O_3$, silica, magnesium oxide (MgO), calcium hydroxide [$Ca(OH)_2$], calcium carbonate [$CaCO_3$], siliceous earth as hyflosupercell and others. In normal phase CC, the adsorption affinity depends on the number of conjugated double bonds, cyclization and the presence of oxygen substituents (Rodriguez-Amaya, 1999). CC has been used for separations of mixtures of carotenes and xanthophylls, aiming for mainly analytical determinations, standard purifications, biological evaluations of carotenoids and the purification of synthesized carotenoids, especially by flash chromatography.

Separations on basic adsorbents such as MgO and $Ca(OH)_2$ are mainly determined by the number and type of double bonds in the carotenoid molecules (Bernhard, 1995). The procedure for isolating and purifying carotenoid standards was established because of the difficulty in obtaining standards commercially. The procedure consists of carotenoid extraction with cold acetone, partition to petroleum ether in a separatory funnel with addition of water, concentration in a rotatory evaporator and chromatographic separation of carotenoids on CC developed with petroleum ether containing increasing percentages of ethyl ether and acetone (Rodriguez-Amaya, 1999). A variety of carotenoid standards (some of them are represented on Figure 10) have been isolated and purified using MgO:Hyflosupercel (1:1) CC developed with 2-8% ethyl ether in petroleum ether and 2-95% acetone in petroleum ether: 98% β-carotene (isolated from carrot), 94% lycopene, 99% β-cryptoxanthin, 91% γ-carotene and 91% rubixanthin (from pitanga) (Porcu e Rodriguez-Amaya, 2008), 91–97% neoxanthin, 95–98% violaxanthin, 97–100% lactucaxanthin, 92–96% lutein, 93% β-cryptoxanthin, 96% zeaxanthin and 90–99% β-carotene (from lettuce, papaya or green corn) (Kimura & Rodriguez-Amaya, 2002; Oliveira & Rodriguez-Amaya, 2007) and 97% trans-ζ-carotene, 99% cis-violaxanthin, 97% trans-violaxanthin, 92% prolycopene and 94% lutein (from passion fruit extracts) (Wondracek et al, 2011). The β-cryptoxanthin elutes before lycopene (Figure 10) in MgO: Hyflosupercel and after lycopene in $Al_2O_3$ column, indicating that the influence of cyclization is greater than that of the presence of hydroxyl substituents in the MgO:Hyflosupercel column  (Rodriguez-Amaya, 1999).

MgO:Hyflosupercel on CC has been also widely used to analysis the main carotenoids of tropical fruits such as *Pouteria campechiana* (Agostini-Costa et al, 2010) and to evaluate the variation in the carotenoid composition of *Malpighia glabra* pulp (Agostini-Costa et al, 2003) and *Eugenia uniflora* fruit (Porcu & Rodriguez-Amaya, 2008). Hydrocarbon carotenes, such as β-carotene, lycopene, α-carotene, ζ-carotene and xanthophylls such as β-cryptoxanthin, violaxanthin and neoxanthin (Figure 10), were extracted from these tropical fruits, separated by MgO: Hyflosupercel on CC developed with ethyl ether  and acetone gradient in petroleum ether, identified by chemical tests and quantified by visible spectroscopy. The chromatographic behavior of carotenoids bears a definite relationship with theirs structure. Although these data cannot be used as the sole criteria for identifying carotenoids, they serve as useful complementary information (Rodriguez-Amaya, 1999). Carotenoids of *Lycium barbarum* L. fruits, a traditional Chinese herb that possesses vital biological properties, were isolated by a column containing MgO: diatomaceous earth. The β-carotene was eluted with n-hexane, β-cryptoxanthin and neoxanthin with ethyl acetate and zeaxanthin with ethyl acetate-ethanol (80:20 v/v). The zeaxanthin fraction was the most effective in scavenging hydroxyl-free radicals (Wang, 2010).

Other basic materials, especially $Ca(OH_2)$, $ZnCO_3$ and $CaCO_3$, are particularly  useful for separating geometrical isomers (Bernhard, 1995). Five lutein derivatives (four isomers and two epoxides) were isolated and separated from inflorescences of *Solidago canadensis* L.  and from flowers of *Chelidonium majus* L., using $CaCO_3$ CC developed with different compositions of toluene-hexane and acetone-hexane. They were identified as lutein-5,6-

epoxide, (9Z)-lutein-5,6-epoxide, (9Z,9'Z)-lutein (neolutein C), (9Z)- and/or (9'Z)-lutein (neolutein B) and (13Z)- and/or (13'Z)-lutein (neolutein A), in addition to (all-E)-lutein, violaxanthin, (9Z)-violaxanthin, flavoxanthin and/or chrysanthemaxanthin (Horvath et al., 2010). Xanthophylls, such as 5,6-diepikarpoxanthin, mutatoxanthin, anteraxanthin, zeaxanthin, β-cryptoxanthin and apo-carotenoids, such as capsorubin, capsochrome, capsanthin 5,6-epoxide and capsanthone, were also isolated from *Asparagus falcatus* using CaCO₃ as adsorbent on CC (Deli et al, 2000). The isomers of β-carotene present in acetone extracts from *Malpighia glabra* pulp (Agostini-Costa et al, 2003) and from *Annona coriaceae* fruit (Agostini et al., 1996) were separated on Ca(OH)₂ CC eluted with 2% ethyl ether in petroleum ether.

Separations on silica and Al₂O₃ are mainly determined by carotenoid polarity. The hydrocarbon carotenes are weakly adsorbed, whereas xanthophylls, containing polar substituents, especially hydroxyl groups, are adsorbed more strongly (Bernhard, 1995). A novel carotenoid (xanthophyll) derivative, lutein-3-acetate, extracted from senescing leaves of rice, was isolated and purified using silica gel TLC (Kusaba, 2009). Phytofluene, β-carotene, ζ-carotene, lycopene, zeinoxanthin, cryptoxanthin and lutein (Figure 10) were separated from ripe and partly ripe *Mormodica charantia* seeds and tomatoes through Al₂O₃ and MgO CC (Rodriguez et al., 1975). The main carotenoid (xanthophyll) in the orange muscle of Yesso scallop was isolated and purified by acetone extraction and silica gel CC. Its structure was identified as that of pectenolone (3,3'-dihydroxy-β,β-caroten-4–one) (Figure 10) (Li et al, 2010).

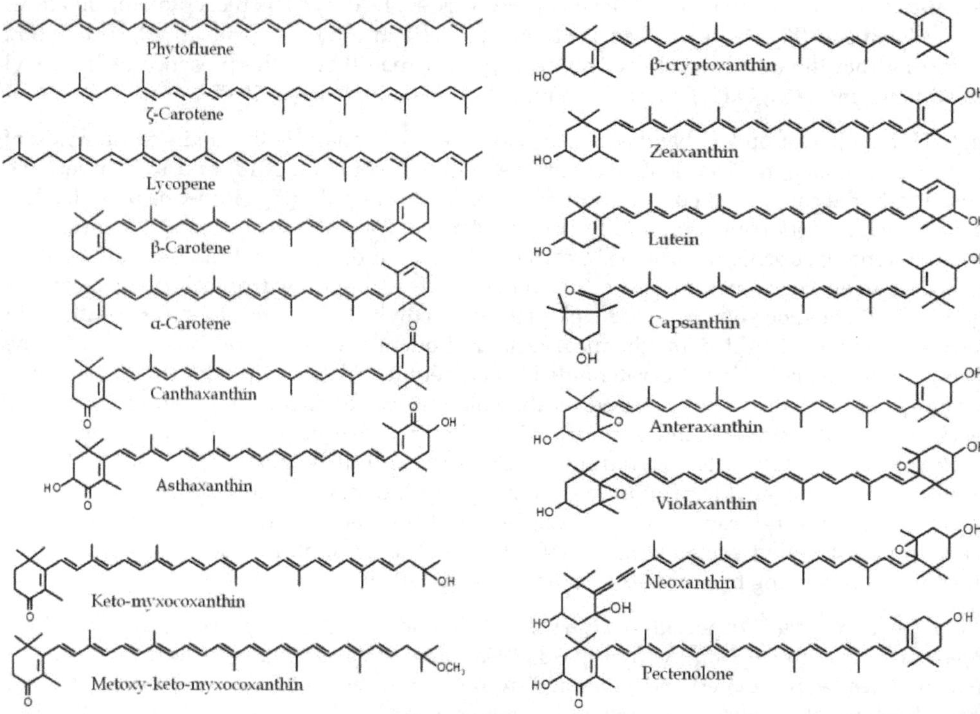

Fig. 10. Main carotenoids discussed in this chapter.

Carotenoid extracts of eight species of lichens were separated by $Al_2O_3$ column and divided into fractions on silica gel TLC and analyzed by ion-pairing in reverse-phase HPLC. Fourteen carotenoids, such as β-carotene, α-carotene, β-cryptoxanthin, lutein, zeaxanthin, canthaxanthin, astaxanthin, violaxanthin and neoxanthin (Figure 10) were separated and identified (Czeczuga et al, 2010). Stereoisomers and epoxy/carbonyl derivatives of β-carotene were separated through $Al_2O_3$ CC developed with 10-60% diethyl ether in $n$-hexane (Marty and Berset, 1990).

CC on silica gel 60 was also used for partial separation of carotenoids and DEAE-Toyopearl 650 M to remove B Chlorophyll and polar lipids during carotenoid isolation from a thermophilic filamentous photosynthetic bacterium *Roseiflexus castenholzii*. Purified fractions provided keto-myxocoxanthin and methoxy-keto-mycoxanthin (Figure 10), besides keto-myxocoxanthin glucoside, keto-myxocoxanthin glucoside ester, myxocoxanthin, myxocoxanthin glucoside ester and others (Takaichi et al., 2001). The major polar carotenoids, myxoxanthophyll and ketomyxoxanthophyll, produced by two cyanobacterias (*Anabaena* sp. and *Nostoc punctiforme*), were also isolated and purified first on a column of silica gel 60. The β-carotene was eluted with hexane, non-polar carotenoids were eluted with acetone/hexane (2:8) and polar carotenoids were eluted with acetone methanol (9:1, v/v). The polar carotenoids (last fraction) were then loaded on a column of DEAE-Toyopearl 650 M and the carotenoids were eluted with acetone/hexane (1:1 v/v) and finally purified on silica gel TLC developed with dichloromethane/ethyl acetate/acetone/methanol (2 : 4 : 2 : 1, v/v) to give myxoxanthophyll and ketomyxoxanthophyll (Takaichi et al., 2005).

## 3. Phenolic compounds

Phenolic compounds are widely distributed in nature. Their chemical structures may vary greatly, including simple phenols ($C_6$), such as hydrobenzoic acid derivatives and catechols, as well as long chain polymers with high molecular weight, such as catechol melanins ($C_6$)$_6$, lignins ($C_6$-$C_3$)$_n$ and condensed tannins ($C_6$-$C_3$-$C_6$)$_n$. Stilbenes ($C_6$-$C_2$-$C_6$) and flavonoids ($C_6$-$C_3$-$C_6$) are phenolic compounds with intermediate molecular weight that present many pharmacological and biological activities. Flavonoids, including anthocyanins, flavonols (such as quercetin and myricetin), isoflavones (such as daidzein and genistein) and others are formed by multiple biosynthetic branches that originate from chalcone.

Phenolic compounds have been widely fractionated in medicinal, aromatic and food plants using CC. Repeated silica gel, sephadex-LH20, RP-18, RP-8, MCI-gel, diaion and toyopearl chromatography columns have been used to fractionate simple phenolics, flavonoids and tannins from kernels and nuts (Zhang et al, 2009; Karamac, 2009), fruits such as apples, *Morus nigra*, *Punica granatum* (Lee et al, 2010; Pawlowska et al, 2008); olive oil (Khanal et al, 2011), tea (Gao et al, 2010; Liu et al, 2009), seeds such as lentils (Amarowicz & Karamac 2003), medicinal species, including *Ulmus davidiana* and *Tridax procumbens* (Jung et al, 2008; Agrawal, 2011); and aromatic plants, including mint and sage (She et al, 2010; Wang et al, 1998).

Separations on silica are mainly determined by polarity, where phenolic compounds containing more hydroxyl groups are adsorbed more strongly. Separations on Sephadex LH-20, a crosslinked dextran-based resin for gel permeation, are mainly determined by molecular sizing of the phenolic compound, outside of adsorption and partition mode.

Phenolic acids (such as ferulic acid and gallic acid), flavonoids (such as flavonol, catechins and anthocyanidins derivatives) and procyanidins (Figure 11) from fruits of wild black berry *Aristotelia chilensis* were obtained using flash and open CC on silica gel and Sephadex LH-20 eluted with hexane, hexane-ethyl acetate (1:1); ethyl acetate-methanol (1:1) and methanol (Cespedes et al., 2010).

Pyrogallol and phenolic acids were purified from defatted extract of *Juglans regia* kernels by repeated CC on Sephadex LH-20 eluted with methanol, followed by silica gel CC developed with chloroform–acetone 40:1, v/v. Ethyl gallate was obtained by recrystallization from methanol. Protocatechuic acid was obtained after CC on Sephadex LH-20 eluted with $CH_2Cl_2$–EtOH, 1:1, v/v and further purification with silica gel CC developed with Chloroform–methanol, 20:1. Gallic acid and 3,4,8,9,10-pentahydroxydibenzo[b,d]pyran-6-one (Figure 11) were separated by silica gel CC and eluted with a mixture of chloroform–methanol (Zhang, 2009). High-molecular-weight tannins (Figure 11) of defatted walnut, hazelnut and almond kernels were also isolated by CC on Sephadex LH-20 gel eluted with 50% (v/v) acetone (Karamac, 2009). The ethanol fraction of *Dipteryx lacunifera* kernels was found to exhibit high radical scavenging activity and was subjected to further fractionation. CC over silica gel with gradient from chloroform to methanol and Sephadex LH-20 eluted with methanol afforded (-)-eriodictyol, (-)-butin, luteolin, 3',4',7-trihydroxyflavone, butein and sulfuretin (Junior et al., 2008).

The total polyphenols from red wine isolated during vinification and storage were isolated by CC and their contribution to wine sensory properties and antioxidant activity was evaluated. Wine samples were evaporated and loaded onto an open CC packed with LiChroprep RP-18; the column was washed with distilled water followed by methanol to recover total polyphenols extract, further separated by HPLC (Sun et al., 2011).

The alkyl esters of protocatechuic acid (Figure 11) were synthesized and the crude product was then purified by CC using petroleum ether/ethyl ether (7:3 to 5:5) as eluent. The increase in the length of ester alkyl chain attached to the catecholic ring had influenced the stabilization of the radicals formed in the oxidation process. Alkyl protocatechuate compounds demonstrated a higher radical-scavenging activity than the natural antioxidant protocatechuic acid. Moreover, the introduction of alkyl groups in the carboxylic acid led to a significant increase in lipophilicity influenced by the antioxidant activity of protocatechuic acid derivatives (Reis et al., 2010).

The virgin olive oil phenol was partitioned successively with hexane and ethyl acetate and fractionated by CC on silica gel and Sephadex LH-20 to give a dialdehydic form of decarboxy-methyl ligstroside aglycone (p-HPEA-EDA) (Figure 11), a phenolic compound that activates AMP-activated protein kinase to inhibit carcinogenesis (Khanal et al., 2011). Phenols and tocopherols were also removed from olive oil by CC over $Al_2O_3$ during to evaluate changes in the phenolic composition of virgin oil during frying. The concentration of hydroxytyrosol (3,4-DHPEA) and its secoiridoid derivatives (3,4-DHPEA-EDA and 3,4-DHPEA-EA) in virgin olive oil decreased rapidly when the oil was repeatedly used for preparing fried food. However, tyrosol (p-HPEA) and its derivatives (p-HPEA-EDA and p-HPEA-EA) in the oil were much more stable during frying operations (Gomez-Alonso et al., 2003).

Phytoalexins are secondary metabolites that plants synthesize for self-defense, and they have shown great promise in chronic disease prevention. The best known example is resveratrol (Figure 11), an induced phytoalexin found in yeast-infected grape skin (Wu et al., 2011). Peanuts also contain several active components including flavonoids, phenolic acids, phytosterols, alkaloids and stilbenes. The latter are characterized by a 1,2-diphenylethylene backbone usually derived from the basic unit of trans-resveratrol (3,5,4'-trihydroxy-stilbene) (Lopes et al, 2011). Peanut seeds, when affected by injuries, pathogenic infections, fungal contamination, insect damage and UV light, could produce phytoalexins such as stilbenoid derivatives (resveratrol, arachidins, 3'-isopentadienyl-3,5,4'-trihydroxystilbene, SB-1, chiricanine A, and arahypins), and pterocarpanoid derivatives (e.g., aracarpene-1 and aracarpene-2). Silica gel column eluted with hexane and hexane ethyl acetate (7:3) was used for fractionation the phytoalexin extract from peanut before analysis by HPLC (Wu et al., 2011). Neutral $Al_2O_3$ - silica gel 60 C18 (1:1 w/w) eluted with 80% ethanol was used as a clean-up column to separate interfering co-eluting with resveratrol from peanut extracts (Potrebko & Resurreccion, 2009).

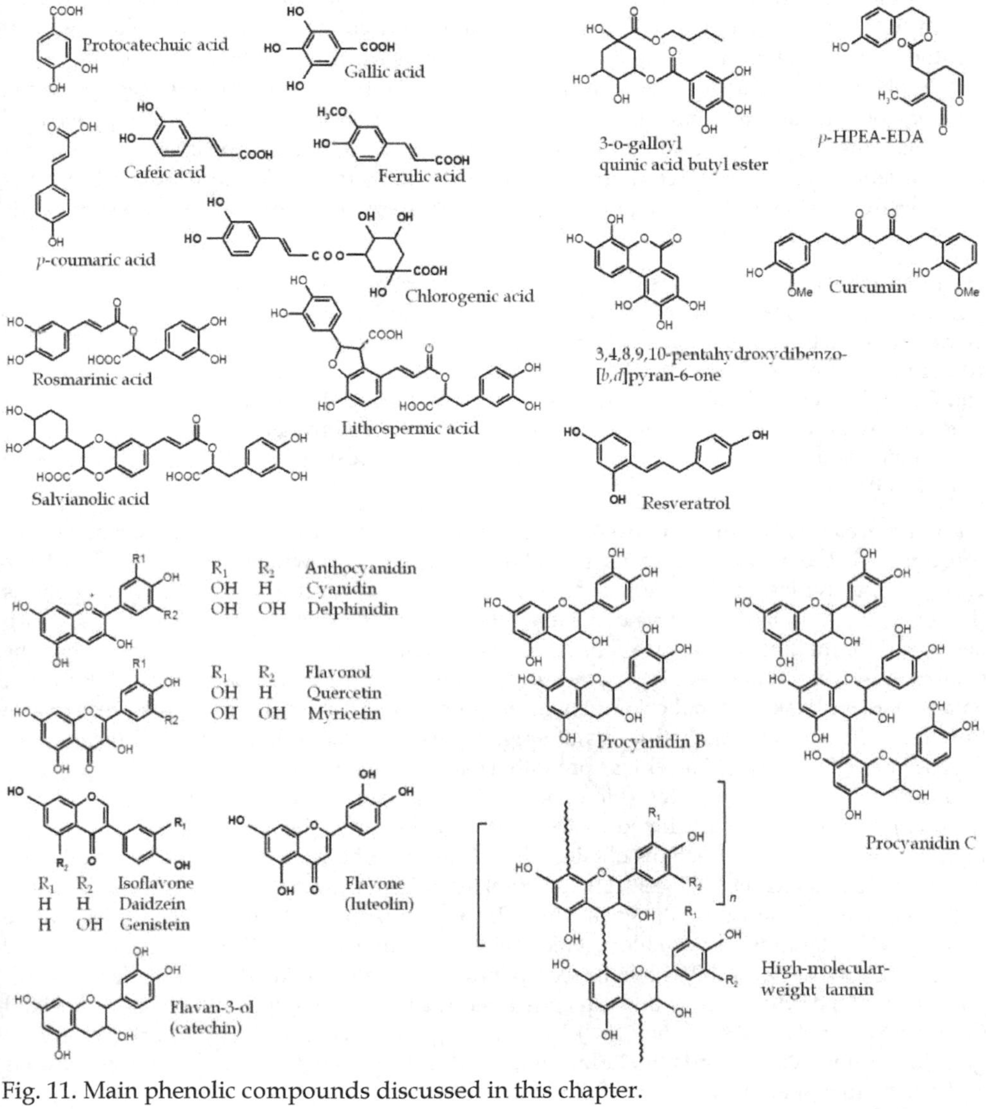

Fig. 11. Main phenolic compounds discussed in this chapter.

A highly efficient column chromatographic extraction of curcumin from *Curcuma longa* was proposed by Zhan et al., 2011. Curcumin (Figure 11) was extracted with minimum use of solvent, minimum volume and high concentration of extraction solution by CC. Turmeric material was loaded into a column with 2-fold of 80% ethanol. After dissolving target compounds, the column was eluted with 80% ethanol. For non-cyclic CCE procedure, 8-fold of eluent was collected, while for cyclic CCE procedure, only the first 2-fold of eluent was collected as extraction solution. A more than 99% extraction rate for curcumin was obtained in both procedures, compared to a 59% extraction rate through the ultrasonic-assisted extraction with 10-fold of 80% ethanol.

A lipid- and essential oil- free infusion of *Cymbopogon citratus* leaves was fractionated on Lichroprep RP-18 column eluted with water and with aqueous methanol solutions. Dry residue was recovered in 50% aqueous ethanol and was fractionated by gel chromatography on a Sephadex LH-20 using ethanol as mobile phase. All the fractionation process provided three major fractions: a tannin rich fraction; a flavonoid rich fraction and two phenolic acid rich fractions. Tannin and phenolic acid were the fractions responsible for the anti-inflammatory effect through inhibition of transcription factor NF-kB, inducible nitric oxide synthase expression and nitric oxide production. These fractions probably had a synergistic effect (Francisco et al., 2011).

Oligomeric procyanidins (Figure 11) in apples were extracted with boiled water and purified on an ADS-17 macroporous resin column to obtain a procyanidin extract. The extract was fractionated according to its degree of polymerization on a Toyopearl TSK HW-40 column eluted with methanol to give procyanidins B2 (epicatechin-(4$\beta$-8)-epicatechin) and C1 (epicatechin-(4$\beta$-8)-epicatechin-(4$\beta$-8)-epicatechin). This method was suitable for the preparation of procyanidin oligomers (from dimers through tetramers in one run) for laboratory research, and is potentially applicable to large-scale production in industry (Xiao et al., 2008).

Thea (Theaceae) is traditionally used for producing tea, one of the most popular beverages consumed in the world due to its polyphenol-rich content, which are reported to have various bioactivities, such as antioxidative, antimicrobial, antitumor and antimutagenesis (Liu et al, 2009). *Camellia sinensis* and some other species from the genus *Camellia* have also been used for making tea and are consumed widely (Gao et al, 2010). More than 96 phenolic compounds were identified in *C. sinensis* tea. The hydrolyzable tannin epigallocatechin gallate was the major phenolic component of green tea and partially fermented teas, while fully fermented black teas had traces of epigallocatechin gallate but contained theaflavins. Glycosylated flavonoids, catechins, proanthocyanidins (condensed tannins) and phenolic acid derivatives were found too (Lin et al., 2008). Aqueous acetonic extract of green tea (*C. crassicolumna*) further partitioned with ethyl acetate and purified by CC led to the identification of various fractions obtained by CC over Sephadex LH-20 eluted with ethanol. Further repeated CC on MCI-gel CHP20P, Sephadex LH-20, and Toyopearl HW-40F, eluted with methanol/$H_2O$ (0:1-1:0) gave five flavan-3-ols, five flavonol glycosides, three hydrolyzable tannins, two chlorogenic acid derivatives and three simple phenolic compounds (Liu et al, 2009). Phenolic compounds were also isolated from the leaves of *C. pachysandra*. The ethyl acetate and aqueous fractions were separately subjected to repeated CC over Diaion HP-20SS, Sephadex LH-20, MCI-gel CHP20P, and Toyopearl HW-40F to give 22 phenolic compounds, including nine hydrolyzable tannins, 11 flavonol glycosides and two simple phenols, without caffein or catechin (Gao et al., 2010).

A new phenolic compound was isolated and purified from butanol fraction of Chinese olive (*Canarium album* L.) fruit through AB-8 adsorption resin CC washed with water to remove impurities and eluted with 90% (v/v) aqueous ethanol to get the phenolic eluents. The phenolic were further separated on a polyamide CC eluted with aqueous ethanol to give several fractions, the ethanol concentration being increased from 0 to 100% in increments of 20%. Fraction obtained from 20% aqueous ethanol was further purified on TSK Toyopearl HW-40 (S) CC developed with aqueous ethanol 0 to 20% to get the purified new compound established as 3-o-galloyl quinic acid butyl ester (Figure 11) (He et al., 2009). Polyamide CC eluted with water and 50%, 70% and 100% aqueous methanol further facilitated GC-MS and HPLC separation of phenolic compounds in *Euphrasia rostkoviana* (Blazics et al., 2008).

*Lycium barbarum* L., a traditional Chinese herb, possesses vital biological properties, such as prevention of cancer and age related macular degeneration. Flavonoids and phenolic acids extracted from fruits of this plant were separated using a Cosmosil 140 C18 OPN column, with phenolic acids being eluted with deionized water and neutral flavonoids with methanol. The flavonoid fraction showed the most pronounced effect in scavenging free radicals, chelating metal ions and reducing power (Wang et al., 2010).

Eight polyphenolic acids were isolated from the aereal part of *Mentha haplocalix* extracted with aqueous acetone 70% at room temperature. Repeated CC on silica gel (developed with chloroform–methanol–$H_2O$, 9:1:0.1 to 7:3:0.5), Sephadex LH-20 and MCI-gel CHP20P and ODS-A, eluted with $H_2O$–Methanol (1:0 to 0:1) afforded rosmarinic acid, cis-salvianolic acid, lithospermic acid (Figure 11), propanoic acid, sodium lithospermate B, magnesium lithospermate B, and lithospermic acid B. Lithospermic acid B, sodium lithospermate B and magnesium lithospermate B displayed stronger activities than the other compounds (She et al., 2010). Ten phenolic compounds were also isolated from a butanol fraction of sage extracts, using repeated CC on silica gel, Lichroprep RP-18 and Sephadex LH20. Among them, the most active antioxidants were found to be rosmarinic acid and luteolin-7-O-$\beta$-glucopyranoside (Wang et al, 1998).

# 4. Alkaloids

Alkaloids are defined as basic compounds synthesized by living organisms containing one or more heterocyclic nitrogen atoms, derived from amino acids (with some exceptions) and pharmacologically active. The class name is directly related to the fact that nearly all alkaloids are basic (alkaline) compounds. Alkaloids constitute a very large group of secondary metabolites, with more than 12,000 substances isolated. A huge variety of structural formulas, coming from different biosynthetic pathways and presenting very diverse pharmacological activities are characteristic of the group (Brielmann et al., 2006).

Archeological evidence has demonstrated the use of alkaloids (plant parts or extracts) since 4000 B.C., and they continue to be very important today (Roberts & Wink, 1998). Poppy (*Papaver somniferum*) and opium have been known and used since antiquity by Sumerians, Arabs, Persians, Egyptians and Greeks. Morphine, obtained from the poppy latex, was the first crude drug isolated. It was named after Morpheus, the god of dreams, one of the sons of Hypnos, the god of sleep, in Greek mythology (Wink, 1998). Morphine is legally used nowadays as an analgesic for severe pain.

Alkaloids are associated with a wide range of pharmacological activities. Many are toxic and can cause death, even in small quantities. Some have antibiotic activities and others interfere with behavior patterns, such as antidepressants (reserpine) and hallucinogens (mescaline). It seems alkaloid function in plants and animals is linked to defense mechanisms. Toxicity is a good weapon to inhibit the action of predators, like herbivores.

## 4.1 Alkaloids from plants

CC is a very important and much used technique in alkaloid isolation and analysis. Usually, after CC, the sample is re-chromatographed either by preparative TLC or HPLC, in order to obtain a pure sample for spectroscopic studies.

Phytochemical investigation in plants from the genus *Kopsia* (Apocynaceae) led to the discovery of different classes of alkaloids, presenting antileishmanial, antimitotic and antitumor activities. From *Kopsia hainanensis*, a native medicinal plant from Hainan, China, two new pentacyclic indole alkaloids have been isolated (Chen et al., 2011). The acidic partition of the methanol extract was neutralized and extracted with chloroform. Alkaloids, as basic substances, can be protonated in acidic media and therefore remain in the aqueous phase. This strategy has been very useful for the isolation of this class of compounds. The pH of the aqueous phase is further adjusted and alkaloids are extracted with a solvent. The concentrated extract was repeatedly submitted to CC over silica gel and eluted with a gradient of chloroform-methanol, alkalinized with a small quantity of triethylamine to give kopsahainanine B. One fraction was re-chromatographed over reversed-phase silica gel and Sephadex LH-20, using methanol-water and pure methanol as eluents, respectively, to give kopsahainanine A (Figure 12).

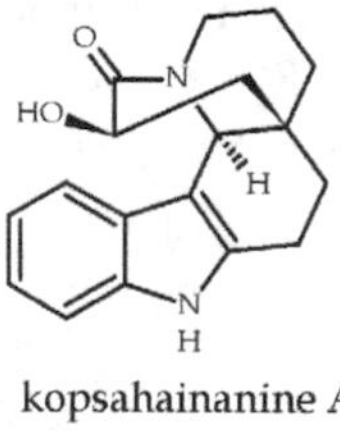

kopsahainanine A 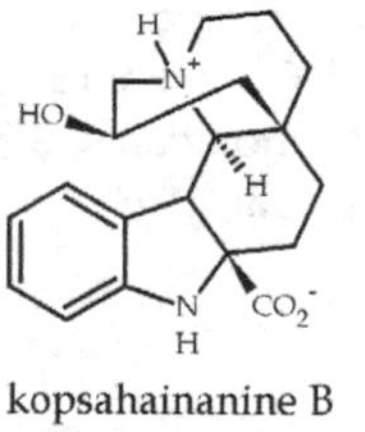                kopsahainanine B

Fig. 12. Pentacyclic indole alkaloids from *Kopsia hainanensis*.

The butanolic fraction of *Lobelia chinensis* Lour. was chromatographed on reversed phase Diaion HP 20 gel, eluted with water-methanol in to six fractions. Fraction 6 was re-chromatographed on Sephadex LH-20 followed by preparative silica gel TLC and 7.3 mg of lobachine (Figure 13), a new alkaloid, were isolated (Kuo et al., 2011).

From *Campylospermum flavum*, a medicinal plant from Cameroon, flavonoids, flavones, chalcones and alkaloids have been isolated. Among them, a new indole alkaloid, flavumindole (Figure 13), was obtained from the methanol extract after partition with ethyl acetate-water and purification on silica gel, Sephadex LH-20 and $RP_{18}$ silica gel, eluted with dichloromethane-methanol mixtures, pure methanol and methanol-water mixtures, respectively (Ndongo et al., 2011). The isolated alkaloid presented a cytotoxicity of 90% according to by brine shrimp (*Artemia salina*) test.

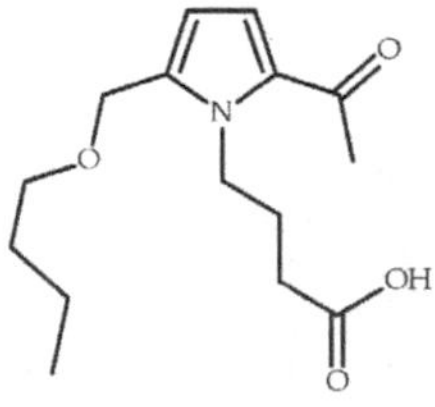

lobachine 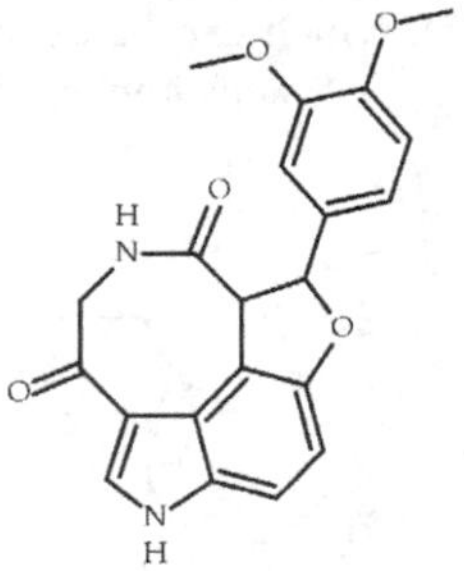                                        flavumindole

Fig. 13. Lobachine, a new alkaloid from *Lobelia chinensis* and flavumindole, from *Campylospermum flavum.*

Nine indole alkaloids were isolated from the roots, bark and leaves of *Tabernaemontana salzmannii* (Apocynaceae). The alkaloids were fully identified by spectroscopic methods and mass spectra by comparison with literature data (Figueiredo et al., 2010). The dichloromethane extract was repeatedly chromatographed over silica gel eluted with dichloromethane-methanol mixtures. Coronaridine, 3-oxo-coronaridine, (19S)-heyneanine, voacangine, isovoacangine, hydroxyisovoacangine, isovoachristine and olivacine (Figure 14) were isolated. For voachalotine, a further chromatographic step was necessary on preparative TLC (silica gel) eluted with dichloromethane-methanol (99:1). After an *in vitro* screening, it was observed that isovoacangine and voacangine alkaloids were able to induce apoptosis cell death in human leukemic cells, line THP-1.

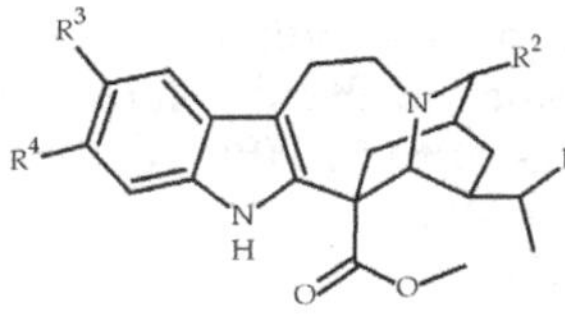

coronaridine: $R^1 = R^2 = R^3 = R^4 = H$
3-oxo-coronaridine: $R^1 = R^3 = R^4 = H; R^2 = OCH_3$
(19S)-heyneamine: $R^1 = OH; R^2 = R^3 = R^4 = H$
voacangine: $R^1 = R^2 = R^3 = H; R^4 = OCH_3$
isovoacangine: $R^1 = R^2 = R^4 = H; R^3 = OCH_3$
isovoachristine: $R^1 = OH; R^2 = R^3 = H; R^4 = OCH_3$
(3S)-hydroxyisovoacangine: $R^1 = R^4 = H; R^2 = OH; R^3 = OCH_3$

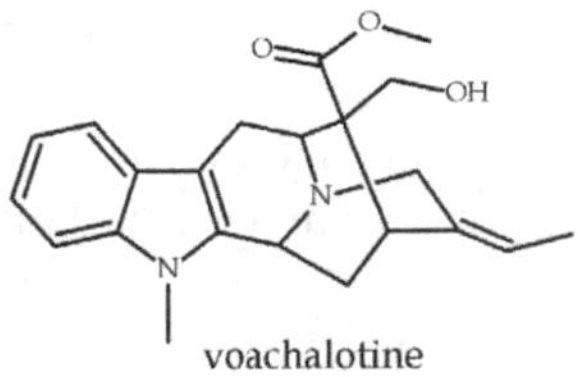

voachalotine

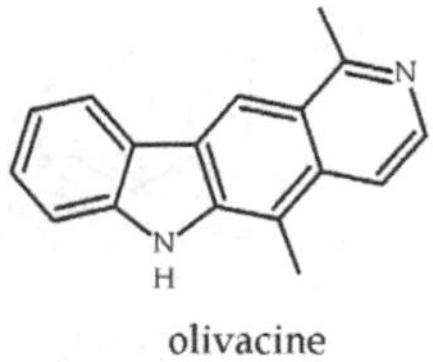

olivacine

Fig. 14. Alkaloids from *Tabernaemontana salzmannii.*

Already having an extremely diverse variety in structural formulas, new and (therefore) unusual carbon skeletons have continued to be found in alkaloids research. That was the case reported by Hitotsuyanagi et al. (2010), in a study with *Stemona sessilifolia*. Plants of the genus *Stemona* are rich sources of alkaloids, and two new ones were isolated by CC of the acidic partition from the methanolic extract of the roots. Sessilifoliamide K and sessilifoliamide L were structurally related, presenting an unusual pyrido[1,2-*a*]azonine skeleton (Figure 15), which can be derived from tuberosteminol-type alkaloids, previously described occurring in *Stemona* genus.

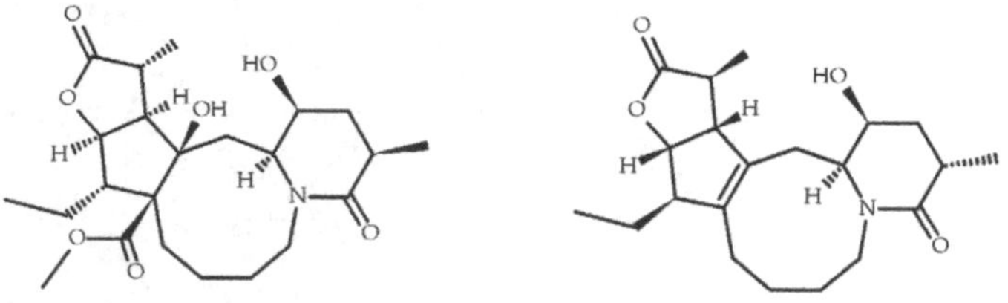

sessilifoliamide K                                    sessilifoliamide L

Fig. 15. New unusual pyrido[1,2-*a*]azinone alkaloids from *Stemona sessilifolia*.

## 4.2 Alkaloids from fungi

Probably the most famous (and infamous) alkaloid producing fungi are the ergots (family Hypocreaceae). Ergotism (holy fire) has killed thousands due to contaminated rye, and even in the last century many cases were reported (Wink, 1998). However, there is a plethora of alkaloid producing-fungi, with different structures and biological activities. Three new alkaloids, lyconadin D, lyconadin E and complanadine E (Figure 16) were isolated from *Lycopodium complanatum* (Lycopodiaceae) by Ishiuchi et al. (2011). The acidified methanolic extract was partitioned with ethyl acetate. The pH of the aqueous phase was adjusted and extracted with chloroform. After solvent evaporation, the chloroform-soluble materials were chromatographed on an amino silica gel column and eluted with hexane-ethyl acetate mixtures, then with a chloroform-methanol-water system. The fractions eluting with pure chloroform were re-chromatographed over silica gel, and lyconadin D and lyconadin E were isolated. Part of the chloroform soluble fraction was eluted with chloroform-methanol over a Sephadex LH-20 column. The fractions were further purified by HPLC ($C_{18}$ column) to yield complanadine E, a dimeric unsymmetrical alkaloid.

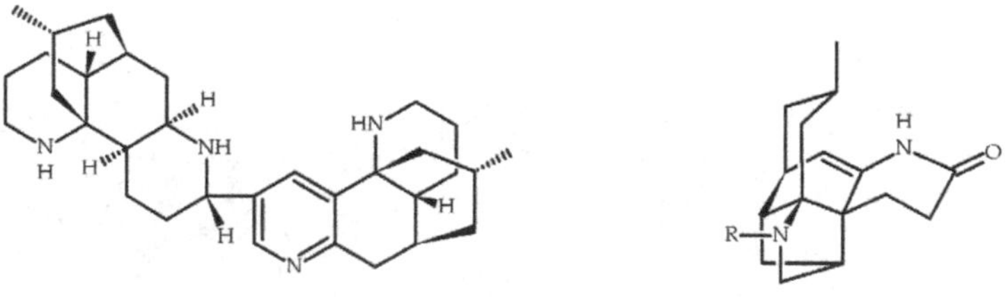

complanadine E                          liconadin D: R = $CH_3$

                                        liconadin E: R = H

Fig. 16. New alkaloids from moss *Lycopodium complanatum*.

The mosses of the genus *Lycopodium* in particular are very prone to produce complex polycyclic alkaloids, with important biological activities, such as inhibition of acetylcholine esterase. From three different species of *Lycopodium*, ten new alkaloids were isolated and characterized (Katakawa et al., 2011). Dihydrolycopoclavamine (Figure 17) was isolated from the extract of *L. serratum* after chromatography on amino silica gel and elution with hexane-chloroform. For *L. clavatum*, after neutralization of the acidic fraction from the methanolic extract, it was submitted to flash chromatography over silica gel and eluted with a chloroform-methanol gradient, then chloroform-methanol-aqueous ammonium hydroxide and pure methanol. One fraction was re-chromatographed on amino silica gel with hexane-chloroform mixtures leading to licopoclavamine A. Another fraction was re-chromatographed on silica gel to give lycopoclavamine B (Figure 17). A similar procedure was applied to *L. squarrosum*, from which lycoposquarrosamine A, acetyllycoposerramine-U, 8-α-hydroxyfawcettimine, 8-β-hydroxyfawcettimine, 8-α-acetoxyfawcettimine, 8-β-acetoxyfawcettimine and lycoflexine N-oxide were isolated. It is quite remarkable that all these compounds were separated solely by means of CC using silica gel, amino silica gel or $Al_2O_3$, and solvent mixtures of hexane, chloroform, methanol and ethyl acetate.

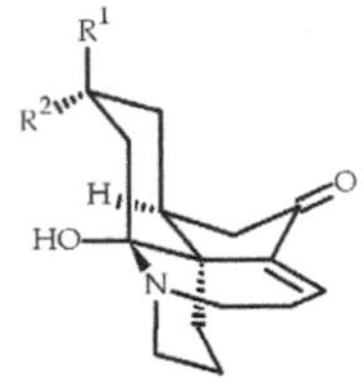

lycopoclavamine A: $R^1 = CH_3$, $R^2 = H$
lycopoclavamine B: $R^1 = OH$, $R^2 = CH_3$

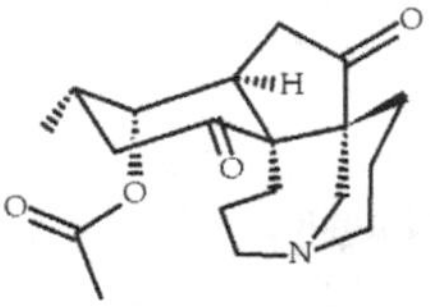

acetyllycoposerramine-U

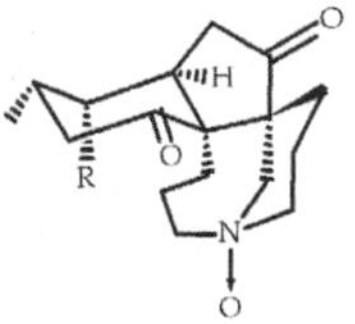

lycopoflexine N-oxide

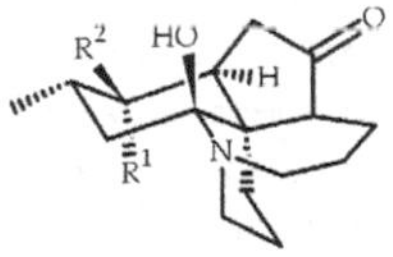

8-α-acetoxyfawcettimine: $R^1 = OCOCH_3$, $R^2 = H$
8-α-hydroxyfawcettimine: $R^1 = OH$, $R^2 = H$
8-β-acetoxyfawcettimine: $R^1 = H$, $R^2 = OCOCH_3$
8-β-hydroxyfawcettimine: $R^1 = H$, $R^2 = OH$

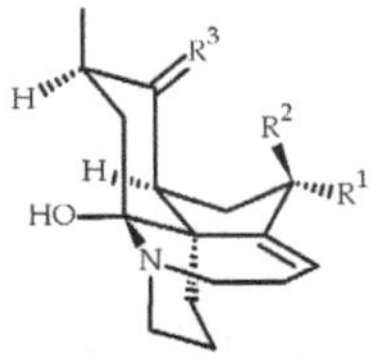

dihydrolycopoclavamine A: $R^1 = OH$, $R^2 = H$, $R^3 = H_2$
lycoposquarrosamine A: $R^1 = H$, $R^2 = OH$, $R^3 = O$

Fig. 17. Alkaloids from Lycopodium clavatum, L. serratum and L. squarrosum.

## 4.3 Marine alkaloids

The marine environment has emerged as a promising source of new substances with medicinal application. Considering only sponges, for example, more than 15,000 compounds have been isolated and characterized, including terpenoids, nucleosides, alkaloids and steroids.

Guanidine alkaloids were isolated from *Monanchora arbuscula* after CC over Sephadex LH-20 with methanol then repeatedly over silica gel and eluted with chloroform-ethyl acetate-methanol mixtures. The compounds (Figure 18) were identified by [1]H and [13]C NMR, IR and MS (Ferreira et al., 2011).

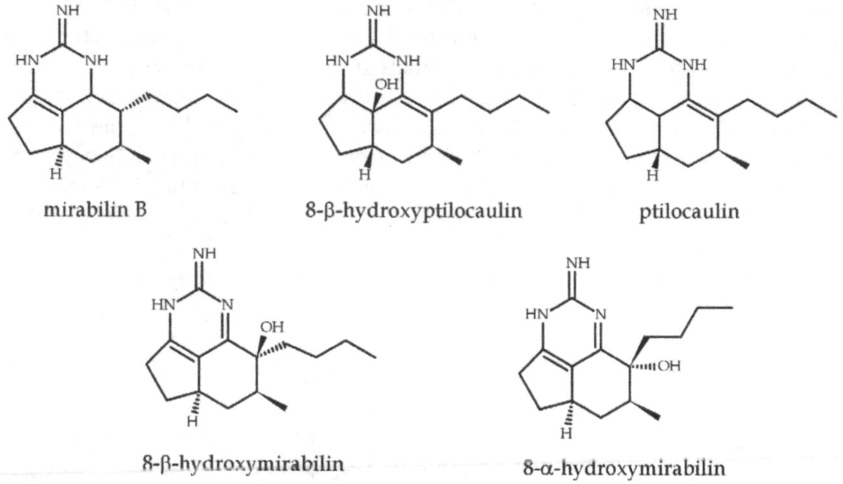

mirabilin B          8-β-hydroxyptilocaulin          ptilocaulin

8-β-hydroxymirabilin          8-α-hydroxymirabilin

Fig. 18. Guanidin alkaloids from sponge *Monanchora arbuscula*.

Two new brominated arginine-derived alkaloids were isolated from the Red Sea sponge *Suberea mollis* (Aplysinellidae), namely subereamine A and subereamine B (Figure 19). Halogenated compounds are common in marine natural products. The subereamines were isolated after CC on Sephadex LH-20 of the ethyl acetate extract, followed by chromatography on silica gel with dichloromethane-methanol mixtures and preparative HPLC on $C_{18}$ column (Shaala et al., 2011).

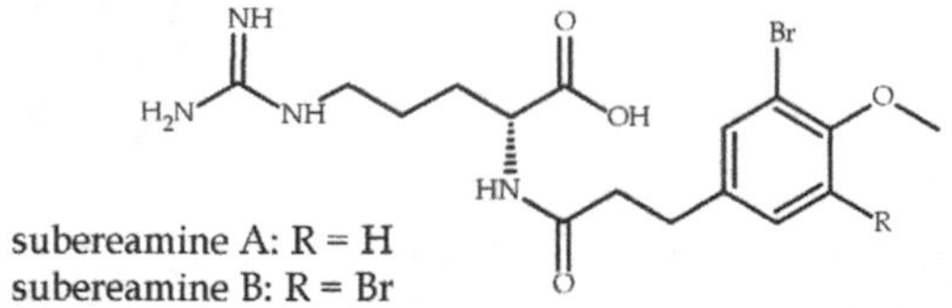

subereamine A: R = H
subereamine B: R = Br

Fig. 19. Brominated alkaloids from sponge *Suberea molis*.

The first natural products containing a 1H-oxazolo [40,50:4,5] benzo[1,2,3-de] [1,6] naphthyridine ring system were isolated from a *Suberites sp.* sponge collected in Okinawa, Japan. Named respectively, nakijinamines C and E (Figure 20), they constitute a group of heteroaromatic alkaloids, hybrids of aaptamine-type and bromoindole alkaloids (Takahashi et al., 2011). Other 23 aaptamine-type alkaloids have been isolated so far, but none contain the hybrid ring system. Another new alkaloid was isolated, namely nakijinamine D, which does not contain the oxazolo ring. The sponges were extracted with methanol and the residue partitioned with ethyl ether and water. The aqueous phase was extracted with butanol and subjected to reversed phase CC ($C_{18}$) followed by repeated HPLC also on a $C_{18}$ column. The alkaloids were isolated initially as racemates. Antifungal activity was recorded for nakijinamines C and E against *Aspergillus niger*.

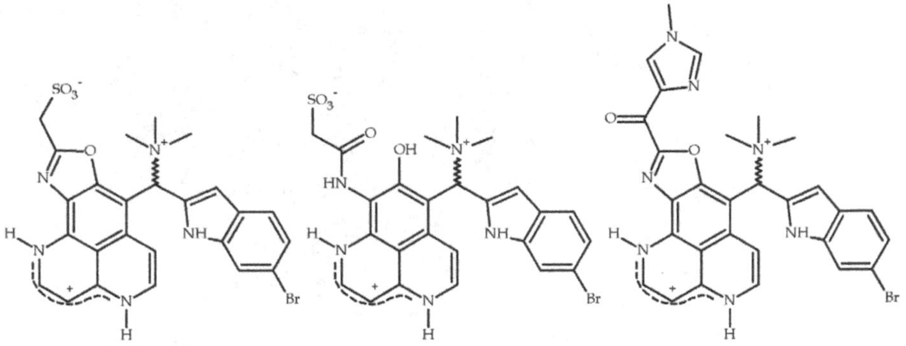

nakijinamine C                     nakijinamine D                     nakijinamine E

Fig. 20. New heteroaromatic alkaloids from *Suberites sp.*

Marine sponges of the genus *Asteropus* have been reported to contain a variety of compounds, such as saponines, sterols and pteridine derivatives. The concentrated methanolic extract was partitioned between water and dichloromethane. The water phase showed toxicity to brine shrimp larvae and was partitioned with butanol. The butanolic extract was submitted to medium pressure liquid chromatography over ODS-A with a methanol gradient. A fraction selected according to the toxicity test result was chromatographed over Sephadex LH-20. The sub-fractions were purified by reversed phase HPLC. The pyroglutamyl alkaloids were new compounds (Figure 21), and therefore fully characterized by spectroscopic methods (Li et al., 2011).

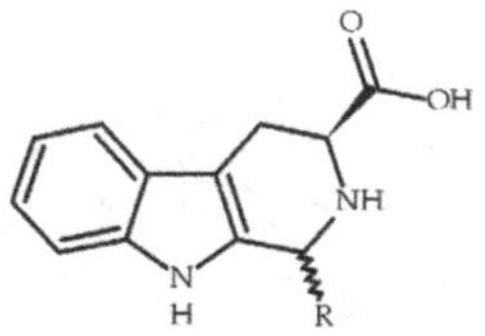

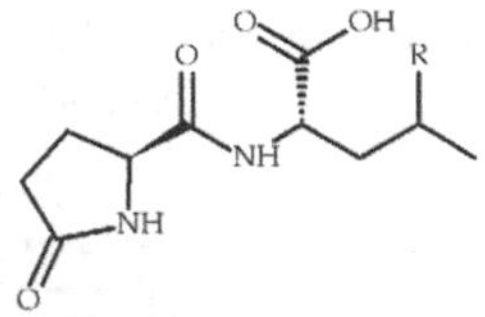

L-p-Glutamyl-L-leucine: R = CH3
L-p-Glutamyl-L-isoleucine: R = H

1,2,3,4-tetrahydro-b-carboline-3-carboxylic acid: R = H
trans-1,2,3,4-tetrahydro-b-carboline-1,3-dicarboxylic acid: R = trans-COOH
cis-1,2,3,4-tetrahydro-b-carboline-1,3-dicarboxylic acid: R = cis-COOH

Fig. 21. Pyroglutamyl dipeptides and tetrahydro-β-carboline alkaloids from *Asteropus* sp.

Carboline alkaloids were also found in tunicates of the genus *Eudistoma*. After purification on Sephadex LH-20, the methanolic extract of *Eudistoma glaucus* was chromatographed on a silica gel column with hexane-ethyl acetate mixtures, and then on another column with amino silica gel, also eluted with the same solvents. Normal phase HPLC led to the isolation of eudistomidin H (Figure 22) and eudistomidin I. Eudistomidins H and I were new compounds containing a unique fused-tetracyclic ring system consisting of a tetrahydro β-carboline ring and a hexahydropyrimidine ring. Using the same stationary phases, but changing the solvent system to chloroform-methanol, eudistomidin K was obtained. For eudistomidin J, the last solvent system was also applied, but without the need for a HPLC separation step (Suzuki et al., 2011).

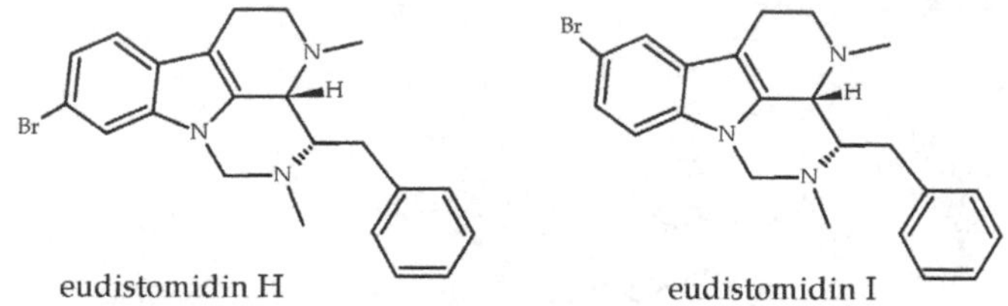

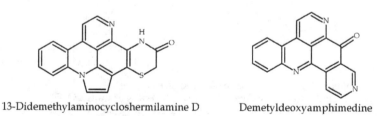

eudistomidin H

eudistomidin I

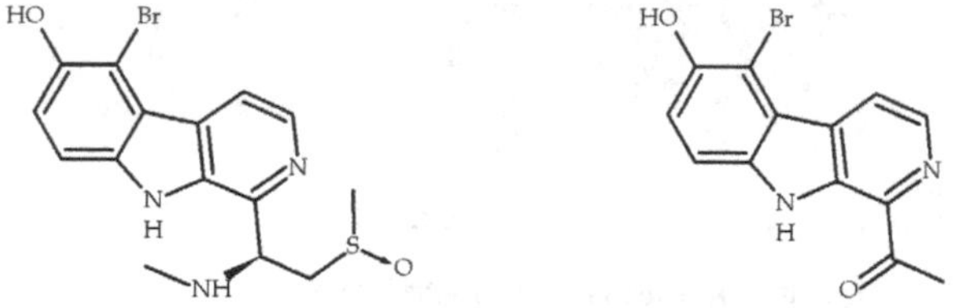

eudistomidin J

eudistomidin K

Fig. 22. Carboline alkaloids from tunicate *Eudistoma glaucus*.

Pyridoacridine alkaloids are responsible for the bright colors observed in sponges and ascidians. Two new pyridoacridine alkaloids, namely 13-didemethylamino-cycloshermilamine D and dimethyl-deoxyamphimedine (Figure 23), were isolated from the purple chromotype of the Western Mediterranean morph *Cystodytes dellechiajei* together with other six known alkaloids of the same kind (Bry et al., 2011). The ascidians were extracted with methanol-dichloromethane-trifluoroacetic acid mixture followed by reverse phase CC. The purple fraction obtained was chromatographed three successive times on reverse phase columns eluted with a methanol-water gradient. After reverse phase semi-preparative HPLC, the two new alkaloids were isolated.

13-Didemethylaminocycloshermilamine D          Demetyldeoxyamphimedine

Fig. 23. New pyridoacridine alkaloids from morph *Cystodytes dellechiajei*.

In this way, the phytochemical investigation of alkaloids has been deeply dependent on CC techniques. Careful and patient repeated CC has led to the isolation of some structurally varied and fascinating compounds, even without the use of modern HPLC methods. To be more efficient and to enhance productivy, of course, combined approaches are the common practice.

It is worth emphasizing the great opportunity that marine organisms offer to researchers looking for biologically active compounds and synthetic models. The real treasures at the bottom of the sea may not come from shipwrecks, but are those which have been lying (or, better, living) there since long before human beings set foot on earth.

# 5. Conclusion

Even in laboratories with good HPLC and GC-MS facilities, the traditional and classical methods of TLC and CC are still widely used, for rapid preliminary screening of extracts, for isolating and purifying bioactive compounds of secondary metabolism for further study, for comparison of samples with standards and for monitoring chemical synthesis or the course of reactions.

# 6. Acknowledgment

Authors thank Embrapa for financial support to the chapter production.

# 7. References

Abe, I. (2007). Enzymatic synthesis of cyclic triterpenes. *Natural Product Reports*, Vol. 24, pp. 1311-1331.

Adams, R.P. (2007). *Identification of Essential Oil Components by Gas Chromatography / Mass Spectrometry* (4th ed.), Allured Publishing Co., ISBN 978-1-932633-21-4, Carol Stream (IL).

Agostini, T.S., Cecchi, H.M. & Godoy, H.T. (1996). Composição de carotenóides no marolo e em produtos de preparo caseiro. *Ciência e Tecnologia de Alimentos*, Vol. 16, pp. 67-71.

Agostini-Costa, T.S., Abreu, L.N. & Rossetti, A.G. (2003). Efeito do congelamento e do tempo de estocagem da polpa de acerola sobre o teor de carotenóides. *Revista Brasileira de Fruticultura*, Vol. 25, pp. 56-58.

Agostini-Costa, T.S., Wondracek, D.C., Lopes, R.M., Vieira, R.F. & Ferreira, F.R. (2010). Composição de carotenoides em canistel (*pouteria campechiana* (kunth) baehni). *Revista Brasileira de Fruticultura*, Vol. 32, pp. 903-906.

Agrawal, S.S. & Talele, G.S. (2011). Bioactivity guided isolation and characterization of the phytoconstituents from the *Tridax procumbens Brazilian Journal of Pharmacognosy*, Vol. 21, pp. 58-62.

Ajit-Simh K.A., Bemberis I., Brown S., De-Mill, B., Deorkar, N., Grund, E., Hays, E., Levine, H.L., Levy, R., Mazzeo, J., Natarajan, V., Nevill, J., Noyes, A., O'Neil, P., Seely, R., Sofer, G. & Wheelwright, S.M. (2003). Column packing for process-scale chromatography: guidelines for reproductibility. *Biopharmaceutical International*, Supp (June), pp. 23-30.

Amarowicz, R. & Karamac, M. (2003). Antioxidant activity of phenolic fractions of lentil (*Lens culznaris*). *Journal of Food Lipids*, Vol. 10, pp. 1-10.

Areche, C., Cejas., P., Thomas, P., San-Martin, A., Astudillo, L., Gutierrez, M. & Loyola, L.A. (2009). Triterpenoids from Azorella trifurcata (Gaertn.) Pers and their effect against the enzyme acetylcholinesterase. *Quimica Nova*, Vol. 32, pp. 2023-2025.

Arroyo, J., Bonilla, P., Ráez, E., Barreda, A. & Huamán, O. Efecto quimioprotector de *Bidens pilosa* en el cáncer de mama inducido en ratas. (2010). *Anales de la Facultad de Medicina*, Vol. 71, pp. 153-159.

Augustin, J.M., Kuzina, V., Andersen, S.B. & Bak, S. (2011). Molecular activities, biosynthesis and evolution of triterpenoid saponins. *Phytochemistry*, Vol. 72, pp. 435-457.

Batista, R., García, P.A., Castro, M.A., Del Corral, J.M.M., Feliciano, A.S. & Oliveira, A.B. (2010). Iso-Kaurenoic acid from *Wedelia paludosa* DC. *Anais da Academia Brasileira de Ciências,* Vol. 82, pp. 823-831.

Begnami, A.F., Duarte, M.C.T., Furletti. V. & Rehder, V.L.G. (2010). Antimicrobial potential of *Coriandrum sativum* L. against different *Candida* species in vitro. *Food Chemistry,* Vol. 118, pp. 74-77.

Bernhard, K. (1995). Chromatography: Part II Column Chromatography. In: Britton, G; Liaaen-Jensen; Pfander, H. (Eds.) *Carotenoids isolation and analysis.* Basel: Birkhauser Verlag, pp. 117-130.

Bicchi, C., Cagliero, C. & Rubiolo, P. (2011). New trends in the analysis of the volatile fraction of matrices of vegetable origin: a short overview. A review. *Flavour and Fragrance Journal,* Vol. 26, pp. 321-325.

Bizzo, H.R., Lopes, D. Abdala, R.V., Pimentel, F.A., Souza, J.A., Pereira, M.V.G., Bergter, L. & Guimarães, E.F. (2001). Sarisan from Leaves of *Piper affinis hispidinervum* C. DC (long pepper). *Flavour and Fragrance Journal,* Vol. 16: 113-115.

Blazics, B., Ludanyi, K., Szarka, S. & Kery, A. (2008). Investigation of *Euphrasia rostkoviana* Hayne Using GC-MS and LC-MS. *Chromatographia Supplement,* Vol. 68, pp. S119-S124.

Brielmann, H.R., Setzer, W.N., Kaufman, P.B., Kirakosyan, A. & Cseke, L.J. (2006). Phytochemicals: The Chemical Components of Plants, In: *Natural Products from Plants* (2nd ed), L.J. Cseke, A. Kirakosyan, P.B. Kaufman, S.L. Warber, J.A. Dule & H.R. Brielmann, pp.1-50, CRC Press, ISBN 978-0-8493-2976-0, Boca Raton.

Britton, G. (1995). Structure and properties of carotenois in relation to function. *The FASEB Journal,* Vol. 9, pp. 1551-1558.

Bry, D., Banaigs, B., Long, C. & Bontemps, N. (2011). New pyridoacridine alkaloids from the purple morph of the ascidian *Cystodytes dellechiajei. Tetrahedron Letters,* Vol. 52, pp. 3041-3044.

Cespedes, C.L., Valdez-Morales, M.V., Avila, J.G., El-Hafidi, M., Alarcon, J. & Paredes-Lopez, O. (2010). Phytochemical profile and the antioxidant activity of Chilean wild black-berry fruits, *Aristotelia chilensis* (Mol) Stuntz (Elaeocarpaceae). *Food Chemistry,* Vol. 119, pp. 886-895.

Chen, J., Chen, J.J., Yao, X. & Gao, K. (2011). Kopsihainanines A and B, two unusual alkaloids from *Kopsia hainanensis.* Organic and Biomolecular Chemistry, Vol. 9, pp. 5334-5336.

Chisholm, M.G., Jell, J.A. & Cass, Jr., D.M. (2003). Characterization of the major odorants found in the peel oil of *Citrus reticulata* Blanco cv. Clementine using gas chromatography–olfactometry. *Flavour and Fragrance Journal,* Vol. 18, pp. 275-281.

Coronel, A.C., Cerda-García-Rojas, C.M., Joseph-Nathan, P. & Catalán, C.A.N. (2006). Chemical composition, seasonal variation and a new sesquiterpene alcohol from the essential oil of *Lippia integrifolia. Flavour and Fragrance Journal,* Vol. 21, pp. 839-847.

Czeczuga, B., Czeczuga-Semeniuk, E. & Semeniuk, A. Chromatic adaptation in lichen phyco- and photobionts. (2010). *Biologia,* Vol. 65, pp. 587-594.

Deli, J., Molnfir, E., Osz, E. & Tdth, G. (2000). Analysis of carotenoids in the fruits of *Asparagus falcatus:* isolation of 5,6-diepikarpoxanthin. *Chromatographia Supplement,* Vol. 51, pp. S-183-S-187.

Delort, E. & Jaquier, A. (2009). Novel terpenyl esters from Australian finger lime (*Citrus australasica*) peel extract. *Flavour and Fragrance Journal*, Vol. 24, pp. 123-132.

Djabou, N., Paolini, J., Desjobert, J.M., Allali, H., Baldovini, N., Costa, J. & Muselli. A. (2010). Qualitative and quantitative analysis of volatile components of *Teucrium massiliense* L. - identification of 6-methyl-3-heptyl acetate as a new natural product. *Flavour and Fragrance Journal*, Vol. 25, pp. 475-487.

Dudareva, N., Pichersky, E. & Gershenzon, J. (2004). Biochemistry of Plant Volatiles. *Plant Physiology*, Vol. 135, pp. 1893-1902.

Duraipandiyan, V., Indwar, F. & Ignacimuthu, S. Antimicrobial activity of confertifolin from *Polygonum hydropiper*. (2010). *Pharmaceutical Biology*, Vol. 48, pp. 187-190.

Ebel, R. (2010). Terpenes from Marine-Derived Fungi. *Marine Drugs*, Vol. 8, pp. 2340-2368.

Ferreira, E.G., Wilke, D.V., Jimenez, P.C., Oliveira, J.R., Pessoa, O.D., Silveira, E.R., Viana, F.A., Pessoa, C., Moraes, M.O., Hadju, E. & Costa-Lotufo, L.V. (2011). Guanidine Alkaloids from *Monanchora arbuscula*: Chemistry and Antitumor Potential. *Chemistry and Biodiversity*, Vol. 8, pp. 1433-1445.

Figueiredo, E.R., Vieira, I.J.C., Souza, J., Braz-Filho, R., Mathias, L., Kanashiro, M.M. & Côrtes, F.H. (2010). Isolamento, identificação e avaliação da atividade antileucêmica de alcaloides indólicos monoterpênicos de *Tabernaemontana salzmannii* (A. DC.), Apocynaceae. *Brazilian Journal of Pharmacognosy*, Vol. 20, pp. 675-681.

Fronza, M., Murillo, R., Slusarczyk, S., Adams, M., Hamburger, M., Heinzmann, B., Laufer, S. & Merfort, I. (2011). In vitro cytotoxic activity of abietane diterpenes from Peltodon longipes as well as Salvia miltiorrhiza and S. sahendica. *Bioorganic & Medicinal Chemistry*, Vol. 19, pp. 4876-4881.

Gao, D.F., Xu, M., Yang, C.R., Xu, M & Zhang, Y.J. (2010). Phenolic Antioxidants from the leaves of Camellia pachyandra Hu. *Journal of Agricultural and Food Chemistry*, Vol. 58, pp. 8820-8824.

Gassenmeier, K. (2006). Identification and quantification of L-menthyl lactate in essential oils from *Mentha arvensis* L. from India and model studies on the formation of L-menthyl lactate during essential oil production. *Flavour and Fragrance Journal*, Vol. 21, pp. 725-730.

Gauvin-Bialecki, A., Aknin, M. & Smadja, J. (2008). 24-O-Ethylmanoalide, a manoalide-related sesterterpene from the marine sponge *Luffariella* cf. variabilis. *Molecules*, Vol. 13, pp. 3184-3191.

Gomez-Alonso, S., Giuseppe Fregapane, G., Salvador, M.D. & Gordon, M.H. (2003). Changes in phenolic composition and antioxidant activity of virgin olive oil during frying. *Journal of Agricultural and Food Chemistry*, Vol. 51, pp. 667-672.

He, Z., Xia, W., Liu, Q. & Chen, J. Identification of a new phenolic compound from Chinese olive (*Canarium album* L.) fruit. (2009). *European Food Research and Technology*, Vol. 228, pp. 339-343.

Hitotsuyanagi, Y., Uemura, G. & Takeya, K. (2010). Sessilifoliamides K and L: new alkaloids from *Stemona sessilifolia*. *Tetrahedron Letters*, Vol. 51, pp. 5694-5696.

Ishiuchi, K., Kubota, T., Ishiyama, H., Hayashi, S., Shibata, T., Mori, K., Obara, Y., Nakanhata, N. & Kobayashi, J. (2011). Lyconadins D and E, and complanadine E, new Lycopodium alkaloids from *Lycopodium complanatum*. *Bioorganic and Medicinal Chemistry*, Vol. 19, pp. 749-753.

Joseph-Nathan, P., Leitão, S.G., Pinto, S.C., Leitão, G.G., Bizzo, H.R., Costa, F.L.P., Amorim, M.B., Martinez, N. Dellacassa, E. & Hernández-Barragán, A. (2010). Structure reassignment and absolute configuration of 9-*epi*-presilphiperfolan-1-ol. *Tetrahedron Letters*, Vol. 51, pp.1963-1965.

Joulain, D. & König, W. (1998). *The Atlas of Spectral Data of Sesquiterpenes Hydrocarbons* (1st ed.), EB-Verlag, ISBN 978-3930826483, Hamburg.

Jung, M.J; Heo, S. & Wang, M.H. (2008). Food Chemistry, Vol. 108, pp. 482–487.

Junior, G.M.V., Sousa, C.M.M., Cavalheiro, A.J., Lago, J.H.G. & Chaves, M.H. (2008). Phenolic Derivatives from fruits of *Dipteryx lacunifera* Ducke and evaluation of their antiradical activities. *Helvetica Chimica Acta*, Vol. 91, pp. 2159-2167.

Karamac, M. (2009). *In-vitro* study on the efficacy of tannin fractions of edible nuts as antioxidants. *European Journal of Lipid Science and Technology*, Vol. 111, pp.1063–1071.

Katakawa, K., Mito, H., Kogure, N., Kitajima, M., Wongseripipatana, S., Arisawa, M. & Takayama, H. (2011). Ten new fawcettimine-related alkaloids from three species of Lycopodium. *Tetrahedron*, Vol. 67, pp. 6561-6567.

Kernan, M.R. & Faulkner, D.J. (1988). Sesterterpene sulfates from a sponge of the family Halichondriidae. *The Journal of Organic Chemistry*, Vol. 53, pp. 4574-4578.

Khanal, P., Oh, W. K., Yun, H. J., Namgoong, G. M., Ahn, S. G., Kwon, S. M., Choi, H. K. & Choi, H. S. (2011). *p*-HPEA-EDA, a phenolic compound of virgin olive oil, activates AMP-activated protein kinase to inhibit carcinogenesis. *Carcinogenesis*, Vol. 32, pp.545-553.

Kimura, M. & Rodriguez-Amaya, D. B. (2002). A scheme for obtaining standards and HPLC quantification of leafy vegetable carotenoids. *Food Chemistry*, Vol. 78, pp. 389–398.

Kuo, P.C., Hwang, T.L., Lin, Y.T., Kuo, Y.C. & Leu, Y.L. (2011). Chemical Constituents from *Lobelia chinensis* and Their Anti-virus and Anti-inflammatory Bioactivities. *Archives of Pharmaceutical Research*, Vol. 34, pp. 715-722.

Kusaba, M., Maoka, T., Morita, R. & Takaichi, S. (2009). A novel carotenoid derivative, lutein 3-acetate, accumulates in senescent leaves of rice. *Plant Cell Physiology*, Vol. 50, p. 1573-1577.

Lanças, FM. (2008). Avanços recentes e tendências futuras das técnicas de separação: uma visão pessoal. *Scientia Chromatographica*, Vol. 0, pp. 17-44.

Lanfranchi, D.A., Laouer, H., Kolli, M.E., Prado, S., Maulay-Bailly, C. & Baldovini, N. (2010). Bioactive Phenylpropanoids from Daucus crinitus Desf. from Algeria. *Journal of Agricultural and Food Chemistry*, Vol. 58, pp. 2174–2179.

Lee, C.J., Chen, L.G., Liang, W. & Wang, A.C.C. (2010). Anti-inflammatory effects of *Punica granatum* Linne in vitro and in vivo. *Food Chemistry*, Vol. 118, pp.315–322.

Li, H., Dant, H.T.,Li, J., Sim, C.J., Hong, J., Kim, D.K. & Jung, J.H. (2011). Pyroglutamyl dipeptides and tetrahydro-β-carboline alkaloids from a marine sponge *Asteropus sp*. *Biochemical Systematics and Ecology*, Vol. 38, pp. 1049-1051.

Li, N., Hua, J., Wanga, S., Cheng, J., Hu, X., Lu, Z., Lin, Z., Zhu, W. & Bao, Z. (2010). Isolation and identification of the main carotenoid pigment from the rare orange muscle of the *Yesso scallop*. *Food Chemistry*, Vol. 118, pp. 616–619.

Lin, L., Chen, P. & Harnly, J. M. (2008). New phenolic components and chromatographic profiles of green and fermented teas. *Journal of Agricultural and Food Chemistry*, Vol. 56, pp. 8130–8140.

Liu, Q., Zhang, Y.J., Yang, C.R. & Xu, M. (2009). Phenolic antioxidants from green tea produced from *Camellia crassicolumna* Var. *multiplex*. *Journal of Agricultural and Food Chemistry*, Vol. 57, pp. 586–590.

Lopes, R. M., Agostini-Costa, T. S., Gimenes, M. A. & Silveira, D. (2011). Chemical composition and biological activities of *Arachis* species. *Journal of Agricultural and Food Chemistry*, Vol. 59, pp. 4321–4330.

Lopez, M.D., Jordan, M.J. & Pascual-Villalobos, M.J. (2008). Toxic compounds in essential oils of coriander, caraway and basil active against stored rice pests. *Journal of Stored Products Research*, Vol. 44, pp. 273-278.

Mahmoud, S.S. & Croteau, R.B. (2002). Strategies for transgenic manipulation of monoterpene biosynthesis in plants. *Trends in Plant Science*, Vol. 7, pp. 366-373.

Maia, J.G.S., Silva, M.L., Luz, A.I.R., Zoghbi, M.G.B. & Ramos, L.S. (1987). Espécies de *Piper* da Amazônia ricas em safrol. *Química Nova*, Vol.10, pp.200-204.

Marty, C. & Berset, C. Factors affecting the thermal degradation of all- trans-β-Carotene. (1990). *Journal of Agricultural and Food Chemistry*, Vol. 38, pp. 1063-1067.

Ndongo, J.T., Shaaban, M., Mbing, J.N., Bikobo, D.N., Atchadé, A.T., Pengyemb, D.E. & Laatsch, H. (2010). Phenolic dimers and an indole alkaloid from *Campylospermum flavum* (Ochnaceae). *Phytochemistry*, Vol. 71, pp. 1872-1878.

Norin, T. & Westfelt, L. (1963). Thin layer, column, and gas liquid chromatography of resin acid esters and related terpenes. *Acta Chemica Scandinavia*, Vol. 17, pp. 1828-1830.

Novak, J., Zitterl-Eglseer, K., Deans, S.G. & Franz, C.M. (2001). Essential oils of different cultivars of *Cannabis sativa* L. and their antimicrobial activity. *Flavour and Fragrance Journal*, Vol. 16, pp. 259-262.

Oliveira, G.P.R. & Rodriguez-Amaya, D.B. (2007). Processed and prepared corn products as sources of lutein and zeaxanthin: compositional variation in the food chain. *Journal of Food Science*, Vol. 72, pp. S79.

Ozen, T.; Demirtas, I. & Aksit, H. (2011). Determination of antioxidant activities of various extracts and essential oil compositions of *Thymus praecox* subsp. *skorpilii* var. *skorpilii*. *Food Chemistry*, Vol. 124, pp. 58–64.

Pawlowska, A.M., Oleszek, W. & Braca, A. (2008). Quali-quantitative Analyses of Flavonoids of *Morus nigra* L. and *Morus alba* L. (Moraceae) Fruits. *Journal of Agricultural and Food Chemistry*, Vol. 56, pp. 3377–3380.

Pichersky, E., Noel, J.P. & Dudareva, N. (2006). Biosynthesis of Plant Volatiles: Nature's Diversity and Ingenuity. *Science*, Vol. 311, pp. 808.

Pinto, S.C., Leitão, G.G., Bizzo, H.R., Martinez, N., Dellacassa, E., Santos, F., Costa, F.L.P., Amorim, M.B. & Leitão, S.G. (2009a). (-)-*epi*-Presilphiperfolan-1-ol, A New Triquinane Sesquiterpene from the Essential Oil of *Anemia tomentosa* var. *anthriscifolia*. *Tetrahedron Letters*, Vol. 50, pp.785-4787.

Pinto, S.C., Leitão, G.G., Oliveira, D.R., Bizzo, H.R., Ramos, D.F., Coelho, T.S., Silva, P.E.A., Lourenço, M.C.S. & Leitão, S.G. (2009b). Chemical Composition and Antimycobacterial Activity from the Essential Oil from *Anemia tomentosa* var. *anthriscifolia*. *Natural Product Communications*, Vol. 4, pp.1675-1678.

Polat, E., Bedir, E., Perrone, A., Piacente, S. & Alankus-Caliskan, O. (2010). Triterpenoid saponins from Astragalus wiedemannianus Fischer. *Phytochemistry*, Vol. 71, pp. 658-662.

Porcu, O.M. & Rodriguez-Amaya, D.B. (2008). Variation in the carotenoid composition of the lycopene-rich Brazilian fruit Eugenia uniflora L. *Plant Foods for Human Nutrition*, Vol. 63, pp. 195–199.

Potrebko, I. & Resurreccion, A.V.A. (2009). Effect of Ultraviolet Doses in Combined Ultraviolet- Ultrasound Treatments on trans-Resveratrol and trans-Piceid Contents in Sliced Peanut Kernels. *Journal of Agricultural and Food Chemistry*, vol.57, pp. 7750-7756.

Quadros, A.P., Chaves, F.C.M., Pinto, S.C., Leitão, S.G. & Bizzo, H.R. (2011). Isolation and Identification of cis-7-hydroxycalamenene From the Essential Oil of *Croton cajucara* Benth. *Journal of Essential Oil Research*, Vol. 23: 20-23.

Reis, B., Martins, M., Barreto, B., Milhazes, N., Garrido, E.M., Silva, P., Garrido, J. & Borges, F. (2010). Structure-property-activity relationship of phenolic acids and derivatives. Protocatechuic acid alkyl esters. *Journal of Agricultural and Food Chemistry*, Vol. 58: 6986-6993.

Roberts, M.F. & Wink, M. (1998). Introduction, In: *Alkaloids: Biochemistry, Ecology and Medicinal Applications*, M.F.Roberts & M. Wink, pp. 1-6, Plenum Press, ISBN 0-306-45465-3, New York.

Rodriguez, D.B., Lee, T.C. & Chichester, C.O. (1975). Comparative Study of the Carotenoid Composition of the Seeds of Ripening *Momordica charantia* and Tomatoes. *Plant Physiology*, Vol. 56: 626-629.

Rodriguez-Amaya, D.B. *A guide to carotenoid analysis*. (1999). Washington: OMINI-ILSI Press, 64p.

Rossi, P.G., Bao, L. Luciani, A., Panighi, J., Desjobert, J.M., Costa,J., Casanova, J., Bolla, J.M. & Berti, L. (2007). (E)-methylisoeugenol and elemicin: antibacterial components of *Daucus carota* L. essential oil against *Campylobacter jejuni*. *Journal of Agricultural and Food Chemistry*, Vol. 55, pp. 7332-7336.

Roze, L.V., Chanda, A. & Linz, J.E. (2011). Compartmentalization and molecular traffic in secondary metabolism: a new undestanding of established cellular processes. *Fungal Genetics and Biology*, Vol. 48, p. 35-48.

Salamci, E., Kordali, S., Kotan, R., Cakir, A. & Kaya, Y. (2007). Chemical compositions, antimicrobial and herbicidal effects of essential oils isolated from Turkish *Tanacetum aucheranum* and *Tanacetum chiliophyllum* var. *chiliophyllum*. *Biochemical Systematics and Ecology*, Vol. 35, pp. 569-581.

Sarr, M., Ngom, S., Kane1, M. O., Wele, A., Diop, D., Sarr, B., Gueye, L., Andriantsitohaina, R. & Diallo, A. S. (2009). *In vitro* vasorelaxation mechanisms of bioactive compounds extracted from *Hibiscus sabdariffa* on rat thoracic aorta. *Nutrition & Metabolism*, Vol. 6, n. 45, p. 1-12.

Shaala, L.A., Bamane, F.H., Badr, J.M. & Youssef, D.T.A. (2011). Brominated Arginine-Derived Alkaloids from the Red Sea Sponge *Suberea mollis*. *Journal of Natural Products*, Vol. 74, pp. 1517-1520.

She, G.M., Xu, C., Liu, B. & Shi, R.B. (2010). Polyphenolic acids from mint (the aerial of *Mentha haplocalyx* Briq.) with DPPH radical scavenging activity. *Journal of Food Science*, Vol. 75, pp. C359-C362.

Shimoda, M., Shigematsu, H., Shiratsuchi, H. & Osajima, Y. (1995). Comparison of the odor concentrates by SDE and adsorptive column method from green tea infusion. *Journal of Agricultural and Food Chemistry*, Vol. 43, pp. 1616-1620.

Shimoda, M., Wu, Y. & Osajima, Y. (1996). Aroma compounds from aqueous solution of haze (*Rhus succedanea*) honey determined by adsorptive column chromatography. *Journal of Agricultural and Food Chemistry*, Vol. 44, pp. 3913-3918.

Sugawara, Y., Hara, C., Aoki, T., Sugimoto, N. & Masujima, T. (2000). Odor distinctiveness between enantiomers of Linalool: difference in perception and responses elicited by sensory test and forehead surface potential wave measurement. *Chemical Senses*, Vol. 25, pp. 77-84.

Sulyok, E., Vasas, A., Rédei, D., Dombi, G. & Hohmann, J. (2009). Isolation and structure determination of new 4, 12-dideoxyphorbol esters from Euphorbia pannonica Host. *Tetrahedron*, Vol. 65, pp. 4013-4016.

Summons, R.E., Bradley, A.S., Jahnke, L.L, & Waldbauer, J.R. (2006). Steroids, triterpenoids and molecular oxygen. *Philosophical Transactions of the Royal Society B: Biological Sciences*, Vol. 361, pp. 951-968.

Sun, B., Neves, A.C., Fernandes, T., Fernandes, A., Mateus, M., Freitas, V., Leandro,M. C. & Spranger, M.I. (2011). Evolution of phenolic composition of red wine during vinification and storage and its contribution to wine sensory properties and antioxidant activity. *Journal of Agricultural and Food Chemistry*, Vol. 59, pp. 6550-6557.

Sutour, S., Bradesi, P., Rocca-Serra, D., Casanova, J. &; Tomi, F. (2008). *Chemical composition and antibacterial activity of theessential oil from Mentha suaveolens ssp. insularis* (Req.) Greuter. *Flavour and Fragrance Journal*, Vol. 23, pp. 107-114.

Suzuki, T., Kubota, T. & Kobayashi, J. (2011). Eudistomidins H-K, new β-carboline alkaloids from the Okinawan marine tunicate Eudistoma glaucus. Bioorganic and Medicinal Chemistry Letters, Vol. 21, pp. 4220-4223.

Takahashi, Y., Kubota, T., Shibazaki, A., Gonoi, T., Fromont, J. & Kobaiashi, J. (2011). Nakijinamines C-E, New Heteroaromatic Alkaloids from the Sponge *Suberites* Species. *Organic Letters*, Vol. 13, pp. 3016-3019.

Takaichi, S., Maoka, T., yamada, M., Matsuura, K., Haikawa, Y. & Hanada, S. (2001). Absence of carotenes and a presence of a tertiary methoxy group in a carotenoid from a thermophilic phylamentous photosynthetic bacterium Roseiflexus castanholzii. *Plant Cell Physiology*, Vol. 42, pp. 1355–1362.

Takaichi, S., Mochimaru, M., Maoka, T. & Katoh, H. (2005). 4, 6 Myxol and 4-Ketomyxol 2'-Fucosides, not Rhamnosides, from Anabaena sp. PCC 7120 and Nostoc punctiforme PCC 73102, and Proposal for the Biosynthetic Pathway of Carotenoids. *Plant Cell Physiology*, Vol. 46, pp. 497–504.

Tantillo, D.J. (2011). Biosynthesis via carbocations: Theoretical studies on terpene formation. *Natural Product Reports*, vol.28, pp. 1035-1053.

Francisco, V., Figueirinha, A., Neves, B.M., Rodríguez, C.G., Lopes, M.C.; Cruz, M.T. & Batista, M.T. *Cymbopogon citratus* as source of new and safe anti-inflammatory drugs: Bio-guided assay using lipopolysaccharide-stimulated macrophages. (2011). *Journal of Ethnopharmacology*, Vol. 133, pp. 818-827.

Vetter, W. & Schröder, M. (2011). Phytanic acid–a tetramethyl branched fatty acid in food. *Lipid Technology*, Vol. 23, pp. 175-178.

Villanueva, H.E. & Setzer, W.N. (2010). Cembrene diterpenoids: conformational studies and molecular docking to tubulin. *Records of Natural Products*, Vol. 4, pp. 115-123.

Wang, C.C., Chang, S.C., Inbaraj, S.B. & Chen, B.H. (2010). Isolation of carotenoids, flavonoids and polysaccharides from *Lycium barbarum* L. and evaluation of antioxidant activity. *Food Chemistry*, Vol. 120, pp. 184–192.

Wang, M., Li, J., Rangarajan, M., Shao, Y., Voie, E.J., Huang, T.C. & Ho, C.T. (1998). Antioxidative Phenolic Compounds from Sage (*Salvia officinalis*). *Journal of Agricultural and Food Chemistry*, Vol. 46, pp. 4869-4873.

Wink, M. (1998). A Short History of Alkaloids, In: *Alkaloids: Biochemistry, Ecology and Medicinal Applications*, M.F. Roberts & M. Wink, pp. 11-44, Plenum Press, ISBN 0-306-45465-3, New York.

Wondracek, D.C., Vieira, R. F., Silva, D.B. & Agostini-Costa, T.S. (2012). Saponification influence in carotenoid determination in cerrado passion fruit. *Química Nova, in prelo*.

Wu, Z., Song, L. &Huang, D. (2011). Food grade fungal stress on germinating peanut seeds induced phytoalexins and enhanced polyphenolic antioxidants. *Journal of Agricultural and Food Chemistry*, Vol. 59, pp. 5993-6003.

Xiao, J.S., Liu, L., Wu, H., Xie, B.J., Yang, E.N. & Sun, Z.D. (2008). Rapid preparation of procyanidins B2 and C1 from granny smith apples by using low pressure CC and identification of their oligomeric procyanidins. *Journal of Agricultural and Food Chemistry*, Vol. 56, pp. 2096–2101.

Ye, J.H., Wang L.X., Chen H., Dong J.J., Lu, J.L., Zheng, X.Q., Wu, M.Y. & Liang, Y.R. (2011). Preparation of tea catechins using polyamide. *Journal of bioscience and bioengineering*, Vol. 111, pp. 232-236.

Zellner, B.D., Dugo, P., Dugo, G. & Mondello, L. (2010). Analysis of Essential Oils, In: *Handbook of Essential Oils*, K.H.C. Baser, G. Buchbauer, pp. 151-184, CRC Press, ISBN 978-1420063158, Boca Raton.

Zhan, P.Y., Zeng, X.H., Zhang, H.M. & Li, H.H. (2011). High-efficient column chromatographic extraction of curcumin from Curcuma longa. *Food Chemistry*, Vol. 129, pp.700–703.

Zhang, Z., Liao, L., Moore, J., Wua, T. & Wang, Z. (2009). Antioxidant phenolic compounds from walnut kernels (*Juglans regia* L.). *Food Chemistry*, Vol. 113, pp. 160–165.

Zhu, B.C.R., Henderson, G., Yu, Y. & Laine, R.A. (2003). Toxicity and Repellency of Patchouli Oil and Patchouli Alcohol against Formosan Subterranean Termites *Coptotermes formosanus* Shiraki (Isoptera: Rhinotermitidae). *Journal of Agricultural and Food Chemistry*, Vol. 51, pp. 4585-4588.

# 10

# Quantification of Antimalarial Quassinoids Neosergeolide and Isobrucein B in Stem and Root Infusions of *Picrolemma sprucei* Hook F. by HPLC-UV Analysis

Rita C. S. Nunomura[1], Ellen C. C. Silva[1],
Sergio M. Nunomura[2], Ana C. F. Amaral[3],
Alaíde S. Barreto[3], Antonio C. Siani[3] and Adrian M. Pohlit[2]
[1]*Department of Chemistry, Amazon Federal University (UFAM), Amazon,*
[2]*Coordenation of Research in Natural Products,*
*Amazon National Institute (INPA), Amazon,*
[3]*Laboratory of Natural Products, Farmanguinhos,*
*Oswaldo Cruz Institute Foundation (FIOCRUZ), Rio de Janeiro,*
*Brazil*

## 1. Introduction

Natural products have been very important to ensure the survival of the man, since the ancient times, especially as remedies to treat different diseases. Today, despite the development of new therapies and new ways of drug development (combinatorial chemistry, ie); natural products continue to play a highly significant role in the drug discovery and development process. (Newman, Cragg, 2007).

Even though fewer drugs have been approved as therapeutical agents lately, nature still inspires the drug development for neglected diseases (malaria, tuberculosis and leishmania) and alternative therapies such as phytotherapy. In both cases, medicinal plants, plants that have been used by the folk medicine for years, are mostly studied. The World Health Organization (WHO) recognized the importance of phytotherapy and the conservation of medicinal plants that stated "the importance of conservation is recognized by WHO and its Member States and is considered to be an essential feature of national programmes on traditional medicines" (Akerele, 1991).

The successful use of some medicinal plants by local population for years, in many cases for centuries, in the treatment of diseases or symptoms associated to some diseases is the basis of the development of drugs or other therapeutical products from them. For instance, artemisinin, a very potent antimalarial, including for drug-resistant malaria strains, was isolated from *Artemisia annua* L., a plant from the traditional Chinese medicine used as remedy for chills and fever for more than 2000 years (Agtmael *et al.*, 1999).

On the other hand, there is an increasing interest for medicines from nature. This interest in products of plant origin is due to several reasons as possible side-effects from synthetic drugs and the awareness that "natural products" are harmless. The world market for phytomedicinal products was estimated in U$ 10 billion in 1997, with an annual growth of 6.5%. In Germany, 50% of phytomedicinal products are sold on medical prescription and the cost being refunded by health insurance. This includes pharmaceutical formulations as plant extracts or purified fractions called phytomedicines or herbal remedies. In many countries, phytomedicines or herbal remedies are controlled as synthetic drugs and they have to fulfill the same criteria of efficacy, safety and quality control (Rates, 2001).

However the quality control of phytomedicines poses a significant challenge due to the complexity of a vegetable extract and column chromatography has proved to be a very helpful and powerful technique. The quality control of *Gingko biloba* L. formulations is good example of this challenge. *Gingko* leaves contains as active compounds flavonoids and terpene lactones (gingkgolides and bilobalide) along with long-chain hydrocarbons, alicyclic acids, cyclic compounds, sterols, carotenoids, among others. Most of the quality control of *Gingko* preparations are based on column chromatography and that was reviewed elsewhere already ( Sticher, 1992; van Beek, 2002).

Column chromatography, especially high performance liquid chromatography (HPLC), has been extensively used in the quality control of plant extracts and phytomedicines formulations, because of its characteristics. The chosen technique must be able to identify the interested compounds (active principles) that are normally not volatile and, in some cases, occur at very small concentrations. Ideally this technique should also be capable to quantify the interested compounds, so one can establishes dosages for the phytomedicine formulation. The required efficiency and selectivity for qualitative and quantitative analysis of the effective components can be achieved by HPLC. Li *et al.* (2011) have recently reviewed the use of different chromatography techniques, such as HPLC, in the quality control of Chinese medicine.

Although, HPLC is a very powerful technique applied in the quality control of medicinal plants, it is necessary to properly identify the active principles of the medicinal plant. This is achieved combining the use of HPLC or other separation technique with a biological test. The search for antimalarials from medicinal plants is one of the most successful examples of this combination as mentioned earlier. In the Amazon region, there are a large number of plants popularly used against malaria or associated symptoms (fever for instance). Milliken (1997) has identified over hundreds of antimalarial plants used by local population in the Amazon region. Many of these plants remained up to now without a study that could confirm their antimalarial activity.

From the fewer plants studied so far, *Picrolemma sprucei* Hook. f., has been studied by our research group. Herein we described the use of HPLC in the quality control of the antimalarial quassinoids, neosergeolide and isobrucein B, the active principles of this species.

*Picrolemma sprucei* Hook. f. (*P. pseudocoffea* Ducke is a commonly cited pseudonym) is a widely distributed and important Amazonian medicinal plant. It is known in the Amazon region by common names which call attention to its resemblance to the coffee plant: *sacha-café* in Peru (Duke & Vasquéz 1994), *caferana* in Brazil (Silva et al. 1977) and *café lane* or *tuukamwi* in French Guiana (Grenand et al. 1987). Infusions of roots, stems, and leaves of *P. sprucei* are traditionally used in different dosages and preparations for the treatment of malaria fevers (Bertani et al. 2005, Vigneron et al. 2005, Milliken 1997), gastrointestinal problems and intestinal worms (Moretti et al. 1982, Duke & Vasquéz 1994). Also, the sale of this plant is sometimes restricted by local vendors due to its use in provoking spontaneous abortions.

Studies on the biological activity of infusions and other derivatives of *P. sprucei* have shown that extracts of this plant have important antimalarial and antihelminthic activities. Bertani *et al.* (2005) reported that a *P. sprucei* leaf infusion inhibited 78 % of *Plasmodium yoelli* rodent malaria growth *in vivo* at a dosage of 95 mg/kg. Furthermore, these same authors reported that of a total of 36 preparations from 25 traditionally used antimalarial plants from French Guiana, *P. sprucei* leaf infusion had the greatest *in vitro* activity against the human malaria parasite *Plasmodium falciparum* (median inhibition concentration, $IC_{50}=1.43$ µg.mL$^{-1}$). These results indicate *P. sprucei* leaf extracts have potential as antimalarials.

In 2006, Nunomura *et al.* showed that water and ethanol extracts of *P. sprucei* at concentrations of 1.3 g.L$^{-1}$ were lethal (90-95 % mortality) *in vitro* towards larvae of the nematoide species *Haemonchus contortus* (Barber Pole Worm), a gastrointestinal nematode parasite found in domestic and wild ruminants. These studies lend support to popular assertions that infusions and other derivatives of *P. sprucei* have important antimalarial and antihelminthic activities.

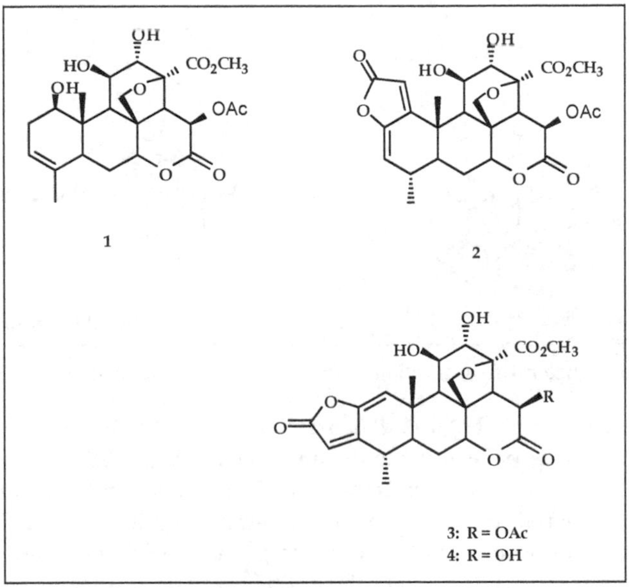

Fig. 1. Quassinoids from *Picrolemma sprucei* Hook.

Two quassinoids have been isolated from *P. sprucei* roots, stems and leaves and identified as isobrucein B (**1**) (Moretti et al. 1982) and neosergeolide (**2**) (Schpector *et al.*1994, Vieira *et al.* 2000). Quassinoid is the name given to any of a number of bitter substances found exclusively in the Simaroubaceae family (Polonsky 1973). Early reports on *P. sprucei* composition from French Guiana (Moretti *et al.* 1982) described the isolation of sergeolide

(3), a structural isomer of 2 and a derivative, 15-deacetylsergeolide (4) (Polonsky *et al.* 1984), from the leaves. Since confirmation of the structure of 2 by x-ray crystallography (Schpector *et al.* 1994) and the systematic application of two-dimensional NMR techniques to the identification of components of *P. sprucei* (Vieira *et al.* 2000, Andrade-Neto *et al.* 2007), neither sergeolide nor its derivative have ever again been described and may be erroneous structures.

Chemically, quassinoids are degraded triterpene compounds which are frequently highly oxygenated. Many quassinoids exhibit a wide range of biological activities *in vitro* and/or *in vivo*, including antitumor, antimalarial, antiviral, anti-inflammatory, antifeedant, insecticidal, amoebicidal, antiulcer and herbicidal activities. For instance, bruceantin (5), brusatol (6), simalikalactone D (7), quassin (8) and glaucarubinone (9) are some of the most well-studied quassinoids and exhibit a wide range of biological activities (Guo *et al.* 2005).

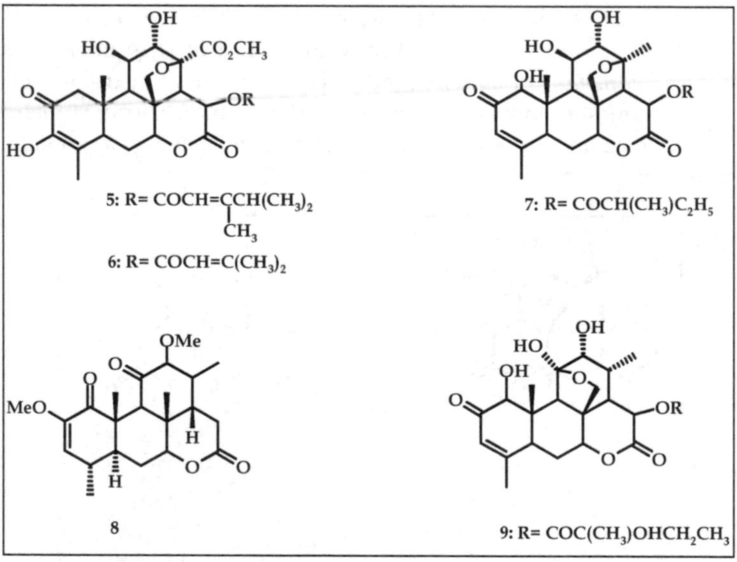

Fig. 2. Quassinoids bruceantin, simalikalactone D, quassin, brusatol and glaucarubinone.

Isobrucein B (Fandeur et al. 1985) and neosergeolide (Andrade-Neto *et al.* 2007) display significant *in vitro* antimalarial activity to the human malaria parasite *P. falciparum*. Recently, the *in vitro* antimalarial activities of isobrucein B and neosergeolide were shown to be comparable to antimalarial drugs quinine and artemisinin (Silva *et al.* 2009). According to this same *in vitro* study, isobrucein B and neosergelide are as cytotoxic or as much as an order of magnitude more cytotoxic than the antitumor drug doxorubicine towards several human tumor strains. Additionally, isobrucein B has been shown to have important antileukemic, antifeedant and leishmanicidal (Moretti *et al.* 1982; Nunomura, 2006).

Bertani *et al.* (2005) conveyed concern about the toxicity of infusions and other preparations based on different parts of *P. sprucei* which is recognized in Amazonian traditional medicine in general. Additionally, these authors were critical of the absence of knowledge of the toxicity of infusions prepared from this species and lack of information available on the

quassinoid composition of these infusions in the study on toxicity published by Fandeur *et
al.* (1985), which focused only on the acute toxicity and antimalarial activity of isolated
quassinoid components o *P. sprucei* and not on toxicity and antimalarial activity of infusions.
Additional studies are needed to prove the *in vivo* efficacy and pharmacological activity of
these infusions as antimalarials with focus on the dose-effect and dose-response to define
the levels of toxicity. The aim of the present study was to develop a method for the
quantification of isobrucein B and neosergeolide in *P. sprucei* root and stem infusions based
on reversed-phase high performance liquid chromatography (HPLC) and ultraviolet
detection (UV).

## 2. Materials and methods

### 2.1 Reagents and solvents

Acetonitrile, HPLC grade, was purchased from Mallinckrodt Baker, Inc. (Xalostoc, Mexico).
The water used in all experiments was purified on a Milli-Q Plus System (Millipore,
Bedford, MA, USA).

### 2.2 Isolation of isobrucein B (1) and neosergeolide (2)

Two collections were performed on the main campus of the University of Amazonas, in
Manaus, Amazonas State, Brazil, in January and July of 1999. Voucher specimens are
deposited at the UFAM Herbarium (Silva 5729 & 5730) and INPA Herbarium (223883).
Identification was performed by Dr. Wayt Thomas as *Picrolemma sprucei* Hook. f. (Wayt
Thomas, personal communication). Roots and stems were cut into small pieces while fresh
and allowed to dry in the shade and were then ground. Air-dried powdered stems (890 g)
were extracted 3 times by maceration in hexanes at room temperature (1 week per
extraction). After concentration, hexane extract (4.79 g) was obtained. Next, the stems were
repeatedly infused in boiling water (Polonsky 1982) which resulted in aqueous solution (20
L). The aqueous solution was concentrated (2.0 L) then continuously extracted with
chloroform (40 h), that after total evaporation yielded chloroform extract (10.8 g).
Chloroform extract was purified on a column of silica gel which was eluted first with
chloroform (100 %), then a gradient of chloroform/methanol 99:1–70:30 (600 mL), 70:30–
50:50 (600 mL), 50:50–25:75 (600 mL), and 25:75–100 % methanol (600 mL) and resulted in
171 collected fractions that were combined based on thin-layer chromatography (TLC)
analysis to yield 11 fractions. Fraction 9 (1.87 g) was purified on a column of silica gel which
was eluted first with 100 % hexane, then 80:20 hexane/chloroform (180 mL), 15:80:5
hexane/chloroform/acetone (800 mL), 10:80:10 hexane/chloroform/acetone (400 mL),
10:70:20 hexane/chloroform/acetone (1440 mL), 10:60:30 hexane/chloroform/acetone (500
mL), 10:50:40 hexane/chloroform/acetone (720 mL), acetone (500 mL), and methanol (500
mL) which resulted in 69 fractions that were combined based on TLC analysis. Combined
fraction 42-50 (360 mg) was re-crystallized from methanol/water and yielded colorless
crystals which were identified as pure **2** (73.9 mg) based on their spectral properties. The
supernatant was re-dissolved in methanol and the insoluble material was washed and
filtered resulting in **1** (62.0 mg), a white solid, which was identified based on its spectral
properties. The isolation of **1** and **2** yielded 0.57% and 0.68 %, respectively. The compounds
**1** and **2** were identified on the basis of their IV, MS and NMR ($^1$H, $^{13}$C, HOMOCOSY,
HMQC, HMBC and NOESY experiments) spectra analysis.

## 2.3 Preparation of root and stem infusions

*P. sprucei* infusions were prepared based on a popular recipe which is used to provoke spontaneous abortions and with which toxic effects are associated according to locals. Stems are the part most commonly used in these remedies. Shade-dried, ground root or stem (9.0 g) was placed in a beaker and boiling deionized water (1.0 mL) was added. The beaker was covered and allowed to stand for 10 min. After this time, the contents of the beaker was filtered hot in a funnel with filter paper which resulted in root and stem infusions. A single infusion was prepared from powdered, dried roots and another from powdered stems obtained from mature plants.

## 2.4 Calculation of extractives

Infusion as prepared above was totally evaporated using rotary evaporation under vacuum and a heated bath (< 50 °C), then freeze-drying. The resulting dry extract was weighed and divided by the mass of plant material used (9.0 g) in the preparation of each infusion and expressed as a percentage (w/w) of extractives.

## 2.5 Preparation of samples of infusions for HPLC analysis

Freeze-dried extracts were dissolved in water to yield final concentrations of stem and root extracts of 445 and 911 mg.L$^{-1}$, respectively.

## 2.6 Preparation of standard solutions of isobrucein B (1) and neosergeolide (2)

Stock solutions of **1** and **2** were prepared at 0.63 g.L$^{-1}$ and 0.50 g.L$^{-1}$, respectively, in methanol. Calibration standards were obtained by appropriate dilution of the stock solutions with methanol. For **1**, the concentrations used in calibration were 100, 50, 25, 10 and 5.0 mg.L$^{-1}$. For **2**, the concentrations used in calibration were 20, 10, 5.0 and 2.5 g.L$^{-1}$. All standard solutions were stored at -20 °C until analysis and protected from light, remaining stable for at least three months.

## 2.7 Apparatus and chromatographic conditions

The liquid chromatography system consisted of an LC-10 Shimadzu, with a SPD-10A UV detector, LC-10AVp quaternary pump, SIL- 10A autosampler and a CBM-10A system controller (Kyoto, Japan). A Supelcosil LC-18 analytical column (250 mm × 4.6 mm i.d., 5 µm particle size) from Supelco (Bellefonte, PA, USA) was used for separation of **1** and **2**. The mobile phase consisted of a gradient of acetonitrile:0.05 % aqueous trifluoroacetic acid delivered at 1.0 mL.min$^{-1}$ as follows: initial (t$_i$=0 min) 10:90, then linear gradient over 20 min to 25:75, and this composition was maintained (isocratic) until the end of each run (t$_f$=30 min). Flow rate was 1 mL.min$^{-1}$. Quantification was performed using the detector set at a wavelength of 254 nm. Injection volume was 50 µL.

## 2.8 Analysis of Infusions by HPLC-UV and calibration curve

Chromatograms of pure **1** and **2** presented retention times of approximately 14 and 25 min, respectively. The peaks corresponding to **1** and **2** were identified in each chromatogram of the infusions with the help of injection of the standard solutions of **1** and **2** or with co-elution (figure 3).

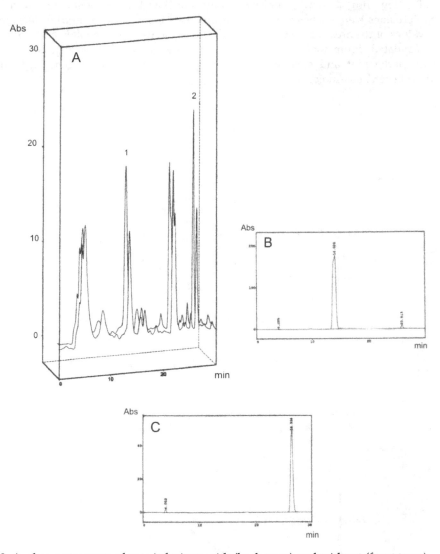

Fig. 3. A: chromatograms of root infusions with (back trace) and without (front trace)
addition of neosergeolide and isobrucein B at 254 nm. B: chromatogram of pure isobrucein B
($t_R$= 14.0 min) at 254 nm. C=chromatogram of neosergeolide ($t_R$= 25.3 min) at 254 nm.

Several injections of standard solution were performed and then average areas were calculated
for each individual concentration injected for isobrucein B (1) and neosergeolide (2). The
calibration curves in the determination of 1 and 2 in *P. sprucei* stem and root infusions (Figure
4A and 4B, respectively) used in the determination of these components in *P. sprucei* stem and
root infusions were obtained by linear regression performed on the average areas versus
standard sample concentrations Y and X, respectively (figure 3) at 254 nm.

After calibration with standard samples of isobrucein B and neosergeolide, *P. sprucei* root and stem infusions were analyzed. Samples of infusions were analyzed in triplicate and the average values of the areas corresponding to the quassinoids neosergeolide and isobrucein B were calculated. From these average areas, the concentration of each quassinoid was calculated in the root and stem infusions using the linear equation generated during calibration of each quassinoid.

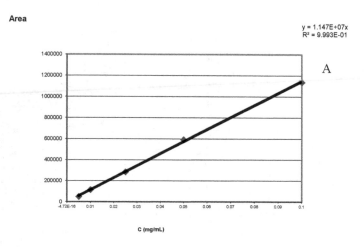

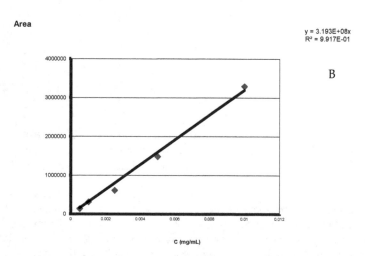

Fig. 4. A: Calibration curve of isobrucein B (**1**); B: Calibration curve of neosergeolide (**2**).

## 3. Results and discussion

The quassinoids isolated from *P. sprucei* were identified by NMR techniques and compared to literature (Moretti, *et al.* 1982, Vieira, *et al.* 2000). The chemical shifts of NMR $^1$H and $^{13}$C of **1** and **2** are presented in tables 1 and 2 respectively.

| Carbon | δ (C) | δ (H) | δ Literature (CDCl₃/Py-5%)[1] | |
|--------|-------|-------|------|------|
| | | | C | H |
| 1 | 81.1 | 4.17 (s) | 81.3 | 4.26 |
| 2 | 197.0 | - | 197.6 | - |
| 3 | 124.3 | 6.11 (q; 2.8; 1.0) | 124.5 | 6.11 |
| 4 | 163.0 | - | 162.6 | - |
| 5 | 51.6 | 2.92 (d; 12.1) | 43.4 | 2.91 |
| 6 | 28.5 | 1.86 (ddd; 14.7; 12.1; 2.6); 2.40 (ddd; 14.7; 2.8; 2.8) | 28.2 | 1.86; 2.41 |
| 7 | 83.1 | 4.75 (d) | 81.7 | 4.75 |
| 8 | 45.5 | - | 45.8 | - |
| 9 | 42.8 | 2.34 (d; 4.0) | 42.4 | 2.38 |
| 10 | 47.5 | - | 47.7 | - |
| 11 | 72.4 | 4.75 (sl) | 74.3 | 4.75 |
| 12 | 75.8 | 4.28 (s) | 75.1 | 4.12 |
| 13 | 80.5 | - | 81.7 | - |
| 14 | 43.5 | 3.04 (d; 12.4) | 52.3 | 3.03 |
| 15 | 66.6 | 6.31 (sl) | 67.8 | 6.30 |
| 16 | 167.0 | - | 167.6 | - |
| 29 | 22.5 | 1.96 (d; 1.0) | 22.4 | 1.95 |
| 19 | 11.6 | 1.18 (s) | 11.3 | 1.18 |
| 30 | 73.3 | 3.75 (dd; 7.7; 2.0); 4.81 (d; 7.7) | 73.0 | 3.75; 4.81 |
| 21 | 172.6 | - | 169.5 | - |
| 1' | 169.0 | - | 170.7 | - |
| 2' | 20.4 | 2.09 (s) | 20.5 | 2.08 |
| OMe (5') | 20.4 | 3.84 (s) | - | 3.83 |
| H-(OH-1) | - | 4.54 (s) | - | - |
| H-(OH-12) | - | 3.25 (s) | - | - |

[1]Moretti et al. (1982).

Table 1. Chemical shifts in NMR $^1$H (500 MHz, CDCl₃) and NMR $^{13}$C (125 MHz, CDCl₃) of isobrucein B (**1**).

| Carbon | δ (C) | δ (H) | δ Literature (CDCl₃/CD₃OD-5%)[1] | |
|---|---|---|---|---|
| | | | C | H |
| 1 | 162.8 | - | 161.89 | - |
| 2 | 149.7 | - | 148.91 | - |
| 3 | 115.9 | 5.71 (dd, 2.0, 2.0) | 116.76 | 5.76 (t, 2.0) |
| 4 | 31.60 | 2.40 (m) | 31.07 | 2.41 (ddq, 1.6, 2.0, 6.8) |
| 5 | 45.70 | 1.82 (d, 2.0) | 45.26 | 1.86 (ddd, 1.6, 12.4, 14.0) |
| 6a | 29.0 | 2.22 (dd, 2.0, 2.0) | 28.80 | 2.27 (dt, 14.0, 2.0) |
| 6b | | 1.76 (dd, 2.0, 2.0) | | 1.73 (dt, 14.0, 2.0) |
| 7 | 83.6 | 4.85 (d, 2.0) | 83.55 | 4.76 (d, 2.0) |
| 8 | 47.5 | - | 46.74 | - |
| 9 | 39.6 | 2.52 (d, 2.0) | 39.06 | 2.34 (brd, 4.0) |
| 10 | 42.7 | - | 42.06 | - |
| 11 | 75.0 | 4.59 (d, 6.0) | 73.60 | 4.56 (d, 4.0) |
| 12 | 77.1 | 4.32 (d, 6.0) | 76.00 | 4.23 (brs) |
| 13 | 82.3 | - | 81.74 | - |
| 14 | 51.1 | 3.29 (ddd, 15.0, 2.0, 2.0, 12.0) | 49.63 | 3.24 (brd, 11.0) |
| 15 | 68.5 | 6.04 (d, 12.0) | 67.00 | 6.09 |
| 16 | 167.5 | - | 168.24 | - |
| 18 | 170.5 | - | 171.25 | - |
| 19 | 18.6 | 1.64 (s) | 18.12 | 1.60 (s) |
| 29 | 19.9 | 1.18 (d, 6.0) | 19.55 | 1.20 (d, 6.8) |
| 30a | 74.2 | 4.78 (d, 8.0) | 73.6 | 4.78 (d, 8.0) |
| 30b | | 3.81 (dd, 8.0, 2.0) | | 3.77 (d, 8.0) |
| 1′ | 169.0 | - | 170.40 | - |
| 2′ | 20.6 | 1.98 (s) | 20.49 | 2.09 (s) |
| 3′ | 171.7 | - | 171.56 | - |
| 4′ | 113.4 | 6.26 (s) | 113.17 | 6.26 (d, 2.0) |
| OMe (5′) | 52.9 | 3.78 (s) | 53.02 | 3,84 (s) |

[1] Vieira, et al. (2000)

Table 2. Chemical shifts in NMR ¹H (200 MHz, CDCl₃) and NMR ¹³C (50 MHz, CDCl₃) of neosergeolide (**2**).

The authenticity of standards is a key-step in quantitative analysis, especially in plant extract analysis. In most cases, authentic standards are not available commercially and this strengths the importance of liquid chromatography. Liquid chromatography enables the isolation of authentic standards at different scales (from microgram until gram scale) and at very high purity that can be used later to perform quantitative analysis. In our study, combining open-column and planar chromatography, we were able to isolate several milligrams of each pure standard, as can be observed at figure 3, that were then used in the quantitative analysis of the quassinoids 1 and 2 in root and stem infusions of *P. sprucei* by HPLC.

The structural authenticity of each standard can be confirmed by the use of modern spectroscopy techniques as MS and NMR. Although these techniques are considered complementary, normally NMR is much more informative. For instance, in our study, HMBC experiments furnished conclusive information that not sergeolide, but neosergeolide was isolated.

As described in the experimental section, samples of stem and root infusions were prepared using approximately 9 g of crushed, dried stems are infused with 1 L of boiling water. HPLC analysis of *P. sprucei* stem and root infusions resulted in the concentrations presented in table 3.

| Quassinoid | Root infusion | | Stem infusion | |
|---|---|---|---|---|
| | mg.L$^{-1}$ | µM | mg.L$^{-1}$ | µM |
| isobrucein B (1) | 32.0 | 67.0 | 14.0 | 29.0 |
| neosergeolide (2) | 0.79 | 1.6 | 0.38 | 0.75 |

Table 3. Concentrations of isobrucein B (1) and neosergeolide (2) in *P. sprucei* stem and root infusions determined by HPLC-UV at 254 nm.

Consistent with the data presented in table 1 the concentrations of both 1 and 2 are at least twice as large in root infusions as in stem infusions. Interestingly, the percentage of extractives of roots during infusion (5.1 %) is twice that of stems (2.5 %) which would seem to be related to the greater concentration of these constituents in the root infusion.

Comparison of root and stem infusions shows that 1 is about 40 times as concentrated as 2 in both stem and root teas on a molar basis. These data suggest that the more relevant active principle in stem and root infusions analyzed is 1.

## 4. Conclusion

The HPLC analysis of infusions (aqueous extracts) of stems and roots of *P. sprucei* revealed higher quantities of isobrucein B than neosergeolide, 40 fold, for both infusions. Considering this information and *in vitro* activity of both compounds, it is very likely that isobrucein B plays more important role for the antimalarial activity than neosergeolide.

More research is needed to describe seasonal, regional and specimen specific variation in *P. sprucei* quassinoid composition which should have a direct influence on the composition of stem and root infusions prepared from samples of different origins. Knowledge of the extent of these variations, especially as they influence quassinoid composition in infusions, is of fundamental importance given the valuable medicinal and dangerous toxic properties of these widely used Amazonian remedies.

The high performance liquid chromatography has proved to be a powerful tool in the plant extract analysis. The possibility to perform qualitative and quantitative analysis, by HPLC, enables the development of new phytoterapeutical products from the Amazon biodiversity.

## 5. Acknowledgment

The autors wish to thank Dr Wanderli Pedro Tadei and CNPq/ FAPEAM - PRONEX, Rede Malária for financial support to the publication of this chapter, Massuo Kato for use of analytical HPLC apparatus and Profs. Norberto P. Lopes and Valquiria P. Jabor for helpful comments regarding this manuscript.

## 6. References

Agtmael, M.A.; Eggelte, T.A.; Boxtel, C.J. (1999). Artemisin drugs in the treatment of malaria: from medicinal herb to registered medication. *Trends in Plant Science*, vol. 20, pp. 199-205, ISSN 1366-2570.

Andrade-Neto, V.F.; Pohlit, A.M.; Pinto, A.C.S.; Silva, E.C.C.; Nogueira K.L.; Melo, M.R.S; Henrique, MC; Amorim, R.C.N.; Silva, L.F.R.; Costa, M.R.F.; Nunomura, R.C.S.; Nunomura, S.M.; Alecrim, W.D.; Alecrim, M.G.; Chaves, F.C.M.; Vieira, P.P.R. (2007). *In vitro* inhibition of *Plasmodium falciparum* by substances isolated from Amazonian antimalarial plants. *Memorias do Instituto Oswaldo Cruz* vol. 102, No. 3, pp. 359-365, ISSN 0074-0276

Bertani, S.; Bourdy, G.; Landau, I.; Robinson, J.C.; Esterre, P.; Deharo, E. (2005). Evaluation of French Guiana traditional antimalarial remedies. *Journal of Ethnopharmacology* Vol. 98, pp. 45-54, ISSN 0378-8741.

Duke, J. A.; Vasquez, R. (1994). *Amazonian ethnobotanical dictionary*, CRC Press, ISBN 0849336643, Florida (USA), 215 pp.

Fandeur, T.; Moretti, C.; Polonsky, J. (1985). *In vitro* and *in vivo* assessment of the antimalarial activity of sergeolide. *Planta Medica,* vol. 51, pp.20-23, ISSN 0032-0943.

Grenand P, Moretti C, Jacquemin H 1987. *Pharmacopées traditionnelles en Guyane, Guyane Française.* Collection Mémoires 108, ORSTOM, Cayenne, French Guiana, 569 pp.

Guo, Z.; Vangapandu, S.; Sindelar, R.W.; Walker, L.A.; Sindelar, R.D. (2005). Biologically active quassinoids and their chemistry: Potentianl leads for drug design. *Current Medicinal Chemistry,* Vol. 12, pp. 173-190, ISSN 1568-0142.

Li, Y.; Wu,T.; Zhu, J.; Wan, L.; Yu, Q.; Li, X.; Cheng, Z.; Guo, C. (2010). Combinative method using HPLC fingerprint and quantitative analyses for qualitaty consistency evaluation of an herbal medicinal preparation produced by different manufaturers.

*Journal of Pharmaceutical and Biomedical Analysis*, vol. 52, pp. 597-602, ISSN 0731-7085.

Li, Y.; Zhao, J.; Yang, B. (2011). Strategies for quality of Chinese medicines. *Journal of Pharmaceutical and Biomedical Analysis*, vol. 55, pp. 802-809, ISSN 0731-7085.

Milliken, W. (1997). *Plants for malaria. Plants for fever: medicinal species in Latin America - a bibliographic survey*, Kew Publishing, United Kingdom, 122 pp.

Moretti, C.; Polonsky, J.; Vuilhorgne, M.; Prange, T. (1982). Isolation and structure of sergeolide, a potent cytotoxic quassinoid from Picrolemma pseudocoffea. *Tetrahedron Letters*, Vol. 23, pp. 647-650, ISSN 0040-4039.

Nunomura, R.C.S.; Silva, E.C.C.; Oliveira D.F.; Garcia, A.M.; Boeloni, J.N.; Nunomura, S.M.; Pohlit, A.M. (2006). In vitro studies of the anthelmintic activity of *Picrolemma sprucei* Hook. f. (Simaroubaceae). *Acta Amazonica*, Vol. 36, pp. 327-330, ISSN 0044-5967.

Rates, S.M.K. (2001). Plants as source of drugs. *Toxicon*, vol. 39, pp. 603-613. ISSN 0041-0101.

Schpector, J. Z.; Castellano, E.E.; Rodrigues, Filho E.; Vieira, I.J.C. (1994). A new quassinoid isolated from *Piclolemma pseudocoffea*. *Acta Crystallographica C*, Vol. 50, pp. 794-797, ISSN 0108-2701.

Silva, E.C.C. (2006). *Isolamento, transformação química e atividade biológica in vitro dos quassinóides neosergeolida e isobruceína B*. Dissertação de Mestrado, Universidade Federal do Amazonas, Manaus, Amazonas, 191 pp.

Silva E.C.C.; Cavalcanti, B.C.; Amorim, R.C.N.; Lucena, J.F.; Quadros, D.S.; Tadei, W.P.; Montenegro, R.C.; Costa-Lotufo, L.V.; Pessoa, C.; Moraes, M.O.; Nunomura, R.C.S.; Nunomura, S.M.; Melo, M.R.S.; Andrade-Neto, V.F.; Silva, L.F.R.; Vieira, P.P.R.; Pohlit, A.M. (2009). Biological activity of neosergeolide and isobrucein B (and two semi-synthetic derivatives) isolated from the Amazonian medicinal plant *Picrolemma sprucei* (Simaroubaceae). *Memorias do Instituto Oswaldo Cruz*, Vol. 104, pp. 48-55, ISSN 0074-0276.

Silva, M.F.; Lisbôa, P.L.B.; Lisbôa, R.C.L.; (1977). *Nomes vulgares de plantas amazônicas*. INPA, Belém, PA, Brasil, 222 pp.

Sticher, O. (1993). Quality of Ginkgo Preparations. *Planta Medica*, 59, pp 2-11, ISSN 0032-0943.

Street, R.A.; Stirk, W.A.; Van Staden, J. (2008). South African traditional medicinal plant trade – Challenges on regulating quality, safety and efficacy. *Journal of Ethnopharmacology*, vol.119, pp. 705-710, ISSN 0378-8741

Van Beek, T.A. (2002). Chemical analysis of *Ginkgo biloba* leaves and extracts. *Journal of Chromatography A*, vol. 967, pp. 21-55, ISSN 0021-9673.

Vieira, I.J.C.; Rodrigues-Filho, E.; Fernandes, J.B.; da Silva, M.F.G.F.; Vieira, P.C. (2000). Complete [1]H and [13]C chemical shift assignments of a new $C_{22}$-quassinoid isolated from *Picrolemma sprucei* Hook. by NMR spectroscopy. *Magnetic Resonance in Chemistry*, Vol. 38, pp. 805-808, ISSN 0749-1581.

Vigneron, M.; Deparis, X.; Deharo, E.; Bourdy, G. (2005). Antimalarial remedies in French Guiana: a knowledge attitudes and practices study. *Journal of Ethnopharmacology*, Vol. 98, pp. 351-360, ISSN 0378-8741.

WHO (1998). Quality control methods for medicinal plant materials. ISBN 92 4 154510 0, Geneva (Switzerland), 127 pp.

# 11

# Purification of Peptides from *Bacillus* Strains with Biological Activity

María Antonieta Gordillo and María Cristina Maldonado
*Institute of Biotechnology, Faculty of Biochemistry, Chemistry and Pharmacy,*
*National University of Tucuman, Tucuman,*
*Argentina*

## 1. Introduction

One of the greatest causes of loss in the food industry is postharvest diseases of fruits and vegetables (Vero Mendez & Mondino, 1999). According to U.S.A. estimations this loss reaches 20 - 25% whereas in developing countries the losses are often more severe due to inadequate storage and transportation facilities (Sharma et al., 2009), but loss has generally been considered to be approximately 10% to 40% depending on packinghouse technology (Vero et al, 2002).

The surface of the fruit or vegetables is covered with fungal spores, bacterial cells and yeasts, which they have acquired from the air during their development on the parent plant, or which they have come in contact during picking or any of the stages of handling the harvested produced. However not every fungal spore or bacterial cell can develop and cause decay in the harvested product, even when conditions suitable for penetration and development are present.

Harvested fruit and vegetables are naturally attacked by own typical pathogens. Fungi are the principal decaying agents in fruit kept in cold storage chambers for long periods (Teixidó et al., 2001).

Fruits of tropical and subtropical origin (mango, papaya, avocado, etc.) are attacked by *Colletotrichum gloesporoides*, which cause actracnose. *Gloesporium musae* attacks banana fruits at the orchard, which becomes active only during the storage.

*Diplodia natalensis*, *Phomopsis citri* or *Dothiorella gregaria* invade the cut stem of tropical and subtropical fruit. *D. natalensis*, *Alternaria citri* and *P. citri* the causal agent of postharvest stem-end rot of citrus fruits.

During and after harvest, the citrus fruits are typically attacked by *Penicillium italicum* (blue mold), *Penicillium digitatum* (green mold), *Geotrichum candidum* (sour rot), *Alternaria citri* (black mold) and *Fusarium* sp. (Wilson & Wisniewski, 1989). *P. digitatum* is an example of a specific fungus that attacks only citrus fruits. *P. italicum* can attack other fruits and vegetables, whereas *P. expansum*, apple and pear pathogen, naturally attack citrus fruits.

*P. expansum, Botrytis cinerea, Gleosporium* spp. *Alternaria alternata* and *Stemphylium botryosum* are typical apple and pear pathogens. The main pathogens of peaches, apricots, nectarines and plums are *Monilia fruticola* and *Rhizopus stolonifer*.

Each harvested fruits and vegetables has its own group of characteristic pathogens to which is susceptible and for which its serves as suitable host.

Strawberries are attacked during the storage by the gray mold fungus (*B. cinerea*) and the soft watery rot fungus (*R. stolonifer*).

*Alternaria alternata* is the major storage decay agent of harvested tomatoes. The main causal agents of soft watery rot in harvested tomatoes are *R. stolonifer, G. candidum* and *Erwinia* spp. (Barkai Golan, 2001).

Postharvest fruit diseases are controlled with careful manipulation practices and synthetic fungicides like 2- 4 thiazalil benzimidazole and imidazole. This method is more widely used against fungal decay because of its low cost and easy application. However, it presents manifold objections since prolonged use generates resistance to synthetic fungicides by major postharvest pathogens (Wilson & Wisniewski, 1994; Fogliata et al., 2001) and increases chemical remainders in fruits with the consequent potentiality engendering iatrogenic diseases (Lingk, 1991). In addition, world trends are moving towards reduced pesticide use in fresh fruit and vegetables. Along with this trend, several physical and biological means have been evaluated as safer alternatives for the use of chemical fungicides. The use of microbial antagonists for the control of postharvest diseases received special attention, and has been extensively investigated (Droby, 2006).

Most of the reported yeast and bacteria antagonists were naturally occurring on fruit surfaces. Microbial biocontrol agents of postharvest diseases have been criticized mainly for not providing as consistent or broad-spectrum control as synthetic fungicides. The "first generation" of biological controls agents for postharvest spoilage relied on the use of single antagonists. Perhaps it is unrealistic for us to expect disease control comparable to synthetic fungicides by the use of single antagonists. It can be expected that enhancing efficacy of biocontrol agents of postharvest diseases to an acceptable level would utilize a combination of different biological and physical means (Droby, 2006).

The mechanism(s) by which microbial antagonists exert their influence on the pathogens has not yet been fully understood. It is important to understand the mode of action of the microbial antagonists because; it will help in developing some additional means and procedures for better results from the known antagonists. It will also help in selecting more effective and desirable antagonists or strains of antagonists (Wilson & Wisniewski, 1989; Wisniewski & Wilson, 1992).

Several modes of action have been suggested to explain the biocontrol activity of microbial antagonist. Still, competition for nutrient and space between the pathogen and the antagonist is considered as the major modes of action by which microbial agents control pathogens causing postharvest decay (Filonow, 1998; Ippolito et al., 2000; Jijakli et al., 2001). In addition, production of antibiotics (antibiosis), direct parasitism, and possibly induced resistance are other modes of action of the microbial antagonists by which they suppress the activity of postharvest pathogens on fruits and vegetables (Janisiewicz et al., 2000; Barkai-Golan, 2001; El-Ghaouth et al., 2004).

## 2. Lipopeptides produced by *Bacillus* strain

Members of the *Bacillus* genus are often considered microbial factories for the production of a vast array of biologically active molecules potentially inhibitory for phytopathogens growth, such as kanosamine or zwittermycin A from *B. cereus*. Their spore-forming ability also makes these bacteria some of the best candidates for developing efficient biopesticide products from a technological point of view. *Bacillus* spores have a high level of resistance to the dryness necessary for formulation into stable products. *Bacillus subtilis* strains are a rich source of antimicrobial peptides with a high potential for biological control applications.

*Bacillus subtilis* has been used for genetic and biochemical studies for several decades, and is regarded as paradigm of Gram-positive endospore-forming bacteria (Moszer et al., 2002). Several hundred wild-type *B. subtilis* strains have been collected, with the potential to produce more than two dozen antibiotics with an amazing variety of structures. All of the genes specifying antibiotic biosynthesis combined amount to 350 kb; however, as no strain possesses them all, an average of about 4–5% of a *B. subtilis* genome is devoted to antibiotic production. Peptide antibiotics, also named lipopeptides, represent the predominant class. They exhibit highly rigid, hydrophobic and/or cyclic structures with unusual constituents like D-amino acids and are generally resistant to hydrolysis by peptidases and proteases (Katz & Demain, 1977). Furthermore, cysteine residues are either oxidized to disulphides and/or are modified to characteristic intramolecular C–S (thioether) linkages, and consequently the peptide antibiotics are insensitive to oxidation (Stein, 2005).

*Bacillus* lipopepdides are synthesized non-ribosomally via large multi-enzymes (non-ribosomal peptide synthetases, NRPSs) (Kowall et al., 1998; Stein, 2005, Finking & Marahiel, 2004). These biosynthetic systems lead to a remarkable heterogeneity among the lipopeptides products generated by *Bacillus* with regards to the type and sequence of amino acid residues, the nature of the peptide cyclization and the nature, length and branching of the fatty acid chain (Ongena & Jacques, 2007).

Lipopeptides are classified into three families depending on their amino acid sequence: iturins, fengycins and surfactins (Fig. 1) (Perez García, et al., 2011). The surfactins are powerful biosurfactants, which show antibacterial activity but no marked fungitoxicity (with some exceptions) (Ongena & Jacques, 2007).

Surfactin, a cyclic lipopeptide is one of the most effective biosurfactants know so far, which was first reported in *B. subtilis* ATCC-21332. Because of its exceptional surfactant activity it is named as surfactin. Surfactin can lower the surface tension from 72 to 27.9 mN/m and have a critical micelle concentration of 0.017 g/L. The surfactin groups of compounds are shown to be a cyclic lipoheptapeptide which contain a β - hidroxy fatty acid in its side chain. Recent studies indicate that surfactin shows potent antiviral, antymicoplasma, antitumoral, anticoagulant activities as well as inhibitors of enzymes. Although, such properties of surfactins qualify them for potential applications in medicine or biotechnology, they have not been exploited extensively till date.

Lichenysin, produced by *Bacillus licheniformis* exhibited similar structure and physicochemical properties to that of surfactin. *B. licheniformis* also produced several other surface active agents which act synergistically and exhibit temperature, pH and salt stability. Lichenysin A produced by *Bacillus licheniformis* strain BAS50, is characterized to

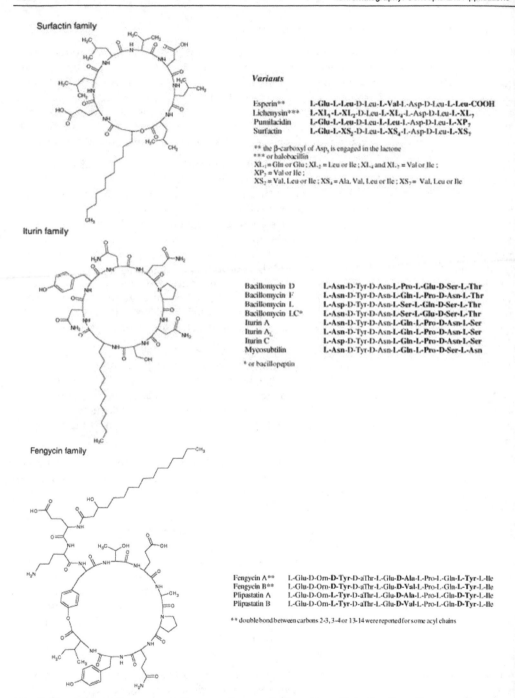

**Surfactin family**

**Variants**

| | |
|---|---|
| Esperin** | L-Glu-L-Leu-D-Leu-L-Val-L-Asp-D-Leu-L-Leu-COOH |
| Lichenysin*** | L-XL₁-L-XL₂-D-Leu-L-XL₄-L-Asp-D-Leu-L-XL₇ |
| Pumilacidin | L-Glu-L-Leu-D-Leu-L-Leu-L-Asp-D-Leu-L-XP₇ |
| Surfactin | L-Glu-L-XS₂-D-Leu-L-XS₄-L-Asp-D-Leu-L-XS₇ |

** the β-carboxyl of Asp, is engaged in the lactone
*** or halobacillin
XL₁= Gln or Glu ; XL₂ = Leu or Ile ; XL₄ and XL₇ = Val or Ile ;
XP₇ = Val or Ile ;
XS₂ = Val, Leu or Ile ; XS₄ = Ala, Val, Leu or Ile ; XS₇= Val, Leu or Ile

**Iturin family**

| | |
|---|---|
| Bacillomycin D | L-Asn-D-Tyr-D-Asn-L-Pro-L-Glu-D-Ser-L-Thr |
| Bacillomycin F | L-Asn-D-Tyr-D-Asn-L-Gln-L-Pro-D-Asn-L-Thr |
| Bacillomycin L | L-Asp-D-Tyr-D-Asn-L-Ser-L-Gln-D-Ser-L-Thr |
| Bacillomycin LC* | L-Asn-D-Tyr-D-Asn-L-Ser-L-Glu-D-Ser-L-Thr |
| Iturin A | L-Asn-D-Tyr-D-Asn-L-Gln-L-Pro-D-Asn-L-Ser |
| Iturin A₁ | L-Asn-D-Tyr-D-Asn-L-Gln-L-Pro-D-Asn-L-Ser |
| Iturin C | L-Asp-D-Tyr-D-Asn-L-Gln-L-Pro-D-Asn-L-Ser |
| Mycosubtilin | L-Asn-D-Tyr-D-Asn-L-Gln-L-Pro-D-Ser-L-Asn |

* or bacillopeptin

**Fengycin family**

| | |
|---|---|
| Fengycin A** | L-Glu-D-Orn-D-Tyr-D-aThr-L-Glu-D-Ala-L-Pro-L-Gln-L-Tyr-L-Ile |
| Fengycin B** | L-Glu-D-Orn-D-Tyr-D-aThr-L-Glu-D-Val-L-Pro-L-Gln-L-Tyr-L-Ile |
| Plipastatin A | L-Glu-D-Orn-L-Tyr-D-aThr-L-Glu-D-Ala-L-Pro-L-Gln-D-Tyr-L-Ile |
| Plipastatin B | L-Glu-D-Orn-L-Tyr-D-aThr-L-Glu-D-Val-L-Pro-L-Gln-D-Tyr-L-Ile |

** double bond between carbons 2-3, 3-4 or 13-14 were reported for some acyl chains

Fig. 1. Structures of representative members and diversity within the three lipopeptides
families synthesized by *Bacillus* species. Boxed structural groups are those that were shown

to be particularly involved in interaction with membranes and/or are supposed to be important for biological activity in addition to the cyclic nature of the molecule (Peypoux, et al, 1999; Bonnatin et al., 2003; Dufour et al., 2005; Ongena & Jacques, 2007).

contain a long chain beta-hydroxy fatty acid molecule. Iturin A, the first compound discovered of the iturin group and its best known member, was isolated from a *Bacillus subtilis* strain taken from the soil in Ituri (Zaire) and its structure was elucidated. The subsequent isolation from other strains of *Bacillus subtilis* of five other lipopeptides such as iturin $A_L$, mycosubtilin, bacillomycin L, D, F and $L_C$ (or bacillopeptin) was reported (Ongena & Jacques, 2007). All have a common pattern of chemical constitution, led to the adoption of the generic name of "iturins" for this group of lipopeptides. The iturin groups of compounds are cyclic lipoheptapeptides which contain a β-amino fatty acid in its side chain. Lipopeptides belonging to the iturin family are potent antifungal agents which can also be used as biopesticides for plant protection.

Fengycin is a lipodecapeptide containing β-hydroxy fatty acid in its side chain and comprises of $C_{15}$ to $C_{17}$ variants which have a characteristic Ala-Val dimorphy at position 6 in the peptide ring. Wang et al. (2004) demonstrated the identification of fengycin homologues produced by *B. subtilis* by using electrospray ionization mass spectrometry (ESI-MS) technique.

These antibiotics are either cyclopeptides (iturins) or macrolactones (fengycins and surfactins) characterized by the presence of L and D amino acids and variable hydrophobic tails. Iturins display strong antifungal action against a wide variety of yeasts and fungi but only limited antibacterial activity. Fengycins also show a strong fungitoxic activity, specifically against filamentous fungi (Ongena & Jacques, 2007). The ability of various *Bacillus* strains to control fungal soil borne, foliar and postharvest diseases has been attributed mostly to iturins and fengycins (Ongena & Jacques, 2007; Romero et al., 2007; Arrebola et al., 2010).

The surfactin family encompasses structural variants but all members are heptapeptides interlinked with a β-hydroxy fatty acid to form a cyclic lactone ring structure (Peypoux et al., 1999). Because of their amphiphilic nature, surfactins can also readily associate and tightly anchor into lipid layers and can thus interfere with biological membrane integrity.

Iturin A and C, bacillomycin D, F, L and LC and mycosubtilin were described as the seven main variants within the iturin family. They are heptapeptides linked to a β-amino fatty acid chain with a length of 14 to 17 carbons. The biological activity of iturins is different to surfactins: they display a strong in vitro antifungal action against a wide variety of yeast and fungi but only limited antibacterial and no antiviral activities (Moyne et al., 2001; Phae et al., 1990). This fungitoxicity of iturins almost certainly relies on their membrane permeabilization properties (Deleu et al., 2003).

However, the underlying mechanisms based on osmotic perturbation owing to the formation of ion-conducting pores and not membrane disruption or solubilization as caused by surfactins (Aranda et al., 2005).

## 3. Purification and identification of lipopeptides

Chemical and structural analysis of lipopeptides is carried out using a broad range of techniques varying from simple colorimetric assays to sophisticated mass spectrometry (MS) and sequencing techniques.

After extraction, the purification procedures of lipopeptides included chromatography methods (TLC, Ion Exchange Chromatography and RP-HPLC). Each step of purification will be monitored by bioassays. The bioassays could be bioautographic methods, dual culture plate, etc.

Generally, identification of the relative percentage of the lipid and protein portions is carried out using simple colorimetric assays, such as Bradford assay for protein determination and spectroscopic methods (FTIR). The molecular mass determination of the compounds of interest may be facilitated by mass spectrometry using assisted laser desorption ionization time of flight mass spectrometry (MALDI-TOF MS).

This should be followed by analysis of the fatty acid portion and determination of the peptide sequence using automated Edman degradation sequencing. This combined approach would provide the necessary information required for complete structural identification.

## 3.1 Purification of lipopeptides

### 3.1.1 Extraction and Thin Layer Chromatography (TLC)

Different solvents are used for extraction of lipopeptides from cell free supernatant. The solvents used are n-hexane, ethyl acetate, petroleum ether, chloroform and methanol, to determine the best solvent for extraction of antifungal compound (Kumar et al., 2009). Yazgan et al., 2001 and Romero et al., 2007 used n-butanol for lipopeptides extraction.

The lipopeptide produced by cultures of *Bacillus mojavensis* strain ROB-2 was used to compare the efficiencies of two purification methods. Method 1 involved acid precipitation using 1 N HCl to adjust the pH of the cell-free culture fluid to 2 (Yakimov et al, 1995; Mc Keen et al., 1986) followed by TLC. Method 2 used ammonium sulfate precipitation (40%) followed by acetone extraction and TLC (Youssef et al., 2005).

Seventy-five percent of the biosurfactant activity remained in the cell-free culture fluid after cell removal. The surface-active fraction obtained from the TLC plate by method 1 had 23% ± 7% of the activity originally presents in the culture, while the surface-active fraction obtained from the TLC plate by method 2 had 63% ± 11% of the activity originally presents in the culture (Youssef et al., 2005).

In order to identify the compounds responsible for antimicrobial activity, extracts of cell-free culture filtrates *Bacillus* strains generally are separated in TLC sheets, using purified iturin A, fengycin, and surfactin as standards. Thin layer chromatography is a simple method, which can be used to detect the presence of lipopeptides while preparative TLC can be used to purify small quantities (Symmank et al., 2002).

Once the butanol layer completely evaporated in vacuum at 40° C or the acid precipitation is neutralized, the residue is dissolved in methanol for further chemical analysis. The methanolic fractions are analyzed using TLC (Razafindralambo et al., 1993) with direct view developed using distilled water spraying. A white spot formed with the same Rf value when the plate was sprayed with water, indicating that the compound is lipophilic.

Cell free supernatant of *Bacillus subtilis* UMAF6614, UMAF6619, UMAF6639 and UMAF8561 and *Bacillus amyloliquefaciens* PPCB004 were evaluated by TLC and the spots with Rf values similar to the standards fengycin (0.09), iturin A (0.3), and surfactin (0.7) (Romero et al., 2007 and Arrebola et al., 2010). To determine which lipopeptides were directly involved in fungal inhibition, the bioautographies were performed using the pathogens as revealing microorganism. It was found that the principal inhibitor was iturin A, which affected all fungi analyzed in this study. Fengycin was also identified as an inhibitor of *Lasiodiplodia theobromae, Botryosphaeria* sp., *C. gloeosporioides, Fusicoccumaromaticum* and *Phomopsis persea.* Surfactine was able to inhibit *L. theobromae,* although slight inhibition of *F. aromaticum* and *P. persea* was also observed (Fig. 2) (Arrebola et al., 2010).

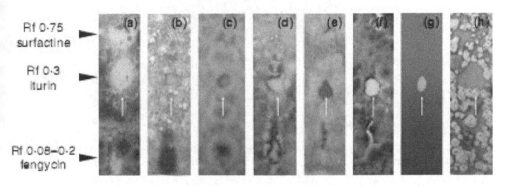

Fig. 2. Bioautographic analysis on thin-layer chromatography plates where lipopeptide extracts from *Bacillus amyloliquefaciens* PPCB004 were separated and the Rf of every lipopeptide are indicated. (a) *Lasiodiplodia theobromae,* (b) *Alternaria citri,* (c) and (d) *Botryosphaeria* sp., (e) *Colletotrichum gloeosporioides,* (f) *Fusicoccum aromaticum,* (g) *Penicillium crustosum,* (h) *Phomopsis persea.* Fungal inhibition zones produced by iturin A activity is indicated with white arrows.

The concentrated extracts from *Bacillus* MZ-7 run on silica TLC plates showed six bands under UV light, having Rf values of 0.1, 0.15, 0.26, 0.37, 0.51 and 0.57. However, a plate bioassay showed two active fractions, those with Rf values of 0.37 and 0.51. The spot with an Rf value of 0.51 was ninhydrin negative and positive to 4,4'-bis (dimehtylamino) diphenylmethane (TDM) reagent. These results indicated the absence of free amino groups and the presence of peptide. The migration and chemical properties of the compound were comparable to surfactin produced by *B. subtilis* strain ATCC 21332. The environmental isolate *B. subtilis* MZ-7 produced more surfactin (170.5 mg/L) than did *B. subtilis* ATCC 21332 (109.5 mg/L) under the same conditions (Mutaz et al., 2007).

Maldonado et al., 2009 run silica gel plates 60 F254 (Merck, 2 mm) and are carried out with a chloroform–methanol–acetic acid (40:4:1) mixture (Batrakov et al., 2003). Plates are developed under UV light at 254 and 365 nm and only one spot of Rf 0.67 is detected (Kumar et al., 2009). The inhibitory activity of the spot was confirmed after TLC by bioautographic assay. Besides the TLC plates were developed with ninhydrin and no spot was observed. Thus, a peptide without free amino groups (cyclic structure) may be presumed (Fig. 3).

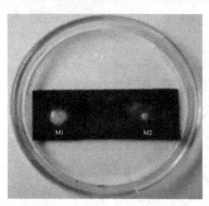

Fig. 3. Bioautographic analysis on thin-layer chromatography plates where lipopeptides extracts from *Bacillus* IBA 33 against *Geotrichum candidum*. $M_1$ and $M_2$ correspond to peaks obtained from DEAE Sephacel chromatography.

### 3.1.2 Ion exchange chromatography and RP-HPLC

The development of efficient HPLC methods generally needs relatively pure biosurfactant samples, which cannot be obtained without tedious isolation and purification operations.

Other technique frequently employed for the characterization and quantification of biosurfactants has been DEAE ion exchange chromatography. This analysis can provide qualitative information about biosurfactants.

The crude extract was applied to a low-pressure chromatography column (Bio-Rad Laboratories, Hercules, CA) filled with DEAE-Sephacel (Pharmacia,Uppsala, Sweden) equilibrated in 50 mM Tris-HCl, pH 7.5. A gradient of 0-0.4 M NaCl in buffer (50 mM, pH 7.5) was run through the column at 0.5 mL/min over a period of 160 min followed by straight 0.4 M NaCl in buffer (50 mM, pH 7.5) for 40 min. The eluent was monitored at 280 nm, and fractions were collected every 5 min (Bechard et al., 1998).

They demonstrated that the antibiotic was retained by DEAE-Sephacel quite strongly but eluted from the matrix in 0.4 M NaCl. Ion exchange chromatography separated the crude antibiotic sample into four peaks. Antibiotic activity corresponds with the fourth and largest peak on the chromatogram. Only 23% of the initial activity was recovered by ion exchange chromatography, and the specific activity actually decreased relative to that in previous purification step. A complete loss of activity was observed when the antibiotic was left in the ion exchange eluent overnight. For this reason the ion exchange protocol was carried out as quickly as possible at 4 °C (Bechard et al., 1998) (Fig.4).

Wang et al., 2004 loaded the crude extract on a DEAE-Sepharose Fast Flow column pre-equilibrated with buffer A (20 mmol/L $Na_2HPO_4$-$NaH_2PO_4$ pH 6.5). After washing the column with buffer A, buffer B (20 mmol/L $Na_2HPO_4$-$NaH_2PO_4$ 1 mol/L $(NH_4)_2SO_4$ pH 6.5) was used to elute the fraction containing fengycin homologues and this fraction was directly loaded on a SOURCE 15 PHE hydrophobic interaction column pre-equilibrated with buffer B. Most impurities were washed by buffer A and fengycin homologues were finally eluted by Milli Q water. After adjusted to 30% acetonitrile/water by adding acetonitrile, this fengycin-containing fraction was ready for ESI mass spectrometry analysis.

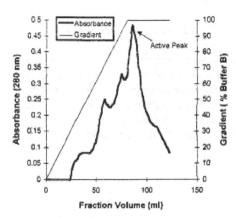

Fig. 4. Elution pattern of the HCl precipitate from DEAE-Sephacel equilibrated in 50 mM Tris buffer, pH 7.5 (buffer A). The peptide was eluted from the column with a gradient of 0.100% buffer B (50 mM tris buffer, pH 7.5, containing 0.4 M NaCl). Flow rate 0.5 mL/min.

Bin Hu et al., 2007 collected the extract, boiled for half an hour, centrifuged at 10.000 g for 10 min and then applied to a DEAE-52 column (16 mm 15 cm) previously equilibrated with 50 mM Tris-HCl buffer (pH 7.5) containing 0.05 M NaCl. Antifungal fractions were obtained from the elution with 0.5–0.7 M NaCl in the same buffer, concentrated by ultrafiltration with a PM10 membrane (Amicon) and placed onto a Sephadex G-100 gel column (15 mm 80 cm). The column was equilibrated with 10 mM ammonium acetate buffer and eluted with the same buffer at a flow rate of 0.5 mL/min. Fractions containing antifungal active compounds were collected, concentrated and lyophilized. Through these purification steps of antifungal active compounds, fractions were determined by the absorbance at 280 nm and the anti *F. moniliforme* fractions underwent further processing.

Maldonado et al. (2009) precipitated the extracted from *Bacillus* IBA 33 with ammonium sulfate 40%, then loaded onto a DEAE-Sephacel (1.5 cm 9.15 cm) column previously equilibrated with 50 mM Tris–HCl buffer pH 7.5 (Bechard et al., 1998). After purification, the chromatogram showed two inhibition peaks in a DEAE-Sephacel column. The first one, exhibited 77% and the second peak had 64% inhibition against *G. candidum* (Fig.5).

HPLC is an excellent method for the separation of lipopeptides (Aguilar, 2004). The most commonly employed technique is reversed phase chromatography, which results in the separation of each lipopeptide structure based on polarity. The separated products are detected by UV absorbance detection and each individual peak can be collected using a fraction collector for further analysis of their structure. Coupling of HPLC with a mass spectrometer provides preliminary information on the molecular mass of each component. Purification with either HPLC-UV or HPLC-MS using different types of column chemistry is also possible.

Gueldner et al., 1988 assayed the crude material dissolved in 50:50 methanol-water, and the solution was chromatographed on a column of C-18 reversed-phase absorbent (Waters Prep Pak 500). Elution with a stepwise gradient of methanol and water (from water up to 80% methanol-water) eluted most of the lipopeptides. Further purification was achieved by droplet counter current chromatography (DCCC) (Buchi B-670) using chloroform-methanol-water (7:13:8) in the descending mode with the chloroform layer as the mobile phase. The main peaks eluted in 300-500 mL of the chloroform layer. The active peptides started to elute from the reversed-phase adsorbent at 65-70% ethanol-water and were completely eluted with 80% methanol - 20% water. Bioassay of the C-18 column fractions with *M. fructicola* showed activity in fractions that contained several peaks as analyzed by HPLC (210 nm). Five HPLC peaks having a profile very similar to those published by Isogai et al. (1982) for *B. subtilis* metabolites and collected from an analytical C-18 column were also active in the *M. fructicola* bioassay. Peak 1 (PKI), corresponding to Iturin A-2.

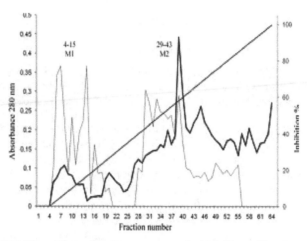

Fig. 5. *Bacillus* sp. IBA 33 antifungal active compounds (AAC) purification step by DEAE-Sephacel chromatography. Dark line: protein (absorbance at 280 nm), light line: AAC (% inhibition).

Mutaz et al., 2007 realized quantitative analysis of surfactin by HPLC active fractions from TLC were further purified by reversed-phase HPLC, using a Thermo Hypersil-Keystone ODS (particle size, 5 μm; column dimensions, 250 by 4.6 mm; Thermo Hypersil, USA). A sample was applied with eluent A (0.1% (vol/vol) trifluoroacetic acid and 20% (vol/vol) acetonitrile) and eluted with segmented gradients of eluent B (0.1% (vol/vol) trifluoroacetic acid and 80% (vol/vol) acetonitrile) as follows: 40% eluent B for 30 min and 40 to 100% eluent B for 10 min. The concentration of surfactin was determined from a calibration curve made by correlating the emulsification index (E 24) with known amounts of surfactin produced by *B. subtilis* ATCC 21332.

Romero, et al., 2007 analyzed first by TLC (Razafindralambo et al. 1993) and afterward by RP-HPLC, using an analytical Zorbax C18 column, 4.6 mm in diameter by 150 mm long (Agilent, Palo Alto, CA, U.S.A.) and solutions of 0.05% trifluoroacetic acid in acetonitrile and in water, with a flow rate of 1 ml/min. The different groups of peaks from butanolic extracts were fractionated by Flash chromatography as described earlier (Deleu et al., 1999; Razafindralambo et al., 1993), followed by (semi)-preparative RP-HPLC using a Vydak C18 column, 22 mm in diameter by 250 mm long (Separations Group, Hesperia, CA, U.S.A.) and the solutions mentioned above with a flow rate of 23 ml/min.

RP-HPLC analysis showed three main groups of peaks were observed at elution times comparable with those observed for standard lipopeptides that correspond to iturin A, fengycin and surfactin (Romero et al., 2007).

Arrebola et al., 2010 performed the analysis by reverse-phase HPLC (RP-HPLC) (Romero et al., 2007) using extracts from *Bacillus subtilis* UMAF6614 and UMAF6639 as standards of lipopeptides production.

The lipopeptide concentration was calculated using the Folin Ciocalteu's reaction (Swain & Hillis, 1959; Harborne, 1984). The results from RP-HPLC analysis showed three main groups of compounds that correspond to bacillomycin, produced by UMAF6614 and iturin A fengycin and surfactin by UMAF6639 (Fig. 6).

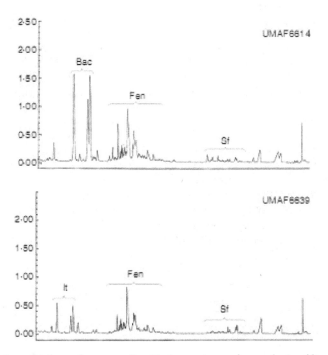

Fig. 6. Reverse-phase high-performance liquid chromatography analysis of lipopeptides produced by *Bacillus subtilis* strains UMAF6614 and UMAF6639. Peaks corresponding to iturin (It), bacillomycin (Bac), fengycin (Fen) and surfactin (Sf) are indicated. Detection is at 214 nm.

## 3.2 Identification of lipopeptides

### 3.2.1 Ultraviolet spectrum

The ultraviolet absorbance spectrum of the antibiotic was measured from the HPLC chromatogram using a Waters 990 photodiode array detector. This device allowed for the simultaneous measurement at all wavelengths from 200 to 600 nm as the antibiotic eluted from the column. The absorbance spectrum for the lipopeptide is measured in acetonitrile/1% acetic acid (68:32) between 200 and 600 nm. The antibiotic shows absorbance maxima at 235, 278, and 285 nm (Fig.7), and there is no appreciable absorbance above 300 nm (Bechard et al., 1998).

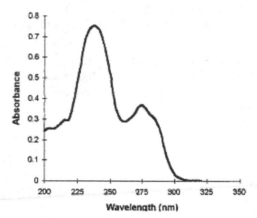

Fig. 7. Ultraviolet spectrum of the peptide antibiotic inacetonitrile/1% acetic acid (68:32).

Maldonado et al, 2009 resuspended the samples ($M_1$ and $M_2$) obtained from *Bacillus* IBA in HEPES buffer 10 mM NaCl 100 mM pH 7.1 to carry out absorption spectra between 250 and 280 nm wavelengths (Beckman DU 7500). They observed the same absorption bands for both samples. Hence, this strain might be a producer of cyclic lipopeptides with antifungal activity belonging to the iturin family. When the wavelength scanning was performed between 250 and 380 nm, they founded absorbance maximums at 280 nm for both samples (Fig. 8 a, b). Furthermore they might infered that the antifungal compounds have tyrosine or tryptophan or both in their composition (Nelson & Cox, 2006). Bechard et al., 1998 reported an absorbance maximum between 210 and 230 nm which they thought was due to the

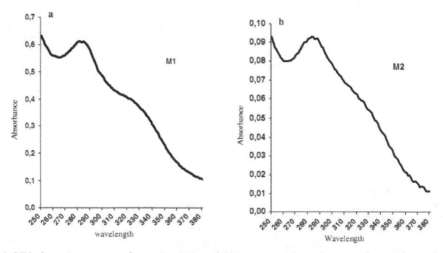

Fig. 8. UV absortion spectra between 250 and 280 nm wavelengths a: peak one, b: peak two.

presence of tyrosine. However, they later discovered that it was a residue of α-aminoacid 4 hydroxyphenilacetic which is structurally similar to tyrosine.

Kumar et al., 2009 dissolved 1 mg extract in 10 ml of methanol and the spectra were recorded at 190-600 nm range. UV spectral data of antibiotic exhibited strong absorption maxima (λ max) at 254, 255 and 277 nm and there was no appreciable absorbance above 300 nm, which was corresponding to characteristic absorption of peptide bond. It is reported that most of peptide antibiotics exhibit absorbance maxima at 210-230 and 270-280 nm (Motta & Brandelli, 2002; Kurusu & Ohba, 1987). A peptide antibiotic cerein, obtained from *Bacillus cereus*, shows UV absorbance peak at 250 and 273 nm.

### 3.2.2 Infrared spectrum (FTIR)

The infrared spectrum of the antibiotic was measured as a potassium bromide pellet. Approximately one-third of the KBr mixture was pressed into a pellet (8 mm diameter) using a non-evacuable Mini Press (Perkin-Elmer Norwalk, CT) and four scans of the sample were taken using a Perkin-Elmer 1600 FTIR spectrophotometer. The FTIR spectrum of the purified antibiotic is shown in (Fig.9). Characteristic absorption valleys at 1,540; 1,650, and 3,300 cm-1 indicate that the antibiotic contains peptide bonds. A lactone ring is suggested by the absorption at 1,740 cm-1 and valleys that result from C- H stretching (2,950; 2,850; 1,460 and 1,400 cm-1) indicate the presence of an aliphatic chain (Bechard et al., 1998).

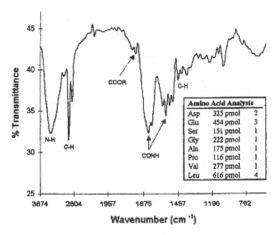

Fig. 9. Fourier transform IR spectrum (KBr pellet) and amino acid composition of the peptide antibiotic.

Romero et al, 2007 performed active extracts from *Bacillus subtilis* strains. The Fourier transform-infrared spectrum (FT-IR) analysis showed bands in the range of 1,630 to 1,680 cm-1, resulting from the stretching mode of the CO-N bond (amide I band), and at 1,570 to 1,515 cm-1, resulting from the deformation mode of the N-H bond combined with C-N stretching mode (amide II band), both indicating the presence of a peptide component and also bands at 2,855 to 2,960 cm-1, resulting from typical CH stretching vibration in the alkyl chain. Also was observed at 1,730 cm-1 due to the lactone carbonyl absorption typical for surfactin and fengycin families of lipopeptides.

Maldonado et al., 2009 showed the FT-IR spectrum of the antifungal compound in $D_2O$, at a concentration of 1 mg/mL. Samples were placed in a liquid cell assembled with $CaF_2$ windows and 0.056 mm lead spacers. The spectrum was taken with a resolution of 2 cm⁻¹ and three regions were observed. The one at 1,650 cm⁻¹ assigned to the vibrational amida I mode which shows the peptide link; another at 1,710–1,740 cm⁻¹ characteristic of carbonyl groups in ester or ketone groups and another at 2,850–2,950 cm⁻¹ bands corresponding to saturated CH links assigned to long chain fatty acids (Fig.10).

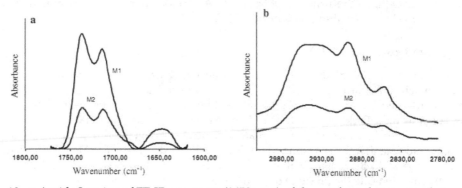

Fig. 10. a: Amida I region of FT-IR spectrum (1650 cm⁻¹) of the antifungal compound at a concentration of 1 mg/mL in $D_2O$. Carbonyl groups (ester or ketone groups) region (1,710–1,740 cm⁻¹) b: Saturated CH links region of FT-IR spectrum (2,850–2,950 cm⁻¹).

Sivapathasekaran et al., 2009 in order to reveal the chemical nature of the biosurfactant, the HPLC purified isoforms from marine *Bacillus* strain (A, B, C and D) were performed by Fourier transform-infrared spectrophotometry (FTIR) analysis. The purified samples were dispersed in spectral-grade KBr (Merck, Darmstadt, Germany) and made into pellets by applying pressure. The spectrum was generated in the range of 400 to 4,000 cm⁻¹ with a resolution of 4 cm⁻¹. The spectral measurement of the compound was carried out in transmittance mode with average of 32 data scans over the entire range of wave numbers (Das & Mukherjee, 2005; Pueyo, 2009).

The IR absorption pattern for fractions A, B, C, and D (Fig. 11) revealed the presence of peptide and carboxyl groups that indicated their lipopeptide nature (Desai & Bannat, 1997; Thaniyavaran et al. 2003; Das & Mukherjee, 2005; Pueyo, 2009). The antimicrobial isoform (fraction A) showed a transmittance valley at 3,274 cm⁻¹ as a result of N-H stretching indicating the peptide groups. Aliphatic chain was indicated by C-H weak stretching vibration observed in the range 2,930 cm⁻¹. The transmittance occurred in 1,658 cm⁻¹ due to the amide I band frequency (C-O stretching in the peptide bond) and transmittance at 1,531 cm⁻¹ range showed C=O bonds. The transmittance at 1,390 cm⁻¹ range may be due to the aliphatic chain of C-H group. Similarly in fraction B, the N-H stretching was observed at 3,477 cm⁻¹ with a broad valley of transmittance. The transmittance at 1,670 cm⁻¹ range resulted in the stretching mode of C-O bond and at 1,143 cm⁻¹ resulted in C-N stretch. The transmittance at 1,203 cm⁻¹ corresponding to C-N stretch and at 1,440 cm⁻¹ and 2,364 cm⁻¹ due to the presence of aliphatic chain (C-H stretching mode) was observed. The absorption at 1,024 cm⁻¹ indicated a C-O stretch. In fraction C, IR spectrum showed absorption at 3,109 cm⁻¹, 1,623 cm⁻¹, and 1,132 cm⁻¹ corresponding to strong absorption band of peptides, resulted from the stretching mode of N–H and C–O bonds, respectively. The presence of CH stretching at 1,398 cm⁻¹ indicated the presence of aliphatic chain. The absorption at 1,198 cm⁻¹ indicated a C–N stretching. The IR spectrum for fraction D revealed absorption corresponding to the presence of peptides at 3,087 cm⁻¹, 1,667 cm⁻¹, and 1,198 cm⁻¹ respectively. The C–H stretching at 1,439 cm⁻¹ and 1,398 cm⁻¹ indicated the aliphatic chain presence. In all the fractions (A–D), the peaks observed at the 800–600 cm⁻¹ range revealed the presence of C–H bend aliphatic chain (Sivapathasekaran et al., 2009).

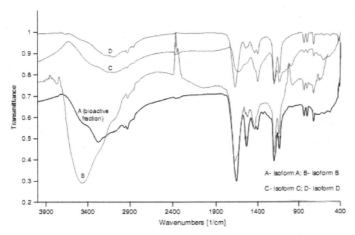

Fig. 11. IR spectrum of HPLC purified surface-active lipopeptide isoforms produced by the marine *Bacillus circulans* DMS-2 (MTCC 8281).

FTIR spectra of antifungal compound had a broad band centering around 3,421.5 cm⁻¹ indicated an amino and hydroxyl group of amino acids (Kumar et al., 2009). Analysis of the spectrum also shows typical absorption bands (1,670.5 and 1,539.8 cm⁻¹) corresponding to N–H stretching of proteins and peptides bonds (Maquelin et al., 2002). Additional absorption valleys 1,418.4 and 1,488.6 cm⁻¹ indicating (C–H) aliphatic side chain may be related with predominance of hydrophobic amino acids such as Val, Leu and Ile or its contains fatty acids in their structure (Bizani et al., 2005).

### 3.2.3 MALDI-TOF mass spectrometry

The lipopeptide molecules are detected, in their protonated form or as Na⁺ or K⁺ adducts, by MALDI-TOF mass spectrometry in the m/z range of 1,400–1,550 Da (Deleu et al., 2008). Several literature reports are available that highlight the analysis and purification of lipopeptide biosurfactants (Desai & Banat, 1997; Maneerat & Phetrong, 2007; Mukherjee et al., 2009; Sen & Swaminathan, 2005). Although methods like ion exchange chromatography (Mukherjee et al., 2006), thin layer chromatography (Desai & Banat, 1997), gel permeation chromatography (Mukherjee et al., 2009) and ultrafiltration (Sen & Swaminathan, 2005; Lin et al., 1998) have been used for the purification of lipopeptide biosurfactants, these techniques have a serious limitation as they do not separate individual isoforms present in the crude lipopeptide mixture.

Mutaz et al., 2007 studied the fractions correlated with surfactin from TLC and reverse phase HPLC using MALDI-TOF-MS. The samples were mixed on the target plate with the matrix solution (α-hydroxycinnaminic acid in acetonitrile-methanol-water, 1:1:1). MALDI-TOF-MS spectra was recorded by using a 337-nm nitrogen laser for desorption and ionization. The mass spectrometer was operated in the refraction mode at an accelerating voltage of 18 kV with an ion flight path that of 0.7 m. The delay time was 375 ns. Matrix-suppression was also used and the mass spectra were averaged over 50 to 100 individual laser shots. The laser intensity was set just above the threshold for ion production. Surfactin isomers were anticipated to have an m/z range of 500–1500. The variance of the m/z of ± 0.8 Da was considered acceptable.

Running MALDI-TOF-MS in refractron mode, was observed a cluster of peaks with mass/charge (m/z) ratios between 1,036 and 1,058, which could be attributed to protonated surfactin isoforms (Fig.12). The peak with a m/z ratio 1,045.86 corresponds to the mass of [M+Na]$^+$ ion of surfactin with a fatty acid chain length of 14 carbon atoms (Huszcza & Burczyk, 2006).

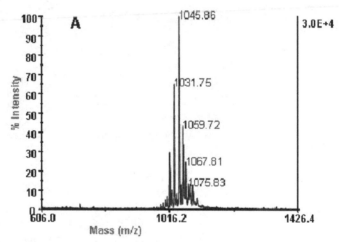

Fig. 12.  Surfactin cluster + Na$^+$obtained by MALDI-TOF-MS.

Romero et al., 2007 confirmed the identification of the antifungal compounds (Lp-a, Lp-b and Lp-c) by scoring the mass spectra contained in the purified fractions using an Ultraflex MALDI-TOF mass spectrometer (Bruker Daltonics, Billerica, MA, U.S.A.) operated in positive ion mode. The samples were prepared according to Williams et al. (2002). Mass spectra were recorded by matrix-assisted laser desorption-ionization time of flight mass spectrometry (MALDI-TOF-MS). The mass spectra of LP-a showed a series of mass number of m/z = 1,030 to 1,074; those of LP-b, m/z = 1,034 to 1,095; and those of LP-c, m/z = 1,435 o 1,499 for UMAF6639 strain (Fig. 13).

Sivapathasekaran et al., 2009 analyzed the HPLC purified isoforms by matrix-assisted laser desorption/ionization time-of-flight analysis (MALDI-TOF) for molecular mass determination. The matrix used for co-crystallization was 2,5-dihydroxybenzoic acid (Sigma, USA). A matrix stock solution was prepared in acetonitrile, methanol, TFA (5:4:1). Equal volume of the purified sample and the matrix were mixed vigorously. After proper mixing,

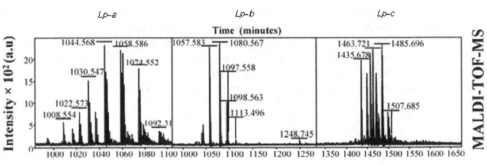

Fig. 13. Mass spectra scored for the purified fractions LP-a, LP-b, and LP-c from *Bacillus subtilis* UMAF6639 strain.

the sample was spotted on the target plate, dried, and was placed inside the sample cabinet of Voyager DE-Pro MALDI-TOF spectrometer (Applied Biosystems Inc, CA, USA). The nitrogen UV laser (337 nm) was used for desorption and ionization and a voltage of 20 kV was maintained to accelerate the molecules. The molecules were separated according to their mass and were detected by the ion detector set in reflector mode (Maneerat & Phetrong, 2007; Leenders et al., 1999). The mass spectral analysis showed that all purified isoforms (fractions A–D) were lipopeptides and belonged to fengycin family (Vater et al., 2002).

The mass spectra of fraction A revealed the presence of a major peak of C16 lipopeptides (1,482 Da and 1,484 Da) in their $Na^+$ adduct form. Similarly, in fraction B showed the presence of a major peak corresponding to protonated C16 lipopeptides (1,464 Da and 1,466 Da). In the surface-active fraction C, C16 lipopeptides were revealed as a major peak in their $H^+$ (1,492 Da) and $Na^+$ (1,514 Da) adduct form, respectively. In the similar manner mass spectra of fraction D, C15 isoform was detected in its protonated form (1,448 Da) and $K^+$adduct (1,480 Da) (Sivapathasekaran et al., 2009).

### 3.2.4 Electrospray ionization/collision induced dissociation (ESI/CID) mass spectrometry (MS)

Fengycin homologues produced by *Bacillus subtilis* JA were analyzed. When each homologue was subjected to ESI/CID analysis, ions representing characteristic fragmentations were detected. These ions can help to identify the homologues; even homologues of the same nominal mass can be discriminated by their ESI/CID spectra. Based on the CID results, fengycin homologues can be correctly assigned (Wang et al., 2004).

They showed the ESI mass spectrum of a group of fengycin homologues purified from the culture of *B. subtilis* JA. Peaks of m/z 1,435.7; 1,449.9; 1463.9; 1477.9; 1,491.8 and 1,506.0 represent different fengycin homologues (Fig.14). Each of these ions was selected as precursor ion for further CID analysis. The results showed the appearance of productions of these precursor ions had regularities: product ions of m/z 966 and 1,080 were found in the CID spectra of precursor ions of m/z 1,435.7; 1,449.9 and 1,463.9; product ions of m/z 994 and 1,108 were found in CID spectra of precursor ions of m/z 1,491.8 and 1,506.0. Product ions at m/z 1,080 and 966 can be explained as neutral loses of (fatty acid –Glu) and (fatty acid –Glu–Orn), respectively, from the N-terminus segment of fengycin A (Fig.15). Ions at m/z 1,063 and 949 found in the spectrum corresponded to neutral loses of ammonia

(−17 Da) from m/z 1,080 and 966. Similarly, product ions with neutral loses of (fatty acid – Glu) and (fatty acid –Glu-Orn) were also found in the CID spectra of precursor ions of m/z 1,491.8 and 1,506.0. But they appeared at m/z 1,108 and 994, exactly 28 Da higher than the corresponding ions of fengycin A (Fig.16). The observation of mass difference reflected the substitution of Ala for Val in the lactone ring and so homologues of m/z 1,491.8 and 1,506.0 belonged to fengycin B. In fact, product ions of m/z 1,209 and 1,237 representing the neutral loses of the fatty acid chain were also detected respectively from the CID spectra of fengycins A and B. But the abundance of these two ions was very low, so sometimes we could not easily find them in the spectra.

An interesting phenomenon was found in the CID result of precursor ion at m/z 1,477.9 (Fig. 17) shows part of the entire CID spectrum). Product ions of m/z 966; 994; 1,080 and 1,108 were all detected. This indicated both fengycins A and B contributed the peak of m/z 1,477.9.

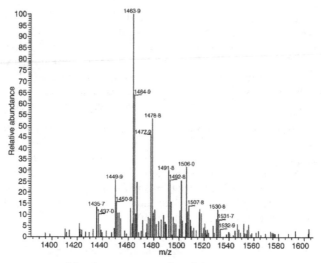

Fig. 14. ESI mass spectrum of the fengycins produced by *B. subtilis*JA.

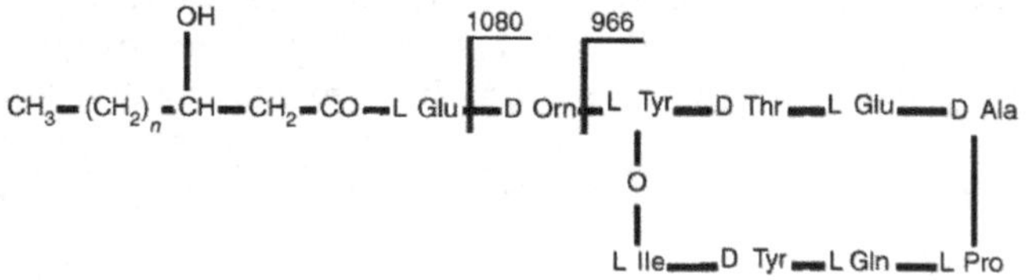

Fig. 15. Structure of fengycin A.

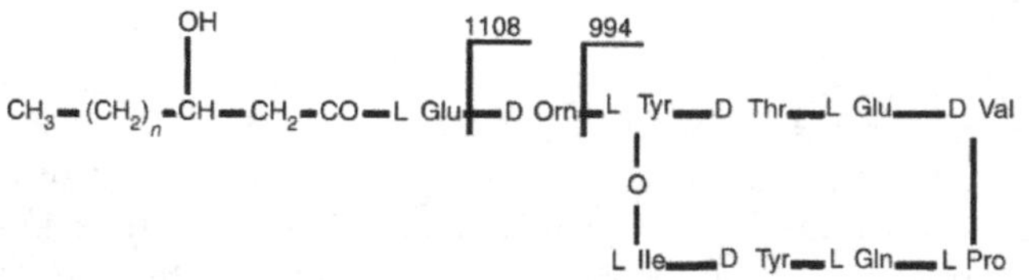

Fig. 16. Structure of fengycin B.

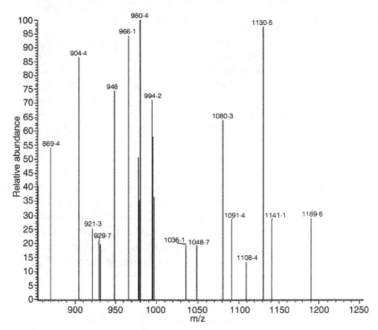

Fig. 17. CID spectrum of precursor ion of m/z 1477.9.

Romero et al, 2007 determined by ESI-MS-MS the amino acid sequences for the peptide moiety. Samples of ring-opened lipopeptides by cleavage of the lactone bond (Williams et al. 2002) were analyzed on an Esquire 3000 Plus ion trap mass spectrometer (Bruker Daltonics) and sequences deduced by comparing the fragmentation spectra with the available databases. When the amino acid compositions were determined, it was found that the fraction LP-a contained Asp, Glu, Val, and Leu in a ratio of 1:1:1:4; fraction LP-b comprised Asp, Ser, Glu, Pro, and Tyr in a ratio of 3:1:1:1:1; and fraction LP-c was composed of Thr, Glu, Pro, and Ala or Val, Ile, Tyr, and Orn in a ratio of 1:3:1:1:1:2:1, corresponding to strain UMAF6639 (Fig. 18).

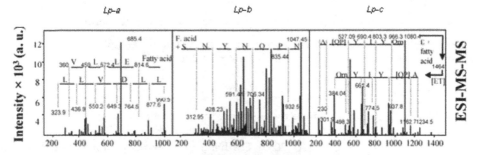

Fig. 18. Mass spectra scored for the purified fractions LP-a, LP-b, and LP-c.
Fragmentation spectra observed by electrospray ionization ion trap mass spectrometry (ESI-MS-MS) and amino acid sequences and fatty acid compositions deduced from the parental peaks 1,008.7 (LP-a, surfactin), 1,065.4 (LP-b, iturin A), and 1,464 (LP-c, fengycin).

### 3.2.5 Edman degradation

Edman degradation is the classic technique for sequencing peptides using chemical methods (Zachara & Gooley, 2000). The method provides an assignment for each residue in the peptide, unlike amino acid analysis which only provides an indication of the ratio of amino acids in the peptide. Edman experiments take place in an oxygen-free environment and involve the modification of the N-terminal residue with phenylisothiocyanate to provide a cleaved phenylthiohydantoin (PTH) amino acid. All of the chemical processes take place on automated sequencers and are followed by a chromatographic step where the retention time of the cleaved PTH amino acid is compared with the retention times of a series of PTH modified amino acid standards to ascertain its identity. The method is relatively slow taking approximately 45 min for each residue, however the amino acid assignment is called with a high degree of confidence and less subject to interpretation, as can be the case with MS/MS methods. The quality of sequence information obtained by the Edman method is subject to the amount of starting material and its purity. For successful sequencing to take place, peptides and proteins must be purified to near homogeneity by chromatographic methods to prevent mixed sequencing during Edman experiments. Lipopeptides needs to be in the open ring form for this type of analysis, which is carried out using mild alkaline hydrolysis.

## 4. Perspectives and conclusions

A great deal of research has been carried out the properly methods for fully characterization of cyclic lipopeptides and their structures.

This review highlights the competitive advantage of efficient purification of surfactin, fengycin and iturin with structural elucidation. Within each family, some structural homologues are seemingly more active than others. In this regard, reverse-phase high-performance liquid chromatography (RP-HPLC) could be extensively used because it is efficient in separation and purification of isoforms (Lin et al., 1998; Thaniyavaran et al., 2003). However the HPLC method, if not optimized, may lead to improper resolution of peaks and longer elution times. Thus, an efficient high-resolution HPLC method is a prerequisite for the purification up to the individual isoform level, required for subsequent commercialization of a particular lipopeptide isoform as a potential therapeutic agent. The products obtained with these methods shows a greater level of purity and profound biological activity (Sivapathasekaran et al., 2009).

Mass spectrometry methods developed to rapidly characterize the lipopeptides nature.

The lipopeptides and moreover the particular isoforms could be exploited as a powerful tools for the selection of useful strains in the context of biocontrol.

## 5. References

Aguilar, M. (2004). Methods in molecular biology. In: *HPLC of Peptides and Proteins; Methods and Protocols*. Aguilar ML, Ed. Totowa, NJ: Humana Press.

Aranda, F.J., Teruel, J.A. & Ortiz, A. (2005). Further aspects on the hemolytic activity of the antibiotic lipopeptide iturin A. *Biochim. Biophys. Acta* 1713: 51–56.

Arrebola, E., Jacobs, R. & Korsten, L. (2010). Iturin A is the principal inhibitor in the biocontrol activity of *Bacillus amyloliquefaciens* PPCB004 against postharvest fungal pathogens. *J. Appl. Microbiol.* 108: 386-395.

Barkai Golan, R. (2001). Each fruit and vegetables and its characteristic pathogens. In: *Postharvest diseases of fruits and vegetables: development and control.* pp 25-26. Ed. Elsevier. ISBN 0-444-50584-9. The Netherlands.

Batrakov, S., Rodionova, T., Esipov, S., Polyakov, N., Sheichenko, V., Shekhovtsova, N., Lukin, S., Panikov, N. & Nikolaev, Y. (2003). A novel lipopeptide, an inhibitor of bacterial adhesion, from the thermophilic and halotolerant subsurface *Bacillus licheniformis* strain 603. *Biochim Biophys Acta* 1634: 107–115.

Bechard, J., Eastwell, K., Sholberg, P., Mazza, G. & Skura, B. (1998). Isolation and Partial Chemical Characterization of an Antimicrobial Peptide Produced by a Strain of *Bacillus subtilis. J. Agric. Food Chem,* 46: 5355-5361.

Bin Hu, L., Zhi Qi Shi, Zhang, T. & Yang, Z. (2007). Fengycin antibiotics isolated from B-FS01 culture inhibit the growth of *Fusarium moniliforme* Sheldon ATCC 38932. *FEMS Microbiol Lett* 272: 91–98.

Bizani, D., Dominguez, A. & Brandelli, A. (2005). Purification and partial chemical characterization of the antimicrobial peptide cerein. *J.Applied Microbiol,* 41: 269.

Bonmatin, J., Olivier, L. & Franoise, P. (2003). Diversity among microbial cyclic lipopeptides: iturins and surfactins. Activity-structure relationships to design new bioactive agents. *Comb. Chem. High Throughput Screen.* 6: 541–556.

Das, K. & Mukherjee A. (2005). Correlation between diverse cyclic lipopeptides production and regulation of growth and substrate utilization by *B. subtilis* strains in a particular habitat. *FEMS Microbiol. Ecol.* 54: 479-489.

Deleu, M., Paquot, M. & Nylander, T. (2008). Effect of fengycin, a lipopeptide produced by *Bacillus subtilis,* on model biomembranes. *Biophys J.* 94: 2667–2679.

Deleu, M., Bouffioux, O., Razafindralambo, H., Paquot, M.,Hbid, C., Thonart, P., Jacques P. & Brasseur,R. (2003). Interaction of surfactin with membranes: a computational approach. *Langmuir* 19: 3377–3385.

Deleu, M., Razafindralambo, H., Popineau, Y., Jacques P., Thonart, P. & Paquot, M. (1999). Interfacial and emulsifying properties of lipopeptides from *Bacillus subtilis. Colloid Surf. A-Physicochem. Eng. Asp.* 152: 3-10.

Desai, J. & Banat, I. (1997). Microbial production of surfactants and their commercial potential. *Microbiol.Mol. Biol. Rev.* 61: 47-64.

Droby, S. (2006). Improving quality and safety of fresh fruit and vegetables after harvest by the use of biocontrol agents and natural materials. *Acta Horticulturae* 709: 45–51.

Dufour, S., Deleu, M., Nott, K., Wathelet, B., Thonart, P. & Paquot, M. (2005). Hemolytic activity of new linear surfactin analogs in relation to their physico-chemical properties. *Biochim. Biophys. Acta* 1726: 87–95.

El-Ghaouth, A., Wilson, C. & Wisniewski, M. (2004). Biologically based alternatives to synthetic fungicides for the postharvest diseases of fruit and vegetables. In: *Diseases of Fruit and Vegetables,* Naqvi, S.A.M.H. (Ed.), pp. (511–535), Kluwer Academic Publishers, The Netherlands.

Filonow, A. (1998). Role of competition for sugars by yeasts in the biocontrol of gray mold of apple. *Biocontrol Sci. Technol.* 8: 243–256.

Finking, R. & Marahiel, M. (2004). Biosynthesis of nonribosomal peptides. *Annu. Rev. Microbiol.* 58: 453–488.

Fogliata, G., Torres, L. & Ploper, L. (2001). Detection of imazalil-resistant strains of *Penicillium digitatum* sacc. In citrus packing-houses of Tucumán Province (Argentina) and their behaviour against current employed and alternative fungides. *Rev Ind Agric Tucumán.* 77: 71–75.

Gueldner, R., Reilly, P., Pusey, L., Costello, C., Arrendale, R., Cox,R., Himmelsbach, D., Crumley, G., & Cutler, H. (1988). Isolation and identification of iturins as antifungalpeptides in biological control of peach brown rot with *Bacillus subtilis*. *J. Agric. Food Chem*. 36: 366-370.

Harborne, J.B. (1984). *Phytochemical Methods: A Guide to Modern Techniques of Plant Analysis*. New York, NY: Chapman & Hall.

Huszcza, E. & Burczyk, B. (2006). Surfactin isoforms from *Bacillus coagulans*. *Zaturforsch [C]*, 61: 727-733.

Ippolito, A., El-Ghaouth, A., Wilson, C. & Wisniewski, M. (2000). Control of postharvest decay of apple fruit by *Aureobasidium pullulans* and induction of defense responses. *Postharvest Biol. Technol*. 19: 265-272.

Isogai, A., Takayama, S., Murakoshi, S. & Suzuki, A. (1982). Structures of β-amino acids in Iturin A. *Tetrahedron Lett*. 23: 3065-3068.

Janisiewicz, W., Tworkoski, T. & Sharer, C. (2000). Characterizing the mechanism of biological control of postharvest diseases on fruit with a simple method to study competition for nutrients. *Phytopathol*. 90: 1196-1200.

Jijakli, M., Grevesse, C. & Lepoivre, P. (2001). Modes of action of biocontrol agents of postharvest diseases: challenges and difficulties. *Bulletin-OILB/SROP* 24: 317-318.

Katz, E. & Demain, A. (1977). The peptide antibiotics of *Bacillus* chemistry, biogenesis and possible functions. *Bacteriol Rev*. 41: 449- 474.

Kowall, M., Vater, J., Kluge, B., Bein, T., Franke, P. & Ziessow, D. (1998). Separation and characterization of surfactin isoforms produced by *Bacillus subtilis* OKB 105. *J Colloid Interface Sci*. 204: 1-8.

Kumar A., Saini, P. & Shrivastava, J. (2009). Production of peptide antifungal antibiotic and biocontrol activity of *Bacillus subtilis*. *Indian J.Experimental Biol*. 47: 57-62.

Kurusu, K. & Ohba, K. (1987). New peptide antibiotics LI-FO3 and characterization. *J. Antibiot*. 40: 1506.

Leenders, F., Stein, T., Kablitz, B., Franke, P. &Vater, J. (1999). Rapid typing of *Bacillus subtilis* strains by their secondary metabolites using matrix-assisted laser desorption/ionization mass spectrometry of intact of cells. *Rapid Commun Mass Spec*.13: 943-949.

Lin, S., Chen, Y. & Lin, Y. (1998). General approach for the development of high-performance liquid chromatography methods for biosurfactant analysis and purification. *J. Chromatogr. A*.825: 149-159.

Lingk, W. (1991). Health risk evaluation of pesticide contaminations in drinking water. *Gesunde Pflangen*. 43: 21-25.

Maldonado, M., Corona, J., Gordillo, M. & Navarro, A. (2009). Isolation and partial characterization of antifungal metabolites produced by *Bacillus sp*. IBA 33. *Curr. Microbiol* 59: 646-650.

Maneerat, S. & Phetrong, K. (2007). Isolation of biosurfactant producing marine bacteria and characteristics of selected biosurfactant. *Songklanakarin J. Sci. Technol*. 29: 781-791.

Maquelin, K., Kirshner, C., Smith, L., Vaden, N., Endtz, H., Naumann, D. & Puppels, G. (2002). Identification of medically relevant microorganisms by vibrational spectroscopy. *J. Microbiol. Meth*. 51: 255.

Mc Keen, C., Reilly, C. & Pusey, P. (1986). Production and partial purification of antifungal substances antagonistic to *Monilia fructicola* from *Bacillus subtilis*. *Phytopathol*. 76: 136-139.

Moszer, I., Jones, L., Moreira, S., Fabry, C. & Danchin, A. (2002). Subti List: the reference database for the *Bacillus subtilis* genome. *Nucleic Acids Res.* 30: 62-65.

Motta, A. & Brandelli, D. (2002). Characterization of an antimicrobial peptide produced by *Brevibacterium linens*. *J. Appl. Microbiol.* 92: 63.

Moyne, A., Shelby, R., Cleveland, T. & Tuzun, S. (2001). Bacillomycin D: an iturin with antifungal activity against *Aspergillus flavus*. *J. Appl. Microbiol* 90: 622–629.

Mukherjee, S., Das, P. & Sen, R. (2006). Towards commercial production of microbial surfactants. *Trends Biotechnol.* 24: 509–515.

Mukherjee, S., Das, P., Sivapathasekaran, C. & Sen, R. (2009). Antimicrobial biosurfactant from marine *Bacillus circulans*: extracellular synthesis and purification. *Lett. Appl. Microbiol.* 48: 281–288.

Mutaz Al-Ajlani, M., Abid Sheikh, M., Ahmad, Z. & Hasnain, S. (2007). Production of surfactin from *Bacillus . subtilis* MZ-7 grown on pharmamedia commercial medium. *Microbial Cell Factories* 6: 17.

Nelson, D. & Cox, M. (2006). Aminoacidos, peptidos y proteínas. In: *Lenhinger Principios de Bioquimica*. Ediciones Omega. pp (75–106). ISBN, 84-282-1410-7. Barcelona, Spain.

Ongena, M. & Jacques P. (2007). *Bacillus* lipopeptides: versatile weapons for plant disease biocontrol. *Trends Microbiol.* 16: 3115-125.

Paquot, M. & Nylander, T. (2008). Effect of fengycin, a lipopeptide produced by *Bacillus subtilis*, on model biomembranes. *Biophys. J.* 94: 2667–2679.

Perez-Garcia, A., Romero, D., & de Vicente, A. (2011).Plant protection and growth stimulation by microorganisms: biotechnological applications of Bacilli in agriculture. *Curr. Opinion Biotech.* 22: 187–193.

Peypoux, F., Bonmatin, J. & Wallach, J. (1999). Recent trends in the biochemistry of surfactin. *Appl. Microbiol. Biotechnol.* 51: 553–563.

Phae, C., Shoda, M. & Kubota, H. (1990). Suppressive effect of *Bacillus subtilis* and its products on phytopathogenic microorganisms. *J. Ferment. Bioeng.* 69: 1-7.

Pueyo, M. Jr, CB RAMC, PdM. (2009). Lipopeptides produced by a soil *Bacillus megaterium* strain. *Microb. Ecol.* 57: 367-378.

Razafindralambo, H., Paquot, M., Hbid, C., Jacques, P., Destain, J. & Thonart, P. (1993). Purification of antifungal lipopeptides by reversed-phase high performance liquid chromatography. *J. Chromatogr.* 639: 81–85.

Romero, D., de Vicente, A., Rakotoaly, R., Dufour, S., Veening, J., Arrebola, E. Cazorla, F., Kuipers,O., Paquot, M. & Pérez-García, A. (2007). The iturin and fengycin families of lipopeptides are key factors in antagonism of *Bacillus subtilis* toward *Podosphaera fusca*. *Mol. Plant- Microbe Interact.* 20: 430–440.

Sen, R. & Swaminathan, T. (2005). Characterization of concentration and purification parameters and operating conditions or the small-scale recovery of surfactin. *Process Biochem.* 40: 2953–2958.

Sharma, R., Singh, D. & Singh, R. (2009). Biological control of postharvest diseases of fruits and vegetables by microbial antagonists: A review. *Biol. Control* 50: 205–221.

Sivapathasekaran, C., Mukherjee, S., Samanta, R. & Sen, R. (2009). High-performance liquid chromatography purification of biosurfactant isoforms produced by a marine bacterium. *Anal. Bioanal. Chem.* 396: 845-854.

Smyth, T., Perfumo, A., McClean, S., Marchant, R. & Banat, I. (2010). Isolation and analysis of lipopeptides and high molecular weight biosurfactants. In: *Handbook of Hydrocarbon and Lipid Microbiology*. K. N. Timmis (ed.). Springer-Verlag Berlin Heidelberg.

Stein, T. (2005). *Bacillus subtilis* antibiotics: structures, syntheses and specific functions. *Mol. Microbiol.* 56: 845–857.

Swain, T. & Hillis, W. (1959). The phenolic constituent of *Prunus domestica* I. The quantitative analysis of phenolic constituents. *J. Sci .Food Agric.* 10: 63–68.

Symmank, H., Frank, P., Saenger, W. & Bernhard, F. (2002). Modification of biologically active peptides: production of a novel lipohexapeptide after engineering of *Bacillus subtilis* surfactin synthetase. *Protein Eng.* 15: 913–921.

Teixidó, N., Usall, J., Palau, L., Asensio, A., Nunes, C. & Viñas, I. (2001). Improving control of green and blue molds of oranges by combining *Pantoea agglomerans* (CPA-2) and sodium bicarbonate. *Eur. J. Plant Pathol.* 107: 685–694.

Thaniyavarn, J., Roongsawang, N., Kameyama, T., Haruki, M., Imanaka, T., Morikawa, M. & Kanaya, S. (2003). Production and characterization of biosurfactant from *Bacillus licheniformis* F2.2. *Biosci. Biotechnol. Biochem.* 67: 1239–2003.

Vater, J., Kablitz, W., Franke, P., Mehta, N. & Cameotra, S. (2002). Matrix-assisted laser desorption ionization- time of flight mass spectrometry of lipopeptide biosurfactant in whole cells and culture filtrates of *Bacillus subtilis* C-1 isolated from petroleum sludge. *Appl. Environ. Microbiol.* 68: 6210–6219.

Vero Mendez, S. & Mondino, P. (1999). Control biológico poscosecha en Uruguay. *Hort. Int.* 7: 26.

Vero, S., Mondino, P., Burgueño, J., Soubes, M. & Wisniewski, M. (2002). Characterization of biocontrol activity of two yeast strains from Uruguay against blue moldof apple. *Postharvest Technol.* 23: 191–198.

Wang, J., Liu, J., Wang, X., Yao, J. & Yu, Z. (2004). Application of electrospray ionization mass spectrometry in rapid typing of fengycin homologues produced by *Bacillus subtilis*. *Lett. Appl. Microbiol.* 39: 98–102.

Williams, B., Hathout, Y. & Fenselau, C. (2002). Structural characterization of lipopeptide biomarkers isolated from *Bacillus globigii*. *J. Mass Spectrom.* 37: 259–264.

Wilson, C. & Wisniewski, M. (1989). Biological control of postharvest diseases of citrus and vegetables an emerging technology. *Ann Phytopathol.* 27: 425–441.

Wilson, C. & Wisniewski, M. (1994). Biological control of postharvest diseases of fruits and vegetables- Theory and Practice. CRC Press Boca Raton. FL.

Wisniewski, M. & Wilson, C. (1992). Biological control of postharvest diseases of fruit and vegetables: recent advances. *Hort Science* 27: 94–98.

Yakimov, M., Timmis, K., Wray, V & Fredrickson,H. (1995). Characterization of a new lipopeptide surfactant produced by thermotolerant and halotolerant subsurface *Bacillus licheniformis* BAS50. *Appl. Environ. Microbiol.* 61: 1706–1713.

Yazgan, A., Ozcengiz, G., & Marahiel, A. (2001). Tn10 insertional mutations of *Bacillus subtilis* that block the biosynthesis of bacilysin. *Biochim. Biophys. Acta-Gene Struct. Expression* 1518: 87-94.

Youssef, N., Duncan, K. & McInerney, M. (2005). Importance of 3-hydroxy fatty acid composition of lipopeptides for biosurfactant activity. *Appl. Environm. Microbiol.* 71: 7690–7695.

Zachara, N. & Gooley, A. (2000). Identification of glycosylation sites in mucin peptides by Edman degradation. *Methods Mol. Biol.* 125: 121–128.

# Permissions

The contributors of this book come from diverse backgrounds, making this book a truly international effort. This book will bring forth new frontiers with its revolutionizing research information and detailed analysis of the nascent developments around the world.

We would like to thank Dr. D. Sasikumar, for lending his expertise to make the book truly unique. He has played a crucial role in the development of this book. Without his invaluable contribution this book wouldn't have been possible. He has made vital efforts to compile up to date information on the varied aspects of this subject to make this book a valuable addition to the collection of many professionals and students.

This book was conceptualized with the vision of imparting up-to-date information and advanced data in this field. To ensure the same, a matchless editorial board was set up. Every individual on the board went through rigorous rounds of assessment to prove their worth. After which they invested a large part of their time researching and compiling the most relevant data for our readers. Conferences and sessions were held from time to time between the editorial board and the contributing authors to present the data in the most comprehensible form. The editorial team has worked tirelessly to provide valuable and valid information to help people across the globe.

Every chapter published in this book has been scrutinized by our experts. Their significance has been extensively debated. The topics covered herein carry significant findings which will fuel the growth of the discipline. They may even be implemented as practical applications or may be referred to as a beginning point for another development. Chapters in this book were first published by InTech; hereby published with permission under the Creative Commons Attribution License or equivalent.

The editorial board has been involved in producing this book since its inception. They have spent rigorous hours researching and exploring the diverse topics which have resulted in the successful publishing of this book. They have passed on their knowledge of decades through this book. To expedite this challenging task, the publisher supported the team at every step. A small team of assistant editors was also appointed to further simplify the editing procedure and attain best results for the readers.

Our editorial team has been hand-picked from every corner of the world. Their multi-ethnicity adds dynamic inputs to the discussions which result in innovative outcomes. These outcomes are then further discussed with the researchers and contributors who give their valuable feedback and opinion regarding the same. The feedback is then collaborated with the researches and they are edited in a comprehensive manner to aid the understanding of the subject.

Apart from the editorial board, the designing team has also invested a significant amount of their time in understanding the subject and creating the most relevant covers. They scrutinized every image to scout for the most suitable representation of the subject and create an appropriate cover for the book.

The publishing team has been involved in this book since its early stages. They were actively engaged in every process, be it collecting the data, connecting with the contributors or procuring relevant information. The team has been an ardent support to the editorial, designing and production team. Their endless efforts to recruit the best for this project, has resulted in the accomplishment of this book. They are a veteran in the field of academics and their pool of knowledge is as vast as their experience in printing. Their expertise and guidance has proved useful at every step. Their uncompromising quality standards have made this book an exceptional effort. Their encouragement from time to time has been an inspiration for everyone.

The publisher and the editorial board hope that this book will prove to be a valuable piece of knowledge for researchers, students, practitioners and scholars across the globe.

# List of Contributors

**Changming Zhang, Zhanggen Huang and Xiaohang Zhang**
State Key Laboratory of Coal Conversion, Institute of Coal Chemistry, Chinese Academy of Sciences, Taiyuan, China

**Hakan Göker, Maksut Coşkun and Gülgün Ayhan-Kılcıgil**
Central Instrumental Analysis II Laboratory, Faculty of Pharmacy, Ankara University, Tandogan, Ankara, Turkey

**Gülçin Saltan Çitoğlu and Özlem Bahadır Acıkara**
Ankara University, Turkey

**Toshiki Mine and Takeshi Yamamoto**
Glycotechnology Business Unit, Japan Tobacco Inc., Japan

**Magid Abdel Ouoba**
Unit of Teaching and Research (UFR) of Health Sciences, University of Ouagadougou, Burkina Faso

**Jean Koudou**
Laboratory of Chemistry of Natural Products, Faculty of Sciences, University of Bangui, Bangui, Central African Republic

**Noya Some**
Institute of Research in Health Sciences, CNRST, Ouagadougou, Burkina Faso

**Sylvin Ouedraogo**
Laboratory of Chemistry of Natural Products, Faculty of Sciences, University of Bangui, Bangui, Central African Republic
Institute of Research in Health Sciences, CNRST, Ouagadougou, Burkina Faso

**Innocent Pierre Guissou**
Unit of Teaching and Research (UFR) of Health Sciences, University of Ouagadougou, Burkina Faso
Institute of Research in Health Sciences, CNRST, Ouagadougou, Burkina Faso

**Vladimir V. Kouznetsov, Carlos E. Puerto Galvis, Leonor Y. Vargas Méndez and Carlos M. Meléndez Gómez**
School of Chemistry, Industrial University of Santander, Bucaramanga, Colombia

**Talita Perez Cantuaria Chierrito, Ananda de Castro Cunha, Regina Aparecida Correia Gonçalves and Arildo José Braz de Oliveira**
Department of Pharmacy, State University of Maringá, Maringá, Paraná State, Brazil

**Luzia Koike**
State University of Campinas, Chemistry Institute, Campinas-SP, Brazil

**Yasser M. Moustafa and Rania E. Morsi**
Egyptian Petroleum Research Institute, Egypt

**Tânia da S. Agostini-Costa, Roberto F. Vieira and Marcos A. Gimenes**
Embrapa Genetic Resources and Biotechnology, Brasília, Brazil

**Humberto R. Bizzo**
Embrapa Food Technology, Rio de Janeiro, Brazil

**Dâmaris Silveira**
Health Sciences Quality, University of Brasilia, Brasília, Brazil

**Rita C. S. Nunomura and Ellen C. C. Silva**
Department of Chemistry, Amazon Federal University (UFAM), Amazon, Brazil

**Sergio M. Nunomura and Adrian M. Pohlit**
Coordenation of Research in Natural Products, Amazon National Institute (INPA), Amazon, Brazil

**Ana C. F. Amaral, Alaíde S. Barreto, Antonio C. Siani**
Laboratory of Natural Products, Farmanguinhos, Oswaldo Cruz Institute Foundation (FIOCRUZ), Rio de Janeiro, Brazil

**María Antonieta Gordillo and María Cristina Maldonado**
Institute of Biotechnology, Faculty of Biochemistry, Chemistry and Pharmacy, National University of Tucuman, Tucuman, Argentina

9 781632 391124